Yearbook on Space Policy 2010/2011

For further volumes:
http://www.springer.com/series/8166

Yearbook on Space Policy

Edited by the
European Space Policy Institute
Director: Peter Hulsroj

Peter Hulsroj • Spyros Pagkratis
Blandina Baranes
Editors

Yearbook on Space Policy 2010/2011

The Forward Look

Springer

Editors
Peter Hulsroj
Spyros Pagkratis
Blandina Baranes
European Space Policy Institute
Vienna, Austria

ISSN 1866-8305
ISBN 978-3-7091-1362-2 ISBN 978 3 7091 1363-9 (eBook)
DOI 10.1007/978-3-7091-1363-9
Springer Wien Heidelberg New York Dordrecht London

Cover illustration: Artist's impression of Herschel. ESA - C. Carreau

Printed on acid-free paper

Springer is part of Springer Science+Business Media (www.springer.com)

Preface Yearbook

Space, the Global Endeavour

The forces that shape and drive human activities in outer space have been changing since the dawn of spaceflight. Where it initially started out as a bipolar and geopolitical competition, space utilisation over time expanded both its scope and the players involved, embracing space exploration, space science and applications that constantly benefit millions of users worldwide.

Recently, we have entered a new phase where the opportunities ahead are great, but the challenges to live up to them equally so. A gradual but steady geopolitical shift – amplified by a financial and economic crisis – points towards a different balance of capabilities and power. This evolution has far-reaching influences on the different dimensions of outer space utilisation. As society requires evermore integrated technology application in a wide array of fields, it triggers the civil side to establish more cooperation to meet these demands. From a military perspective, the geopolitical shift introduces the challenge of a widening multi-actor system and more global interdependencies. In addition, existing issues, like the sustainable use of outer space, require ever more attention in order to be addressed in a timely and efficient manner. These processes sometimes push the current operation mode of the system to its limits, and – what is more – they will necessitate and drive substantial change in the medium to long term. It is for these reasons that the thematic title of this Yearbook highlights the global dimension of the space endeavour.

As in previous editions, this yearbook is comprised of three parts. The first part seeks to set out a comprehensive overview of the economic, political, technological and institutional trends that affect space activities. It was prepared in-house at ESPI and while its perspective is European, it also provides a comparative analysis of space efforts around the world.

The second part of the Yearbook brings together the views of nine distinguished experts in the space field. These experts touch topics or events which stirred the space sector in 2010, thus reflecting on the establishment of the UK Space Agency and the formulation of a new German space strategy, but also on the usability of space data, intergovernmental cooperation and issues of sustainability such as balanced development and space debris.

An important milestone in the preparation of the Yearbook was again ESPI's Autumn Conference, where authors met for an exchange on the drafts of their contributions. Having taken place in Vienna in September 2011, it provided a forum for constructive exchange and coordination of the contributions.

The third part of the Yearbook carries forward the character of the Yearbook as an archive of space activities. Again prepared in-house by ESPI, a bibliography, chronology and data about institutions is provided where readers of the now five volumes of the Yearbook can identify statistical development and evolutions.

In closing, I would like to thank the contributors of the articles in Part II for their engagement in this publication, as well as the ESPI staff that has been instrumental in the Yearbook's production.

Peter Hulsroj

Contents

List of Figures

List of Tables

Acronyms

A

ABS	Asia Broadcast Satellite
ACE	Advanced Composition Explorer
ACES	Atomic Clock Ensemble in Space
ACLS	Advanced Closed-Loop System
AEB	Agência Espacial Barasileira (Brazilian Space Agency)
AIA	Atmospheric Imaging Assembly
AIS	Automatic Identification System
ALHAT	Automated Landing and Hazard Avoidance Technology
ALOS	Advanced Land Observing Satellite system
AMS	Alpha Magnetic Spectrometer
APL	John Hopkins University Applied Physics laboratory
APSCO	Asia-Pacific Space Corporation Organisation
APRSAF	Asia-Pacific Regional Space Agency Forum
ARV	Advanced Re-entry Vehicle
ASAT	Anti Satellite
ASI	Agenzia Spatiale Italiana (Italian Space Agency)
ATV	Automated Transfer Vehicle

B

BAA	Broad Area Announcement
BGAN	Broadband Global Area Network
BIS	UK's Department for Business, Innovation and Skills
BNSC	British National Space Centre
BSS	Boeing Satellite Systems

C

CAGR	Compound Annual Growth Rate
CASC	China Aerospace Corporation
CBERS	China-Brazil Earth Resources Satellite
CCD	Charged Couple Device
CD	Conference on Disarmament
CDTI	Centre for the Development of Industrial Technology
CEOS	Committee on Earth Observation Satellites
CESPO	Centre d'Etudes Spatiales de la BIOsphère
CFAS	Federal Commission for Space Affairs
CFSP	Common Foreign and Security Policy
CGWIC	Chinese Great Wall Industry Corporation
CIS	Commonwealth of Independent States
CME	Coronal Mass Ejection
CNES	Centre Spatiale d'Etudes Spatiales (French Space Agency)
CNSA	China National Space Administration
CONCORD	European Commission's Annual Competitiveness Assessment Conferences
COPUOS	Committee on the Peaceful Uses of Outer Space
COSPAR	Committee for Space Research
COTS	Commercial Orbital Transportation Services
CPIP	Common Pre-Frontier Intelligence Picture
CRuSR	Commercial Reusable Suborbital Research Programme
CSA	Canadian Space Agency
CSDP	Common Security and Defence Policy

D

DARPA	US Defence Advanced Research Projects Agency
DARS	Digital Radio Audio Satellite
DBS	Direct Broadcast Service
DLR	Deutsches Zentrum für Luft- und Raumfahrt (German Space Agency)
DoD	US Department of Defence
DRDO	Defence Research and Development Organisation
DSI	German SOFIA Institute

E

EAC	European Astronaut Centre
EADS	European Aeronautic Defence and Space Company
EBITDA	Earnings Before Interest, Taxes, Depreciation, and Amortization
ECLS	Environmental Control and Life Support
EEA	European Environment Agency

EGNOS	European Geostationary Navigation Overlay Service
EGP	Eurobot Ground Prototype
EJSM	Europa Jupiter System Mission
ELINT	Electronic Signals Intelligence
ELV	European Launch Vehicle
EMSA	European Maritime Safety Agency
EO	Earth Observation
ERA	European Research Area
ESA	European Space Agency
ESDP	European Security and Defence Policy
ESP	European Space Policy
ESPI	European Space Policy Institute
ESTEC	European Space Technology and Research Centre
EU	European Union
EUMETSAT	European Organisation for the Exploitation of Meteorological Satellites
EUR	Euro (Currency)
EUROSUR	European Border Surveillance System
EUSC	European Union Satellite Centre
EVE	EUV Variability Experiment

F

FAA	Federal Aviation Administration
FCC	Federal Communications Commission
FCT	Foundation for Science and Technology
FIFA	Fédération Internationale de Football Association
FSS	Fixed Satellite Services

G

GA	General Assembly
GATS	General Agreement on Trade in Services
GCI	GEOSS Common Infrastructure
GDP	Gross Domestic Product
GEO	Geostationary Orbit
GEO	Group of Earth Observation
GEODSS	Ground-Based Electro-Optical Space Surveillance System
GEOSS	Global Earth Observation System of Systems
GERD	Gross Domestic Expenditure on Research and Development
GIS	Geospatial Intelligence System
GLONASS	Global Navigation Satellite System
GMES	Global Monitoring for Environment and Security
GMT	Greenwich Mean Time

GNSS	Global Navigation Satellite System
GPS	Global Positioning System
GRACE	Gravity Recovery and Climate Experiment Satellite Mission
GSC	GMES Space Component
GSLV	Geosynchronous Satellite Launch Vehicle
GSM	Global System for Mobile Communications
GTO	Geostationary Transfer Orbit

H

HDTV	High Definition Television
HDU	Habitat Demonstration Unit
HMI	Helioseismic and Magnetic Imager
HPGP	High-Performance Green Propulsion system
HST	Hubble Space Telescope
HTV	H-II Transfer Vehicle

I

IADC	Inter-Agency Space Debris Coordination Committee
IAI	Israel Aerospace Industry
IBM	Integrated Border Management
ICG	International Committee on Global Navigation Satellite Systems
ICT	Information and Communication Technology
IDF	Israeli Defence Force
IG	Information Gathering Satellites
IGSO	Inclined Geosynchronous Orbit
ILA	Internationale Luftfahrtausstellung (Berlin Air Show)
ILN	International Lunar Network
ILS	International Launch Services
IMOD	Israeli Ministry of Defence
INSAT	Indian National Satellite System
IOV	In-Orbit Validation
IPRs	Intellectual Property Rights
IRNSS	Indian Regional Navigation Satellite System
ISA	Israeli Space Agency
ISECG	International Space Exploration Coordination Group
ISIC	International Space Innovation Centre
ISO	International Organisation for Standardization
ISRO	Indian Space Research Organisation
ISRU	In-Situ Resource Utilisation Technologies
ISS	International Space Station
ITAR	International Traffic in Arms Regulations
ITU	International Telecommunication Union

J

JAXA	Japan Aerospace Exploration Agency
JDA	Japanese Defence Agency
JWST	James Webb Space Telescope

K

KARI	Korea Aerospace Research Institute
KSC	Kennedy Space Center

L

LCO	Lifeline Connectivity Obligation
LEO	Low Earth Orbit
LES	Living with a Star Programme
LLDCs	Least-Developed Countries
LSS	Large Space Simulator

M

MDGs	Millenium Development Goals
MDRS	Mars Desert Research Station
MEO	Medium Earth Orbit
MER	Mears Exploration Rover
MMO	Mercury Magnetospheric Orbiter
MoU	Memorandum of Understanding
MPCV	Multi-Purpose Crew Vehicle
MPO	Mercury Planetary Orbiter
MRO	Mars Reconnaissance Orbiter
MS	Member State
MSL	Mars Science Laboratory
MTG	Meteosat Third Generation Programme

N

NASA	National Aeronautics and Space Administration
NATO	North Atlantic Treaty Organisation
NCC	National Coordination Centre
NDPBs	Non-departmental Public Bodies
NEO	Near Earth Object
NERC	Natural Environment Research Council
NGA	US National Geospatial-Intelligence Agency
NGL	Next-Generation Launcher
NICT	National Institute of Information and Communications Technology
NLR	National Aerospace Laboratory of the Netherlands

NOAA	National Oceanic and Atmospheric Administration
NPOESS	National Polar-Orbiting Operational Environmental Satellite System
NRC	US National Research Council
NRO	US National Reconnaissance Office
NSAU	State Space Agency of Ukraine
NSC	Norwegian Space Centre

O

OHB	Orbitale Hochtechnologie Bremen
ONERA	Office Nationale d'Etudes et de Recherches Aérospatiales
OOSA	Office of Outer Space Affairs
OST	Outer Space Treaty

Ö

ÖWF	Österreichisches Weltraumforum (Austrian Space Forum)

P

PMM	Permanent Multipurpose Module
PND	Portable Navigation Device
PNT	Positioning-Navigation-Timing
PPP	Public Private Partnership
PSLV	Polar Satellite Launch Vehicle
PSO	Polish Space Office
PWR	Pratt & Whitney Rocketdyne

Q

QZSS	Quasi-Zenith Satellite System

R

R&D	Research and Development
RATS	Desert Research and Technology Studies
ROSA	Romanian Space Agency

S

SAC	Space Activities Commission
SAP	United Nations Programme on Space Applications
SAR	Synthetic Aperture Radar
SDC	Space Data Centre
SDMP	Space Debris Mitigation Plan

SDO	Solar Dynamics Observatory mission
SEV	Space Exploration Vehicle
SHF	Super High Frequency
SIGINT	Signal Intelligence
SIS	Space Infrastructure Services
SMDC	Space Missile and Defence Command
SME	Small and Medium Enterprise
SNSB	Swedish National Space Board
SOHO	Solar and Heliospheric Observatory
SPAC	Satellite Positioning Research and Application Centre
SPRN	Russian Satellite Network, Russian Abbreviation for Satellite for Warning on Rocket Attack
SRON	Dutch Space Research Organization
SSA	Space Situational Awareness
SSTL	Surrey Satellite Technology Limited
STEREO	Solar Terrestrial Relations Observatory
STFC	Science and Technology Facilities Council
STSS	Space Tracking and Surveillance System
SWOT	Surface Water Ocean Topography Mission

T

TCBM	Transparency and Confidence-Building Measures
TFEU	Treaty on the Functioning of the European Union
TRAI	Telecom Regulatory Authority of India
TRIPS	Agreement on Trade-Related Aspects of Intellectual Property Rights
TSB	Technology Strategy Board

U

UARS	Upper Atmosphere Research Satellite
UAS	Unmanned Aircraft System
UHF	Ultra High Frequency
UK	United Kingdom
UKTI	UK Trade and Investment
ULA	United Launch Alliance
UN	United Nations
UNCOPUOS	United Nations Committee on the Peaceful Uses of Outer Space
UNESCO	United Nations Educational, Scientific and Cultural Organization
UNGA	United Nations General Assembly
UNIDIR	United Nations Institute for Disarmament Research
UNIWG	United Nations Geographic Information Working Group
UNOSAT	United Nations Operational Satellite Applications Programme

UNSDI	United Nations Spatial Data Infrastructure
UN-SPIDER	United Nations Platform for Space-Based Information for Disaster Mangement and Emergency Response
US	United States
USAF	US Air Force
USAT	Ultra Small Aperture Terminals
USD	US Dollar (Currency)
USNSSS	US National Security Space Strategy
USRA	Universities Space Research Association

V

VHF	Very High Frequency
VLS	Veículo Lançador de Satélites
VSAT	Very Small Aperture Terminals

W

WARC	World Administrative Radio Conference
WRS	World Radiocommunications Seminar
WSIS	World Summit on the Information Society
WTO	World Trade Organisation

Part I
The Year in Space 2010/2011

Chapter 1
European Space Activities in the Global Context

Spyros Pagkratis

1.1 Global Political and Economic Trends

1.1.1 Global Economic Outlook

In 2010 and 2011 the symptoms of the 2008 financial crisis were still being broadly felt worldwide. Although global financial expansion had already resumed in 2009, its pace remained slow and uncertain. The key element of the recovery has been the uneven pace of its impact on developed and emerging countries, with the latter achieving a much faster and bullish return to growth, as was predicted in the "Space Policies Issues and Trends in 2009/2010" report. Although global economic activity was expected to rise by roughly 5% in 2010 and 4.2% in 2011, advanced economies were projected to expand by only 2.2–2.7%, whereas emerging countries' growth was expected to be up to three times higher.[1]

Furthermore, while emerging economies seem to have overcome the worst of the crisis with the help of suitable fiscal policies, advanced economies are still under considerable fiscal stress, especially in Europe. The growing difference between emerging and developed economies has increased the systemic danger of a new crisis by exacerbating global financial imbalances and encouraging excessive capital volatility and especially flows from advanced to emerging economies. This unstable situation is mostly fuelled by the different approaches chosen by developed and emerging economies in order to counter the consequences of the crisis. The former have adopted policies of fiscal consolidation and monetary contraction in order to hedge the risks from the 2008 bailouts, which have resulted

[1] International Monetary Fund. World Economic Outlook: recovery, Risk and Rebalancing. Washington DC: IMF, Oct. 2010.

S. Pagkratis (✉)
ESPI, Vienna, Austria
e-mail: spyros.pagkratis@espi.or.at

P. Hulsroj et al. (eds.), *Yearbook on Space Policy 2010/2011*,
Yearbook on Space Policy, DOI 10.1007/978-3-7091-1363-9_1,

in rising unemployment, crippled domestic demand and increased exports. The latter are following policies that favour domestic consumption, which was expected to rise by 8% in 2010.[2]

The main reason behind the different approaches adopted was the poor condition of sovereign debt and bank balance sheets in developed economies compared to emerging economies. Another reason was the limited margin for further consumption growth in developed economies: starting from much higher consumption rates, they would not have been able to profit from further expansion as much as emerging economies, even with targeted fiscal measures. In the years immediately preceding the crisis, emerging and developed economies had been following completely different paths to economic development. The former chose a fiscally solid, industrial output and export oriented posture, while the latter a fiscally more precarious, financial services' and domestic consumption oriented approach. In the time of global economic expansion, the two different approaches were able to cooperate and complement each other in a mutually beneficial way. It is clear however that in the aftermath of the crisis and the recovery efforts that followed they are set upon two very different and rapidly diverging paths of economic development.

Advanced economies still have a number of challenges lying ahead, with the most important being to carefully balance the necessary sovereign and banking fiscal stabilisation measures with the need to restrain unemployment and improve household finances and consumption. First of all, financial sector policies and practices need to be improved, addressing the financial and banking imbalances that caused the recent crisis. This applies particularly to the banking sector, where bank consolidation is still in process, as well as to mitigating the effects of the sovereign debt crisis that followed the financial melt down. On the other hand however, in the presence of very low interest rates and tax policies that favour production rather than consumption, such measures also increase the risk of deflation. In the short term, such measures seem to be creating excessive unemployment, further complicating efforts to restore demand to pre-crisis levels. In the medium term, they could pose a threat to medium and small enterprises that rely mostly on domestic demand rather than exports. In the longer term, finally, they could lead to further stagnation and deflation as exports to emerging economies will drop as a result of their rapidly improving position vis-à-vis the advanced economies.

The challenges that lay ahead for emerging economies are very different by comparison. Most countries in that category have so far opted to boost their domestic consumption in order to offset the drop in export volume caused by declining international consumer demand created by the crisis. In a similar fashion, they have increased spending on retooling and refurbishing their industrial infrastructure, in an effort to increase liquidity in their economies and limit unemployment. However, such efforts on their own cannot restore output to pre-crisis levels, especially in the case of countries that have relied heavily on manufactured goods demand from advanced economies to boost their growth. In addition to these measures, carefully balanced monetary and exchange rate policies would have to

[2] Ibid.

be put in place in order to avert inflationary risks and maintain credit growth at reasonable levels.

Finally, a key plank during the past 12 months has been the increased capital volatility and flow from advanced to emerging markets. At this point in time, excessive capital outflows from established markets could potentially destabilise the global financial system and compromise its recovery, as they have a direct negative impact on market confidence, credit availability and interest rate levels in advanced economies, further obstructing their path to recovery. At the same time, this increased volatility and speculation in capital markets is also increasing inflationary risks in emerging economies. In essence, this imbalance in global capital and credit flows could increase the pro-cyclical movement of the global economy, significantly complicating the recovery effort worldwide.[3]

In the face of these challenges, the space economy and especially the commercial space services sector can play a role in boosting global economic growth, without further fuelling the aforementioned imbalances. Being inherently global by its very nature, it has the potential to repatriate capital flows to advanced economies while at the same time distributing beneficial services to emerging countries. In fact, space services can instigate growth in both cases, without creating the negative effects from capital flows mentioned above. Space infrastructure creates jobs both in advanced and emerging economies; it encourages global synergies and enhances international cooperation; and its operating costs are evenly distributed among its users, who can nevertheless reap the full advantages of its use. As space infrastructure is not based in any territory, it does not have any of the disadvantages related to more traditional international investments or services' exports, especially for the importing countries. Furthermore, it allows space services' providers to enjoy operational and financial flexibility that constitute a decisive advantage in view of the crisis's consequences for global economic integration. At the same time, it can benefit both from the fiscal consolidation measures applied in high income economies (that encourage outsourcing services to the private sector), and from the booming growth in emerging countries.

1.1.2 Political Developments

1.1.2.1 Security

From a security perspective, 2010 and 2011 were marked by the geopolitical events that took place in the Middle East. Earlier protests that took place in Egypt and Tunisia were followed on 16 February 2011 by civil protests against the Libyan leader Moammar Gadhafi and his government that had been running the country since a coup d'état in 1969. Gadhafi confronted the civil protests and declared his

[3] The World Bank. Global Economic Prospects: Navigating Strong Currents. Washington DC: World Bank, Jan. 2011.

intention not to leave the country.[4] Amid violent repression of the protesters and government attacks against civilians, the death toll began to rise. This continued violence forced the U.N. Security Council to intervene.[5] It approved a no-fly zone resolution for Libya and gave permission for the use of any means necessary to protect civilians. There were no opposing votes at the 15-member council, but China, Russia, Germany, India and Brazil abstained. The prolonged civil unrest and armed conflict that followed had repercussions in many fields, including global oil prices and the stock market. OPEC was forced to intervene in order to stabilize petroleum prices, as Libya's production fell to less than 400,000 barrels per day, or approximately ¼ of pre-crisis production levels, with the increased risk of a complete production halt, according to Libya National Oil officials.

Prolonged combat between government and opposition forces created significant refugee flows to neighbouring countries. Satellite imagery was one of the means applied by international and humanitarian organisations to identify and mitigate the consequences of this event by setting up appropriate humanitarian aid operations. The use of satellite imagery and maps during the crisis was particularly successful on a United Nations level, with the exploitation of the UN Operational Satellite Applications Programme (UNOSAT).[6] Finally, on 22 March 2011, international military operations started in western Libya against Gadhafi's regime among protests from Russia, China and certain Arab League countries that originally supported the operation.

1.1.2.2 Environment

On March 11, 2011 a tsunami spawned by the fifth-largest earthquake ever recorded slammed Japan's eastern coast. The magnitude 8.9 offshore earthquake unleashed a 7 m tsunami that was followed by more than 50 aftershocks that shook Japan for hours, causing major damage in broad areas of northern Japan.[7] The incident's death toll climbed to over 20,000 people.[8] Following the earthquake, a catastrophic failure occurred at the Fukushima nuclear power station that was closest to the earthquake's epicentre and was also struck by the ensuing tsunami. A state of

---

[4] "The Libya War of 2011."STRATFOR Global Intelligence, 21 March 2011, http://www.stratfor.com/analysis/20110319-libyan-war-2011?utm_source=SpecialReport&utm_medium=email&utm_campaign=110319a&utm_content= readmore&elq=3c7acd0ccb4540ad8f71e78079162462

[5] "U.N. Security Council Approves No-Fly Zone in Libya." CNN, 18 March 2011, http://edition.cnn.com/2011/WORLD/africa/03/17/libya.civil.war/index.html

[6] Godoy, Julio. "Satellite Technology to Help the Displaced." Terraviva, 15 March 2011, http://www.ipsterraviva.net/UN/news.asp?idnews=54851 and http://allafrica.com/stories/201103151462.html

[7] "Japan Hit By Tsunami after Massive Earthquake." 11 March 2011, http://www.bbc.co.uk/news/world-asia-pacific-12709850

[8] "Japão confirma 22 mil mortos e desaparecidos." Publico, 22 March 2011, http://www.publico.pt/Mundo/japao-confirma-22-mil-mortos-e-desaparecidos_1486147.

emergency was declared and the entire region evacuated. As a consequence of the damages caused to the nuclear plant, large amounts of radioactive particles were released into the atmosphere, making this the worst nuclear incident since Chernobyl.[9] The situation inside the plant came close to a total core meltdown, a danger that was averted by the narrowest of margins and only after considerable self-sacrificing efforts by fire fighting and damage control crews. A rise in seawater contamination levels near the Fukushima facility was also reported, with iodine levels reaching 27.1 times higher than normal and caesium levels 2.5 times higher than normal.[10] In the process of identifying and preventing further damage from the spreading radiation, weather and oceanographic satellites offered considerable assistance in identifying the speed and pattern of contamination, based on prevailing winds and sea currents in the region. The availability of such information in near real time was pivotal in the operational planning of evacuation and humanitarian assistance operations in the immediate aftermath of a disaster, the full environmental impact of which will reveal itself only in the long term.

1.1.2.3 Energy

Increased volatility in energy sector prices was a persistent trend throughout the reporting period that continued from the first half of 2010. In spite of the fact that consumption levels still lagged behind pre-crisis levels, investing in oil and other fossil energy resources was a constant trend since the 2008 financial crisis, as investors increasingly took refuge in commodity positions to hedge against increased market volatility and decline. However, the perception of oil as a secure asset class in the midst of financial turmoil led to considerable fluctuations in the oil market as well, including a considerable decoupling of oil prices from real demand and supply conditions. The unforeseen geopolitical events in the Middle East at the end of 2010 and the beginning of 2011 mentioned above further exacerbated the situation. It should be noted that despite the progress made in alternative power sources, fossil fuel is expected to remain dominant in the global energy consumption mix. This long term trend is mainly due to rising demand in the transportation sector, particularly in emerging Asian markets that are responsible for 75% of the projected consumption growth in the next 20 years. On the other hand, the rising

---

[9] "Concentracões elevadas de iodo radioactivo detectadas no pacífico." Publico, 21 March 2011, http://www.publico.pt/Mundo/concentracoes-elevadas-de-iodo-radioactivo-detectadas-no-pacifico_1485890

[10] "Japanese Government Confirms Meltdown." STRATFOR Global Intelligence, 12 March 2011, http://www.stratfor.com/analysis/20110312-japanese-government-confirms-meltdown?utm_source=redalert&utm_medium=email&utm_campaign=110312%286%29&utm_content=readmore&elq=96da7bd6198c44f9a29a05659469a594

potential of gas utilisation might also lead to a change in the internal balance of fossil fuel consumption, from crude oil to gas use.[11]

Compared to the years immediately preceding the crisis, demand in 2010 witnessed modest growth over 2009, although consumption was still low compared to pre-crisis times. On the other hand, oil extraction costs continued to rise as well, requiring considerable upstream investments to increase production, especially within the OPEC countries. In the absence of such investments, crude oil production would remain at roughly the present levels in the medium term, something which may not be sufficient if global economic activity returns to pre-crisis levels, which at the same time could instigate further price speculation should additional geopolitical disturbances occur. On the other hand, if recovery proves to be slower than anticipated, demand could remain flat in the medium term, depriving fossil fuel extraction companies of the resources necessary to invest in modernising production.[12] In conclusion, the current situation is full of challenges for the oil industry that might become trapped in a vicious circle of stagnating demand, coupled with rising investment costs for future exploitation projects.

The combination of low demand and rising infrastructure modernisation costs has particularly hit the refining industry in advanced economies that, since the outbreak of the financial crisis, has been consistently demonstrating increasing excess capacity. This trend is expected to continue as refining capacity in emerging countries (especially Asia) increases, severely limiting the medium-term profitability of European and North American competitors. In general, demand in advanced economies is considered to have reached its historical peak, a fact that, when combined with the aforementioned medium term trends, might move the centre of downstream oil services from the Atlantic to the Pacific region.[13] This trend could be even accelerated in the medium term, if economic recovery remains sluggish in advanced economies and new climate change and energy policies are implemented, especially in the U.S. Carbon related legislation in particular, which is still in a formative stage, could potentially further reduce demand growth and increase competition for markets.

Whatever may be the long term prospects for oil consumption, increased gas supply and demand will most likely preserve the capital importance of fossil fuels in the global energy mix. This trend is expected to confirm the importance of satellite services for the operation of global energy flows. Earth observation satellites will continue to provide useful support for geological surveys related to fossil fuel exploitation, navigation and communication spacecraft will ensure the reliability of fuel transport, and environmental monitoring spacecraft will help monitor and mitigate pollution issues. In the long term, the increased importance of gas supply might increase the role of terrestrial energy flows via pipelines,

[11] Organisation of the Petroleum Exporting Countries. World Oil Outlook. Vienna: OPEC, Sept. 2010.

[12] Ibid.

[13] Ibid.

compared to the current preponderance of maritime transport. In such an event, relevant satellite services might swiftly change from maritime to pipeline operations' support, causing a slow change in the business portfolio of commercial satellite services. The beginnings of this development, however, did not yet appear during the reporting period.

1.1.2.4 Resources

In spite of the fragile economic recovery witnessed on average worldwide, world trade and commodity prices showed considerable increases in 2010. This development was principally driven by emerging-market economies that had already reached their previous pre-crisis peak of April 2008, with almost all emerging economies outside the European region returning to their pre-crisis growth levels in 2010. While developing economies led the global recovery, the situation in developed countries stagnated and showed increasing signs of resembling patterns leading up to the crisis, especially in Europe, sparking fears of a double-tip recession. Only countries with a strong reliance on exports such as Japan or Germany were able to make modest and precarious progress, which in the case of the former was jeopardised by the results of the March 2011 earthquake detailed above.[14]

The rise in primary commodity prices that had started in 2009 continued in 2010, particularly in metals, minerals and energy products. However, when put into perspective, it should be noted that these commodities had experienced the steepest fall during the previous 2 years of crisis. Agricultural commodities' appreciation was particularly felt, rising by almost 50% in 2009 from its low in 2008 and continuing upwards in 2010. Strong demand from emerging countries and financial investor behaviour were the main reasons behind this trend, especially after food shortages as a result of poor crops worldwide in 2008. In spite of the improvement in 2010, food security remains a pressing issue for the developing world.[15]

1.1.2.5 Knowledge

The economic and social philosophy of the European Union is particularly influenced by free market and knowledge based growth, in the form of developing innovating new technologies. By placing innovation at the core of its economic policy as a key driver behind European competitiveness, the EU aspires to secure long-term growth and prosperity for its citizens. This policy is particularly noticeable in the organisation of the European Research Area, but it is also shared by all

[14] United Nations Conference on Trade and Development. Trade and Development Report, 2010. Geneva: UNCTD 2010.

[15] Ibid.

European Institutions, including ESA. At the same time, the benefits of innovation are spread over a great variety of economic and social activities, including agriculture, environment, energy, security etc. The benefits of innovation in space technologies are both direct (when creating specific services) and indirect (when creating new technologies with applications outside the space domain). Due to the cutting edge nature of space technology and the considerable upfront investments involved in its development, state and institutional actors continue to play a pivotal role in space related innovation. However, as the global commercial market for space-related services increases, private sector interest in developing space technologies is expected to grow. In addition to this, companies outside the narrow space industry are increasingly taking an interest in technologies that can have immediate secondary applications outside the space domain.

In a sense, space technology development and innovation become increasingly evident, as their broader applications and benefits catch the public eye. Some examples of secondary applications are technologies related to alternative spacecraft propulsion systems, especially nuclear ones. During the reporting period, the issue of developing bigger and more efficient nuclear propulsion systems gained momentum, especially in the U.S. Space Nuclear Power Systems are generally considered pivotal for future human solar system exploration missions,[16] where solar energy might not prove sufficiently efficient or reliable. Despite being a highly contested issue given the environmental hazards it entails, nuclear propulsion research continues, with the aim of increasing the robustness and safety of these systems. Although nuclear propulsion research is specific to space exploration missions, it also has positive effects on terrestrial nuclear energy applications through the advancement of safety features and the miniaturisation of relevant components, in the same way that solar panel technologies have profited from their use in the space industry.

Further technology spin-offs from space oriented technological solutions can have significant industrial applications.[17] These include cutting edge fields such as nanotechnologies, the application potential of which is enormous.[18] Recent breakthroughs in these technologies include their use for space propulsion, navigation and sensor manufacturing, such as ceramic diskless turbines and micro-electronic optical cameras.[19] The potential of such systems is not limited to space applications, but extends to areas such as medicine (artificial muscles and valves), the automobile industry (replacing several electrical motorised actuators and

[16] FP8 hearing on Space Research, Jean-Pierre Roux, AREVA, 8 Dec. 2010, Brussels, Belgium.

[17] "NASA Technologie SpinOff 2010." NASA, 2010, http://www.sti.nasa.gov/tto/Spinoff2010/pdf/Spinoff2010.pdf

[18] FP8 hearing on Space Research, Jeroen Rotteveel, Chris Verhoeven, Mark Bentum, ISIS – Innovative Solutions In space; Delft University of Technology; University of Twente, 8 Dec. 2010, Brussels.

[19] FP8 hearing on Space Research, Edwin Gevorkyan and Valery Manchesko, CermetULtd, University of Twente, 8 Dec. 2010, Brussels.

valves) and aeronautics (replacement of electrical motors and generation of smart structures).

Robotics and laser technologies also have considerable applications outside space, for example in developing underwater robotic equipment that could be crucial for the future exploitation of sea and underground resources. To this list can be added the invention of the optical frequency comb[20] and more generally laser technology that was initially mainly developed for space activities and is now regularly used on Earth. After funding a space related programme, it is necessary to follow up by fully exploiting its industrial potential in other areas. For this purpose, an efficient and innovation-oriented industrial base able to combine academic and research expertise with sound business management is indispensable. Finally, innovative approaches always benefit from an open market and globalised approach that targets the widest possible audience. In this sense, any market barriers raised against the full exploitation of innovative technologies could seriously jeopardise such programmes in their development. Market distortions such as export controls or other protectionist measures have a negative influence on willingness to assume the technological and investment risks associated with adopting innovative technologies. In this context, the revision of the ITAR export control process related to space material in the U.S. could boost space inspired innovation in that country as well as worldwide, even if it lessens the competitive advantages in the market place which the ITAR regulations have given parts of European space industry.

Space activities also facilitate scientific research in the narrower sense of promoting knowledge of our environment, be it on a molecular or planetary scale. With respect to the former, astrobiology[21] research on the International Space Station is highly valuable for research related to applications on Earth (in medicine, biology etc.), and is considered a unique (and consequently indispensable) platform for performing experiments in the physical and life sciences. Zero gravity experiments on the ISS greatly advance our understanding of natural processes such as fluid behaviour, material and plasma physics, plant and cell biology, and even human physiology and psychology. It also helps researchers understand the effects of gravity itself on the human body, facilitating the elaboration of adequate treatments for wide spread muscle and bone illnesses on Earth. Even in psychology,[22] space activities help to expand our knowledge of crucial social and psychological mechanisms related to individual isolation or space confinement in a much more thorough and complete fashion than on Earth. Through experiments in orbit and appropriate space related analogues on Earth (such as the widely publicised Mars500 experiment) useful conclusions on human behaviour

[20] FP8 hearing on Space Research, Steve Lecomte, Centre Suisse d'Electronique et de Microtechnique SA, Switzerland, 8 Dec. 2010, Brussels.

[21] FP8 hearing on Space Research, Prof. Charles Cockell, Open university, 8 Dec. 2010, Brussels.

[22] FP8 hearing on Space Research, Richard Aked, Space Applications Services NV, 8 Dec. 2010, Brussels.

can be drawn and exploited in the study of similar conditions, for example in submarines and off-shore stations.

Environmental studies are a scientific field particularly influenced by space technologies and applications. As a global problem, air pollution and the related climate change consequences are best monitored and fully assessed from space. Relevant climate and environment monitoring technologies have greatly profited from space programmes, both in their development and operational maturity and assessment phases. Space provides the tools for European decision makers to measure and assess the consequences of global climate change, promoting adequate European policies to mitigate them.[23] Apart from improving our understanding of global environmental challenges, space R&D also instigates and further develops ways of combating them through the development and use of greener technologies, such as fuel-efficient propulsion systems.

1.1.2.6 Mobility

Despite renewed demand in emerging economies and especially in the Pacific region, maritime transport remained stagnant in 2010, continuing the downward trend of 2009. The considerable increase in size of the world merchant fleet, a direct result of the 1.3 record new deliveries figure in 2009, resulted in declining productivity for shipping companies. Excessive tonnage supply and lower tonnage carrying capacity trends were expected to continue well into 2011. The only notable exception to the overall picture has been China, which pursued its spiralling growth both in supply and demand in shipping services. Chinese companies are the fastest growing worldwide and the country is home to the most important container and crane manufacturers. As Chinese containerised exports represent 25% of the global total, this trend is expected to continue. In 2009 alone, China overtook Germany as the third largest shipping country, Japan as the second ship-building country and India as the busiest ship-recycling country.[24]

1.1.3 Main Science and Technology Indicators Relevant for Space Activities

Space activities imply large investments in cutting-edge technologies, which are often difficult to make in market conditions and require a more centralised public sector led effort. Global space industry and business operations have consequently

---

[23] FP8 hearing on Space Research, Per Høeg, Technical University of Denmark, DTU Space, 8 Dec. 2010, Brussels.

[24] United Nations Conference on Trade and Development. Review of Maritime Transport 2010. Geneva: UNCTD 2010.

created a web of private and public technology development schemes that interact and complement each other in various levels and to different extents. This is particularly true on a European Union level, where knowledge and innovation are considered a key driver of future economic and societal development, not only within but also beyond the space domain. This is also reflected on the creation and development of the European Research Area (ERA), in which the space sector occupies an important place. Technological development and innovation is in fact a quantifiable activity, based on its input and output. As has been explained in previous editions of the ESPI Yearbook, the term "input" describes the level of human resources (both tangible and intangible) dedicated to research and development in a certain geographic area and for a given research topic. The term "output", on the other hand, is the tangible result of these inputs, consequently measuring the added value of innovation in the process of transforming resources into concrete scientific knowledge and technological inventions. Since the 1960s, this input–output balance has been applied to the space sector, in an effort to quantify a global investment return ratio for space activities that are by definition of large scale and long-term impact.

1.1.3.1 Science and Technology Inputs

According to the aforementioned definitions, science and technology inputs on a European level are measured on a constant basis against the benefits they provide. In 2010 and 2011, the global economic crisis did not seem to catch up with investment levels. On the contrary, there were considerable investments in R&D within this time frame, generally in line with recent historic levels. The fact that R&D in Europe continued its long term trend in 2010 and 2011 can be seen as a sign of innovation's resilience as a strategic aspect of European development beyond short term ups and downs. In addition to this, it can be noted that innovation investments might in fact have benefited from the crisis, inasmuch as the latter instigated an increase in public spending as a counter to the economic slowdown. From this point of view, and given the predominantly public nature of R&D (particularly space related) in Europe, it can be said that the supporters of innovation policies rather increased their influence.

Although total R&D expenditure on a European scale had not been fully quantified at the time this book's content was finalised, data regarding the last two quarters of 2010 allow us to make the following observations. According to Eurostat, Gross Domestic Expenditure on R&D (GERD) stood at approximately €245 billion by the end of 2010, or 3.8% higher than the previous year (in current prices). When measuring GERD against European GDP levels (which helps put things into perspective) it appears to have declined to 2% in 2010. This short term trend could be explained by the relatively higher increase of European GDP compared to GERD in 2010. This performance was clearly inferior to Japan's and the United States' indicators during the same period, in accordance with long term trends in this field. It should be noted, however, that Japanese expenditure on

R&D has been growing continuously over the past decade, in spite of the fact that the country has struggled to recover from a significant financial and economic downturn throughout this period. This observation might be good news for Europe as well, as it demonstrates how investing in innovation can actually grow in times of financial turmoil, and be even seen as a remedy against it. On a national level, the European countries with the highest investment levels of GDP on innovation in 2010 were Finland (3.87%), Sweden (3.42%) and Denmark (3.06%), while eight European Member States reported expenditures under 1%, mostly in South and Eastern Europe.[25]

In addition to this, a further breakdown of statistical evidence underscores the great significance of private R&D investment in Europe, alongside public funded projects. Indeed, countries with higher GERD levels also showed a generally high level of innovation investment within the private business sector (which was especially the case with Scandinavian countries, the Netherlands, Germany and Austria). Countries with larger GDPs (such as France and Germany) tended to field equally large government led investments in R&D, mostly through higher education research.

Private sector funded R&D accounted for nearly 55% of the European total in 2010, as compared to 78.2% in Japan and 67.3% in the United States in 2008. Government expenditures in this field also varied significantly from one European region to the other, with the highest levels observed in Central and Eastern European countries.[26] The great variety of European approaches to R&D funding illustrates the challenges in planning appropriate policies on a European level.

1.1.3.2 Science and Technology Outputs

In contrast to R&D input, the actual results of investments in innovation (or output) are more difficult to measure and assess, given the complexity of the market dynamics that affect the private sector. It is widely understood that innovation offers a competitive edge to companies that are willing to invest and – more importantly – follow up and capitalise on their R&D projects. Clear-cut indicators, such as the number of patents lodged by country, industrial sector and individual companies, are of only indicative value as their effect on financial performance is also heavily dependent on external market competition and general economic sentiment factors. According to relevant sources, in 2010 approximately 1,400 companies invested more than $30 million on R&D worldwide. Among them, companies with the best performance reported that 27% of their sales were directly related to products introduced over the previous 3 years. For companies that are already investing heavily on R&D (and are therefore presumably operating in a highly competitive

[25] European Commission: Eurostat. Statistics Explained: R&D Expenditure. Accessed on 14 January 2012. http://epp.eurostat.ec.europa.eu/statistics_explained/index.php/R_%26_D_expenditure.

[26] Ibid.

environment) there seems to be a linear link between investment and market performance. However, this trend is highly dependent on the nature of the market they operate in, and cannot be considered as a holding true in all industrial sectors. Indeed, from a global point of view it is much more difficult to establish a direct and consequential link between R&D and company performance, making such estimates only meaningful on a case by case basis and on a sector level at best.[27]

From a cross-sectoral point of view, the uneven benefits of corporate innovation on a European scale are further demonstrated in the European Commission's annual competitiveness assessment conferences CONCORD. Evidence presented during the 2011 meetings confirmed both the existence of concrete market benefits for innovative companies, as well as their globally non-linear nature that makes their marginal efficiency difficult to measure. In fact, it appears that innovation-fuelled corporate performance greatly depends on how technology driven the entire market sector is. From this point of view, one should expect the space sector to be one of the most influenced by R&D, a conclusion that is indeed empirically confirmed. In general, the more high-tech a sector is, the greater and broader investment returns one can expect from innovation. By contrast, innovation in low-tech industry segments seems to focus primarily on improving production efficiency and not directly creating new products or even new markets. This situation, in turn, creates the need for carefully planned and finely tuned innovation policies on a European scale that would be able to meet a great variety of cases. In fact, any horizontal measures taken on a European level to improve competitiveness through innovation would have to be accompanied by appropriate custom-made policies for the different market segments, corporate sizes and geographical regions of Europe.[28]

1.2 Worldwide Space Policies and Strategies

1.2.1 The United Nations System

Various institutions within or associated with the United Nations are relevant for space policy. In this section, the UN General Assembly (UNGA), UNGA Committees and other UN bodies and organs are discussed regarding space activities.

[27] European Commission Joint Research Centre. Institute for Prospective Technological Studies. The 2010 Survey on R&D Investment Business Trends. Seville, 2011.

European Commission Joint Research Centre. Institute for Prospective Technological Studies. The 2011 EU Industrial R&D Investment Scoreboard. Seville, 2011.

[28] European Commission Joint Research Centre, Institute for Prospective Technological Studies. CONCORD-2011. 3rd European Conference on Corporate R&D: The Dynamics of Europe's Industrial Structure and the Growth of Innovative Firms. Seville, 2012.

1.2.1.1 United Nations General Assembly (UNGA)

As in its 2009–2010 session, the United Nations General Assembly passed a number of annual Resolutions pertaining to the use and exploration of Outer Space. More precisely, in the course of the UNGA's 65th session (2010–2011) the GA adopted four Resolutions on space activities. On 13 January 2011 it adopted Resolution 65/44 on the "Prevention of an Arms Race in Outer Space". In the resolution, the GA reaffirmed the importance of transparency and confidence building measures (TCBM) to avoid an arms race in space, stating that relying on legal measures alone would be inefficient. Indeed, TCBMs were seen as an integral part of broader agreements on the prevention of such an arms race. The GA did in fact recall that the existing legal framework for outer space does not guarantee the prevention of an arms race and asked states, especially the major space faring nations, to negotiate further. The Conference on Disarmament (CD) was seen as the relevant multilateral disarmament forum. The Resolution also called for establishing an Ad Hoc Committee on the Prevention of an Arms Race in Outer Space within the CD. In general, it also acknowledged the complementary nature of multilateral and bilateral efforts in this area.[29]

The GA also adopted on the same day its annual Resolution on "Transparency and Confidence-building Measures in Outer Space Activities" (65/68). This Resolution too was nearly identical to the one tabled in 2009, noting that an arms race in space would constitute a significant danger to peace and security and inviting UN Member States to continue submitting proposals on TCBM to the Secretary General.[30] Next, the chain of annually adopted UNGA Resolutions on space activities was completed by Resolution 65/97 on "International Cooperation in the Peaceful Uses of Outer Space", adopted on 20 January 2011. As in the previous year, the resolution referred to all the central aspects and challenges of the peaceful use of outer space and recalled the crucial importance of international cooperation to tackle all relevant issues. Furthermore, it provided an overview of the necessary steps that should be taken in this regard, including conferences, sessions of relevant entities and noteworthy progress in implementing corresponding programmes.[31] The fourth resolution, "International Day of Human Space Flight", Resolution 65/271 adopted on 21 June 2011, sets 12 April as the International Day of Human Space Flight to celebrate each year at the international level the beginning of the space era for mankind.[32]

---

[29] UN GA Resolution. 65/44. Prevention of an arms race in outer space. Available at: <http://www.un.org/ga/search/view_doc.asp?symbol=A/RES/65/44>.

[30] UN GA Resolution. 65/68. Transparency and confidence-building measures in outer space activities. Available at: <http://www.un.org/ga/search/view_doc.asp?symbol=A/RES/65/68>.

[31] UN GA Resolution. 65/97. International cooperation in the peaceful uses of outer space. Available at: <http://www.un.org/ga/search/view_doc.asp?symbol=A/RES/65/97>.

[32] UN GA Resolution. 65/271. International Day of Human Space Flight. Available at: <http://www.un.org/ga/search/view_doc.asp?symbol=A/RES/65/271>.

1.2.1.2 UNGA Committees

UNGA has established several committees that are involved in space policy and associated matters. Some of them are discussed here.

Disarmament and International Security Committee

The resolutions on the prevention of an arms race in outer space and on transparency and confidence-building measures had been introduced in the Disarmament and International Security Committee, also referred to as the First Committee, beforehand. The debates were marked by enduring differences between the U.S. on the one hand and Russia and China on the other.

Committee on the Peaceful Uses of Outer Space (COPUOS)

UNCOPUOS activities during the reporting period included the Committee's plenary session on 9–18 June 2010 and the annual sessions of its technical and legal subcommittees, on 6–17 February 2011 and 28 March-8 April 2011 respectively. At the scientific and technical subcommittee, discussions continued on a broad range of space related issues, including the use of nuclear power sources in outer space, possible dangers emanating from near-Earth objects, the space debris problem and the challenges it poses to the long-term sustainability of outer space activities, space-based applications useful for disaster management support, recent developments in global navigation satellite systems, as well as space safety issues. The Subcommittee received and considered information provided by the Member States on their activities in all these fields. The proceedings of the legal subcommittee were dominated by the issue of outer space delimitation, on which several national points of view were presented, leading to a fruitful exchange of views. Finally, during its plenary session the committee endorsed the findings and recommendations of its subcommittees and proposed that discussions on the promotion of peaceful uses of outer space should continue in the following year. Other key planks of the proceedings included the promotion of regional and interregional cooperation in space activities (especially in the SE Asia and Pacific region, reflected in APSCO and APRSAF activities), improvement of the international legal framework pertaining to space activities, and equal and fair access to benefits from space by all countries regardless of their technological level and in a manner respectful of their sovereignty.[33]

---

[33] UNGA. Report of the Committee on the Peaceful Uses of Outer Space - Fifty-third session (9-18 June 2010). Available at: <http://www.oosa.unvienna.org/pdf/gadocs/A_65_20E.pdf>.

1.2.1.3 Other UN Bodies and Organs Monitoring Outer Space Activities

Beyond the UN General Assembly and its Committees, there are other UN bodies, programmes and organs related to space activities. In the following, the ITU (a specialised agency of the UN), UN-SPIDER; the UN Programme on Space Applications, the International Committee on Global Navigation Satellite Systems (ICG), the United Nations Spatial Data Infrastructure (UNSDI), the Conference on Disarmament (CD) and UNIDIR are discussed.

International Telecommunication Union (ITU)

The International Telecommunication Union (ITU) held its annual Council on 2 October 2010 in Guadalajara, Mexico. In addition, it held its annual World Radiocommunication Seminar (WRS) in Geneva on 6–10 December 2010. During its meetings, a significant number of space related issues were advanced, including radio spectrum allocation and orbital slots procedures, emergency communications standardisation, and the coordination of satellite networks. As in previous years, the meeting provided a forum to exchange views on the associated technical, procedural and operational aspects of satellite communications, and particularly technical issues related tp technological advancements, such as the rapid growth of space and terrestrial based digital television services worldwide. The next World Radiocommunication Conference is scheduled in Geneva on 11–21 October 2011.

UN-SPIDER

Several workshops and regional meetings were organised in the framework of the United Nations Platform for Space-based Information for Disaster Management and Emergency Response (UN-SPIDER). This platform was set up by UNGA in 2006 with the aim of providing universal access to all types of space-based information and services relevant to disaster management support. In this context, it should be noted that the International Charter on Space and Major Disasters was activated several times by the Office of Outer Space Affairs (OOSA) at the request of other UN entities. Key geopolitical events that activated the charter included the Libyan crisis and the Japan earthquake, which were discussed in detail above.

UN Programme on Space Applications (SAP)

The UN Programme on Space Applications (SAP) is concerned with cooperation in space science and technology. Several activities were carried out under its auspices in the reporting period which dealt with, for instance: Technology Contribution to Infection Surveillance and to the Health-related MDG Goals; Integrated Space

Technologies and Space-based information for Analysis and Prediction of Climate Change; Integrated Applications of Global Navigation Satellite Systems; and Integrated Space Technology Applications for Socioeconomic Benefits.

International Committee on Global Navigation Satellite Systems (ICG)

The aim of the International Committee on Global Navigation Satellite Systems (ICG) is to promote cooperation in matters of satellite navigation. OOSA serves as the Executive Secretariat of the ICG and the associated Providers' Forum. The fifth meeting of the ICG took place in Turin, Italy, on 18–22 October 2010. It was co-organised by the European Commission and Italy, and it attracted participants from industry, government, non-governmental officials and academia to review and discuss developments in global navigation systems.

The ICG work plan was organised into four working groups: on compatibility and interoperability, on enhancement of performance of GNSS services, on information dissemination and capacity building, and on reference frames, timing and applications. The issues of interoperability and interference detection and mitigation were the key topics discussed during the committee's sessions, while further progress was made in the organisation of the committee's work, and especially on the establishment of its information centres worldwide. The next ICG meeting is scheduled to take place in Tokyo, Japan.

United Nations Spatial Data Infrastructure (UNSDI)

The United Nations Geographic Information Working Group (UNGIWG) held its eleventh annual meeting in Geneva, Switzerland, on 14–18 March 2010. The UNSDI is understood as a comprehensive, decentralised geospatial information network to facilitate decision-making.

Conference on Disarmament (CD)

The Conference on Disarmament (CD) is the key multilateral disarmament and arms control negotiating forum within the international community. The stalemate in its work regarding space security has been ongoing. In the course of the 2010 and 2011 sessions, the prevention of an arms race in outer space was again a central topic.

United Nations Institute for Disarmament Research (UNIDIR)

Several projects of the United Nations Institute for Disarmament Research (UNIDIR) deal with space security, directly or indirectly. Among other things,

UNIDIR intends to review former proposals and to propose new options for breaking the deadlock in space weaponisation matters at the Conference on Disarmament (CD).

1.2.2 The Group on Earth Observation

The Group of Earth Observation (GEO) is a voluntary partnership of governments and international organisations whose task is to coordinate efforts to build a system of systems, GEOSS. In 2010 and 2011 GEO had a fruitful year focused on improving its governance mechanisms and operational capabilities. The key events of the period in question as well as all recent developments are presented in the relevant chapter of the second part of this Yearbook.

1.2.3 Europe

1.2.3.1 European Space Agency

In 2010 and 2011, the European Space Agency continued to cope with the effects of the financial turmoil in Europe, while maintaining the pace of progress towards the completion of Galileo and GMES, the two key European satellite programmes it is developing on behalf of the European Union. The resilience of the Agency's programmes in the face of difficult financial conditions was reaffirmed on June 8 by ESA Director General Jean-Jacques Dordain. In his statement, he said that despite the growing debt crisis in the ESA's 18 member states, he expected to maintain all programmes previously agreed. Notwithstanding ESA's governments' approval in 2008 of €10 billion in spending for the coming years, many doubts emerged, such as the ability of European nations to service their debt without major spending cuts. Spain, Italy and Britain were facing severe government budget cuts, to mention a few. In order to keep its contributions constant, Spain authorised the ESA to take out loans on its behalf should this be necessary, while Italy, reaffirmed its commitment to the ExoMars exploration mission whilst being asked to contribute more to maintaining ISS operations.[34]

At the same time, ESA announced the third 4-year term for Jean-Jacques Dordain as Director General, extending his tenure to 2015. Dordain identified several challenges for the agency. Among these were the development of the

[34] De Selding, Peter B. "Dordain: Even With Gov't Cuts, ESA Programs Should Maintain Funding." Space News, 14 June 2010: 5.

agency's cooperation with the European Commission and the future role of ESA in the European institutional architecture.[35]

Another issue that drew attention during the reporting period was the distribution and access policy that would govern the use of Earth observation data from the future GMES EO satellite constellation. The idea of offering free and open access to almost all data from the Sentinel spacecraft was not shared by the private sector whose business models could be jeopardised by the presence of free-access products. As an indication of the private sector's reaction, the comments of Pedro Duque, managing director of Deimos Imaging of Spain, during the 2010 Living Planet Symposium in Norway, can be noted. In his remarks, he claimed that open access to GMES products would compromise the prospects of a competitive and commercially viable European Earth observation data services industry. Apart from this company, also Spot Image of France, Infoterra of Britain and Germany and e-Geos in Italy, could see their business plans affected. Deimos launched in July 2009 its own small Earth observation satellite (Deimos 1) that is integrated in the Disaster Monitoring Constellation. This satellite aims to provide services to the agricultural market, one of the areas of activity where GMES products could be distributed freely. ESA's mission manager for the Soil Moisture and Ocean Salinity satellite confirmed during the same event the notable demand growth for such data services over recent years, which indicate the dynamics of this commercial market in Europe. The idea of at least dropping access fees for basic GMES service was championed by the Group on Earth Observation. At that time, the European Commission had not yet delivered its final decision on data-access.[36]

Increased competition from U.S. commercial Earth observation services companies was also mentioned by industry officials as a threat to the creation of a viable EO commercial market in Europe. Indeed, industry officials noted the major indirect civil and defence public funding enjoyed by their competitors through the grant of long-term guaranteed services contracts by U.S. government authorities. Of particular concern was the new long-term Enhanced View contracting vehicle adopted by the U.S. National Geospatial- Intelligence Agency (NGA) to ensure the availability of high-resolution imaging data and products for military and government use. Another perceived threat to European competitiveness were ideas circulated within the E.U. to lower launch costs of the Galileo satellite navigation system and other E.U. space programmes by opening commercial launch contract bids to international suppliers, as required by standard E.U. open market rules. In response to such views, numerous European Commission and industry officials argued that space activities are not in fact conducted in a true

[35] "ESA Chief Dordain Elected to Third Term." Space News, 21 June 2010: 3.

[36] "ESA Data Access Policy Draws Mixed Reviews." Space News 12 July 2010: 12.

commercial context and that industry policy considerations should temper E.U. competition rules, at least in the absence of reciprocal measures from third countries.[37]

Notwithstanding this, satellite data access has been debated for many years and has been discussed in the context of the GMES programme. The ESA and the European Parliament have endorsed the idea of free access to that data, at least for imagery with a ground resolution under 10 m, but strong objections from industry officials have persisted. Others have maintained that free access would in fact stimulate the broader use of Earth observation data, ultimately creating a larger market segment for private companies offering processed EO products and added value services. As an example, proponents of this point of view have pointed to the example of the U.S. satellite Landsat that saw downloads of its images surge after they became freely available to the scientific community in 2008.[38]

In the field of access to space, Arianespace was set to gain a capital injection from ESA in early 2011. The company announced that it was taking measures needed to counter competition from lower-cost launch providers and to help defray the extra burden of operating two new launch vehicles, the European version of the Soyuz rocket and the new Vega light launcher. At the time of the announcement, twelve launches were planned for 2011: six of the Ariane 5 ECA rocket, five of the Soyuz and the inaugural Vega mission.[39]

1.2.3.2 European Union

During the reporting period, the European Union continued the development of its flagship Galileo GNSS and GMES space programmes. Developments in Galileo are presented in a separate section. Regarding GMES, the ESA and the European Commission programme managers met in June 2010 to discuss the short-term problems in the programme that had already cost ESA and the European Commission over €2.2 billion. Budgets showed an over €500 million gap that required immediate attention. Apart from funding issues related to the programme's deployment, additional resources needed to be found to stimulate services, augment existing ground-based facilities and pay for the system's early operations. Total annual operating and maintenance costs were estimated at that time at €600 million. ESA had funded 2/3 of the programme and the remaining 1/3 was contributed by the European Commission. Finally, another issue that needed to be resolved was the system's interoperability with other regional and global Earth observation

[37] Taverna, Michael A. "Industry Says EU Open Competition Rules Should Not Apply To Space" Aviation Week & Space Technology 1 Nov 2010: 36.

[38] De Selding, Peter B. "European Officials Embrace Open Data Policy for GMES Satellites." Space News, 5 July 2010: 5.

[39] Taverna, Michael A. "New Ariane Financing Likely To Be Decided In March." Aviation Week & Space Technology 10 Jan 2011: 37.

programmes, especially in terms of ground segment and final product compatibility.[40]

In the same context, the Head of the GMES Space office at ESA, Mr. Aschbacher, offered in July 2010 a cost estimate for the programme. He affirmed that ESA governments had spent €790 million for the first segment of GMES satellites, while the second segment's final budget would amount to €830 million. Of these costs, a total of €630 million had already been paid for through FP7 funding (the European Commission's Seventh Framework Programme for Research), while the programme's total budget was then expected to reach €2.2 billion. Aschbacher also predicted that the European Commission would need to find €560 million of additional funding for the GMES for 2011–2014 in view of the Agency's eventual scale down of funding for the programme, after the completion of the space component's development phase. In fact, ESA's role in GMES after that point (in 2014) was expected to be limited to research and development activities related to the constellation's operational use and performance that should average no more that €170 million annually.[41]

At the same time, other European institutions were elaborating plans for GMES's operational utilisation. The European Environment Agency (EEA), for example, was calculating the cost of designing, installing and maintaining a network of ground-based sensors to work as part of GMES Earth observation system. For this purpose, EEA had created a separate office within its administrative structure to coordinate its involvement in the programme, which consists in ensuring that GMES data are collected, archived and distributed in an easily accessible and practical manner, incorporating not only space based imagery, but also data collected by a large number of terrestrial sensors. The amount of GMES data bandwidth that would have to be downloaded could reach as much as 27 petabytes by 2020. ESA's Earth observation strategy division had also previously acknowledged the challenge for ESA related to securing funding for maintaining relevant data on a long-term basis. In this context, the Agency also undertook long-term cooperation steps with the Canadian Space Agency.[42]

In a separate development, the European Commission, following its Seventh Framework Programme for Research and Technological Development, planned to award a total of €8 million in 2011, for the development of cubesats and other small satellites. To form an eligible consortium, each cubesat development company would have to include at least three independent European research organisations, such as universities or private non-profit organizations, from at least three European Union Member States. The proposals, which should had been submitted by 25 November 2010, would include space-based research experiments, environmental and climate change monitoring, as well as biological and microgravity research.

[40] De Selding, Peter B. "European Earth Observation program Faces Big Funding Shortfall." Space News, 5 July 2010: 5.

[41] "GMES Chief Tallies Tab for Earth-monitoring System." Space News, 5 July 2010: 15.

[42] "GMES Ground Component Poses Its Own Challenges". Space News 12 July 2010: 12.

The cubesat initiative was part of a broader €99 million European Commission space research plan.[43]

On 22 September 2010, a cooperation agreement was signed between Norway and the European Union allowing the former to cooperate in the Galileo System by providing hardware equipment and facilities for two ground stations, one on Svalband Island in the Arctic and another at the Norwegian Antarctic base, coupled with €70 million of funding.[44]

Finally, in December 2010 European leaders discussed the possibility of a partnership with the U.S. to build a global space-based maritime surveillance network. This network, known as C-Sigma, would allow the early detection and identification of vessels engaged in piracy, contraband, illegal immigration and other sea-based threats. C-Sigma would in fact involve the development of the necessary IT tools and compatible ground equipment to combine data from the already existing automatic identification system (AIS) receivers onboard commercial vessels with imagery from commercial and government Earth observation satellites. The result was expected to be a highly integrated near real-time operational picture of areas most affected by piracy and other illegal maritime activities, offering useful intelligence and operational decision-making support to local and international authorities on the ground.[45]

1.2.3.3 EUMETSAT

One of the key developments regarding EUMETSAT programmes was the progress made in the development of the organisation's next generation geo-stationary weather satellites. In June 2010, ESA was waiting for final approval to build 6 meteorological satellites, with a $1.5 billion contract. The Meteosat Third Generation (MTG) programme industrial leadership was then hanging in the balance, with serious problems emerging from disagreements on the satellite's industrial manufacturing shares between Germany and France. After long-lasting acrimony that started in 2008 a compromise was finally reached in mid-2010. According to this compromise, Thales Alenia and OHB Technology of Bremen, would be the prime contractor with Astrium added as a subcontractor Their proposal was selected on the grounds of its more competitive price that foresaw a €1.18 billion deal. The addition of Astrium to the programme resulted in a €60 million increase in the contract's final price, but still within the initial budget.[46]

---

[43] "Europe Solicits Ideas for Cubesat Launch Options". Space News 9 Aug. 2010: 8.

[44] "Norway Pledges 70 M Euros As It joins Galileo Programme". Space News 27 Sept. 2011: 8.

[45] Taverna, Michael A. "Network Would Combine AIS And Imaging Satellite Capabilities." Aviation Week & Space Technology 20 Dec 2010: 86.

[46] De Selding, Peter B. "European Weather Satellite Contract Dispute Appears Resolved." Space News, 28 June 2010: 5.

1.2.3.4 National Governments

France

The French aerospace research agency (Onera) considered that the time is right for a big leap forward for air transportation. Its belief is that projecting 40 years away is the right timeframe to consider next generation aircraft and propulsion systems, air traffic management, energy resources and environmental constraints. This position was clearly stated in the "Research Paths for a Viable Air Transport Industry in 2050", in which Onera offered useful insights, saying that we should immediately present fresh conceptual and technological options for a 40 year time line.[47]

The French National Accounting Office had doubts on the vast French plan to sell off assets and outsource services to help offset a drop in military procurement budgets. The Office is analyzing the proposal to sell and lease back the Syracuse 3 secure satellite communications system and raised some questions, for example, as to whether the Syracuse sale is strictly intended to improve operating efficiencies or is a purely financial mechanism to escape budget restrictions.[48]

A French solar-weather spacecraft, together with two Swedish formation-flying satellites, was placed into low Earth orbit, on June 15, by a Russian-Ukrainian Dnepr rocket. This launch was delayed several times because of a dispute between Russia and Kazakhstan over the issue of rocket debris falling on Kazakh territory. France's Picard satellite was developed by the French space agency, CNES, and will operate from a 725 km orbit for at least 2 years, in order to study solar activity. The two Swedish Prisma satellites, Mango and Tango, were released together and remained a single unit until August 3, when they were separated and started working and testing formation flying. Mango weighs 140 kg and will spend 10 months manoeuvring around the 40 kg Tango. The goal is to test technologies for future, larger missions in Europe that will require precise positioning of satellite groups. CNES also had a role in the Prisma mission, proving the formation flying in-orbit ranging demonstration subsystem.[49]

The $325 million generated by the French government bond issue initially directed for satellite broadband technologies and a Ka-band satellite, could instead be invested in locally managed subsidies for consumer broadband terminals. France's State Secretary for development of the digital economy said that the government had the intention to spend that amount of money for broadband development. No mention was made of the Megasat advanced Ka-band broadband project. A governmental decision was expected after consultation with the

[47] Sparaco, Pierre. "Onera Offers 2050 Technology Road Map." Aviation Week & Space Technology 22 Nov 2010: 41.

[48] Morring, Frank Jr. "French Watchdog Agency Criticizes Syracuse 3 Sell-off." Aviation Week & Space Technology 14 Feb 2011: 22.

[49] "Dnepr Rocket Lofts Trio Of European Satellites." Space News, 21 June 2010: 9.

Parliament, in order to select what kinds of broadband technologies and capabilities would be targeted for the investment.[50]

Finally, the French space agency, CNES, was expecting to award contracts valued at more than $900 million to develop new next-generation rockets, two Earth-observation satellites and satellite-delivered broadband. The funds were allocated to three different project categories. One third of the money was allocated to finance the demonstration and early design work on a next-generation launch vehicle, to replace today's heavy-lift Ariane5. A second category funded two environmental satellites. The first satellite to be built in partnership with Germany, dubbed Merlin, is a methane-measuring spacecraft. The second satellite – Surface Water Ocean Topography – SWOT – is a joint French-U.S. mission, in which CNES is going to invest €170 million. The third category was to extend broadband access to regions of France.[51]

Germany

The most important development in the field of German space policy in 2010 and 2011 was the announcement of the country's new space strategy. The 30 page document offers a detailed analysis of the strategic environment in which German space activities are expected to evolve in the next decade, describing both the general strategic orientations and the specific objectives that the new policy would have to achieve in order to meet them. As far as the first are concerned, the new strategy adopts a utilitarian approach aimed at highlighting the specific benefits that space activities can achieve for society, insists on the principle of sustainability, maintaining a long term view on the development of space infrastructure, and adheres to the importance and further development of international cooperation in space.[52] As concerns the second, the new strategy describes in detail 10 key policy objectives that need to be addressed in the medium term. These include maintaining and expanding Germany's scientific expertise and industrial base related to space technologies and ensuring the continued growth and excellence of the country's space sector; creating a unified legal instrument for space activities and encouraging further European integration in this respect; incorporating space applications into the country's overall security strategy; shaping the distribution of roles in the European space sector; ensuring European independence in space technologies; maintaining European flight capabilities and considering an unmanned mission to the Moon; and finally ensuring the sustainability of German space activities through promoting international cooperation. In conclusion, the new German strategy is a

[50] "France Rethinks Plans for Megasat Broadband Satellite". Space News 2 Aug. 2010: 3.

[51] "French Bond Issue To Fund Rocket, Satellite Projects." Space News, 7 June 2010: 3.

[52] Federal Ministry of Economics and Technology. Making Germany's Space Sector Fit for the Future: The Space Strategy of the German Federal Government. Berlin: BMWI. November 2010: 9.

balanced document that seeks to uphold the country's leading position in European space activities, both from an institutional as well as industrial view point. A more detailed analysis of the new policy and its implications for the country's space activities can also be found in the second part of this Yearbook.

On May 26, the U.S. space agency's telescope – SOFIA – became operational. The first in-flight photos were captured, namely multicoloured infrared images of Jupiter and the galaxy M82. SOFIA is an extensively modified Boeing 747P jet aircraft with a 2.5 m infrared reflecting telescope and is expected to provide completely new astronomical science observations and great observatory-class astronomical science. NASA's Dryden Flight Research Centre, a joint venture of NASA and the German Aerospace Centre, is overseeing the telescope's operations. The control is done by researchers flying along in a separate cabin inside the SOFIA aircraft, for a period of 8 h, flying at 800 km per hour. The first infrared images of Jupiter and M82 were revealed. In a Jupiter snapshot the gas giant planet's internal heat bleeding through holes in its thick bands of clouds was visible. The infrared look at the M82 galaxy allowed astronomers to peer through the galaxy's interstellar dust clouds and spot several starburst knots, where infant stars are forming by the tens of thousands. The 10 scientist inaugural crew included researchers from NASA, the Universities Space Research Association (USRA), the German SOFIA Institute (DSI) and Cornell University.[53]

After 20 years of absence, the German Government was considering the development of a new series of communications satellites to cover both public and private demand. This decision would contest French dominance in manufacturing large telecommunications satellites. This development project would be aimed at creating the necessary technological spin-offs that might trigger a larger presence of German communication satellites manufacturers on a global scale, especially in view of increasing competition from emerging industrial competitors like China and India. Relevant ESA programmes, such as ARTES, could also be exploited in this direction. In addition to this, German authorities increased funding for the development of national satcom technology demonstrators, such as the Heinrich Hertz spacecraft.[54]

At the same time, the country was aiming to expand its considerable expertise in Earth observation satellites into the manufacturing of spacecrafts equipped with optical observation instruments as well. As part this strategy, German authorities were looking for commercial and government partnerships to develop such high-resolution optical Earth observation satellites. During the ILA 2010 Berlin Air show, DLR officials announced that such spacecrafts were in the agency's work plan. DLR's executive Chairman, Johann-Dietrich Woerner confirmed that DLR and Astrium GmbH of Ottobrunn co-financed work on the KompSat-3 optical EO satellite for South Korea as a first step in this process. The satellite, which at that time was set for a late 2011 launch, was designed to provide 70 m. ground

[53] "SOFIA Airborne Telescope Returns First Scientific Images" Space News, 7 June 2010: 9.

[54] "Germany Seeks Niche in Telecom Satellite Market." Space News, 14 June 2010: 12.

resolution. According to the same official, work on the development of a high-resolution spacecraft would continue based on this model. The relevant project, known as Hi-Ros would offer Germany the possibility to integrate optical and RADAR imagery products within the same commercial platform, consequently improving the commercial appeal of domestic satellite data products and market penetration of the country's EO industry. At the same time, Mr. Woerner affirmed that the Hi-ROS system would only have civil and commercial applications, indirectly reaffirming Germany's commitment to European multinational cooperation programmes for dual-use and defence related EO satellite development programmes.[55]

In a related development, the ComsatBw-2 dedicated military communications satellite was declared operational on July 9, 2010, and its control was passed over to the German military. ComsatBw-2 carries four SHF and five UHF channels, was designed to operate for 15 years and its architecture is nearly identical to the already in service Comsat Bw-1. ND Satcom of Germany was responsible for the development of its ground segment and Astrium for the space segment, although final integration was performed by Thales Alenia Space.[56]

Also during the Air Berlin Show, NASA Deputy Administrator Lori Garver and DLR executive chairman Johann- Dietrich Woerner, signed an agreement to extend the GRACE satellite mission for 2 years. The Gravity Recovery and Climate Experiment (GRACE) satellites operate together in low Earth Orbit, with a spatial separation of 220 km. They measure small changes in the gravity field caused by mass changes in different areas of the Earth. DLR's share of GRACE operating costs is around €1.75 million, including data download and distribution.[57]

Finally, research on in-orbit servicing technologies in order to extend satellites' operational life and develop orbital manoeuvring know-how continued to be a key plank of the country's national space programme in 2010 and 2011. Relevant work was continued by DLR on the study of in orbit refuelling, orbital rendezvous and docking techniques.[58]

Italy

During the reporting period, Italy was another European country that announced a new strategic plan for space, extending its space programme funding and visibility for another 10 years. The new strategic plan improved ASI's ability to reinforce existing initiatives and established new priorities. Among these priorities are: expanding Italy's radar Earth-observation know-how as well as science missions, and developing a new liquid upper stage and improved solid propulsion motors for

[55] "German Government Eyes High-Resolution Satellites." Space News, 14 June 2010: 13.
[56] "German Comsat Bw-2 Declared Operational". Space News 19 July 2010: 9.
[57] "United States and Germany Agree to Extend GRACE Mission." Space News, 14 June 2010: 13.
[58] "Germany Still Committed to In-space Satellite Servicing". Space News 26 July 2010: 12.

the Vega light launcher. Another important priority is the reinforcement of ties with other space agencies. The strategy document also discussed the country's involvement in extending the operational life of the International Space Station (ISS) and the new opportunities for flying Italian instruments on NASA scientific missions.[59] At the same time, a new bilateral agreement between Italy and the U.K., centred on cooperation in space related R&D, was under consideration.[60]

In addition to public initiatives, private industry and public-private financing schemes were also expected to be central in implementing the new Italian space strategy, while allowing for priority to national missions. Among the key national projects explicitly mentioned in the document are the creation of a domestic high-speed broadband satellite system, the expansion of remote sensing capability into the hyper spectral optical domain and the broadening of the country's know-how in space launcher development beyond the field of rocket motors and boosters and into integrated space launch systems of a greater size than Vega.[61]

The United Kingdom

Like Italy, the U.K. focused its attention on inter-agency and industry collaboration for the new space strategy. One first step was to conclude a new bilateral agreement with Russia, focused on commercial aspects, such as reducing Russian import duties and promoting the use of Britain's small satellite technology. The U.K.'s top priorities included examining how satellites could help improve broadband access in rural areas and, in particular, seeing if it is in need of its own operational Earth-observation network. The initiative is supported by a $7.6 million Earth-observation clearinghouse facility.[62]

To this effect, a consortium led by Astrium was selected by the British government to create an Earth observation satellite operations and data processing centre in Britain – EO Hub. It was scheduled for March 2011 and cost $7.5 million. ESA said that building a facility in Britain could stimulate space investment in the country, which could reflect on future British contributions to the agency. The EO Hub was installed at the International Space Innovation Centre as part of an effort to make the Centre a link between regional space capabilities. On July 20, Astrium announced that its Infoterra division bought download and processed data

[59] Italian Space Agency. Strategic Vision 2010–2020. Rome: ASI, 2010.

[60] Nativi, Andy and Taverna, Michael A. "Rome, London Seek ways To Compensate For Limited Space Budgets." Aviation Week & Space Technology 26 Jul 2010: 44.

[61] Taverna, Michael A. "Italy's Plan To Favor National Missions Over ESA." Aviation Week & Space Technology 3 Jan 2011: 28.

[62] Nativi, Andy and Taverna, Michael A. "Rome, London Seek ways To Compensate For Limited Space Budgets." Aviation Week & Space Technology 26 Jul 2010: 44.

from the ERS-2 and Envisat radar Earth observation satellite for $9.7 million. Under this contract, Infoterra is responsible for managing downloading and processing images from ESA's Swarm satellites.[63]

1.2.4 The United States

One of the key policy issues where reform was underway in 2010 and 2011 was the issue of export control regulations. Export control is expected to have a direct effect on commercial space activities, since most related hardware is affected. Indeed, shortly after the new Space Policy's announcement, the White House disclosed that it aimed to create an independent agency responsible for licensing exports of military and dual-use technologies, or a "single licensing agency". The agency is to be governed by a cabinet-level board of directors reporting to the President. U.S. National Security Advisor, James L. Jones did not give a date for the creation of the agency, but it will be part of the third and final phase of the Administration's effort to reform the export control regime. The Administration also expressed its intention to create a tiered munitions list in order to allow the U.S government to prioritize export controls and process license application in a more efficient manner.[64]

U.S. manufacturers depend on the domestic market and figures confirm the tendency towards lifting export controls as internal demand may vary significantly from year to year.[65] In 2009 70.6% of the sales of the U.S. aerospace sector were to the domestics market (NASA and other agencies with the Department of Defence orders), in 2010 it was 72%. The increase of general sales revenues from the aerospace industry was small between 2009 and 2010 (about 0.9%), while domestic market orders rose by 5.35% in 2010. These figures indicate that the domestic market plays a crucial role in the U.S. aerospace industry which is not of course new, but shows clearly that public expenditures must rise in an almost mandatory and increasing pace to cover the weakness of the market and assure sales growth. Employment figures support strongly this observation because they show a significant and constant decrease of the number of employees.[66] The U.S. space industry is therefore at a turning point since the recent political decisions. It will be strongly affected by the decision to privatize a part of the NASA's activity and the will to reform the set of U.S. export controls.[67]

---

[63] "Astrium-led Group to Build Satellite Center in Britain". Space News 26 July 2010: 8.

[64] "White House Calls for Export Control Agency." Space News, 5 July 2010: 3.

[65] Aerospace Industry Association. "Aerospace Industry Sales by Customer 1997–2011." 2011 http://www.aia-aerospace.org/economics/year_end_review_and_forecast.

[66] Aerospace Industry Association. "Aerospace Related Employment 2002–2010." 2011 http://www.aia-aerospace.org/economics/year_end_review_and_forecast.

[67] Antonie, Boessenkool. "DoD: U.S. Space Industry May Lose Edge." Defense News 25 May 2010 http://www.defensenews.com/story.php?i=4641686.

Another issue that stood out in 2010 and 2011 was the continuing advance of outsourcing public space services to the private sector, including defence related ones. This issue generated considerable friction between the executive and legislative branches of the U.S. government, particularly due to the short term job losses in the public sector it could entail. An example is a provision of the Senate Defence bill that would restrict the kinds of commercial satellite imagery the Pentagon could purchase, which was objected to by the Obama Administration. Indeed, the Administration's Defence Authorization Bill included a provision to bar the Pentagon entry into new imagery purchase after 2010, with the exception of the 1.5 m telescopes.[68] Such a restriction could negatively affect commercial data providers, limiting innovation in commercial technologies and increasing risks in future government contracts for commercial data services. Industry officials claimed that the timing of such a development could not be worse, with contracts already awarded to DigitalGlobe of Longmont, Colo. and GeoEye of Dulles, with a total worth of $7.3 billion over 10 years, allowing them to finance the development of a new generation of 1.1 m aperture satellites for 2012.

1.2.5 Russia

In 2010 and 2011 Russian space activities continued on the same path of rapid development as in the previous year. The key element of the Russian space programme remains the rapid restoration of the country's space infrastructure across the board, to levels that would eventually rise to capabilities that existed prior to the 1990s. The principal characteristics of this effort remain its strong high level political backing and budgetary commitment on one hand, and the willingness to pursue the simultaneous development of almost all of the country's space capabilities rather than prioritising among them, on the other. Nevertheless, there are some programmes that clearly attract more funding than others, which would imply that they are at least perceived as relatively more important, or closer to operational maturity. These include the completion of the Glonass GNSS, the creation of a new spaceport on Russian territory, the renewal of the country's launcher fleet, and the further development of space related scientific research.

Indeed, the restoration of the Glonass GNSS constellation to its full operational capacity remained the Russian space programme's top priority in 2010 and 2011. Between 2006 and 2010 related expenditures witnessed a fivefold increase, to reach $925 million, or nearly one third of the country's space agency total budget. This impressive budgetary commitment led Russian to the first place among all space faring countries in GNSS spending, surpassing the U.S. for the first time. The constellation was expected to reach full operational deployment in 2010, but the loss of three spacecraft during launch in December postponed this to 2011. At

[68] "White House Objects to Imagery Provision in Bill". Space News 27 Sept. 2011: 3.

the same time, the Russian space agency is developing the new generation of GNSS satellites Glonass K, which are expected to improve the system's accuracy and resilience dramatically, bringing it to comparable levels to the GPS and Galileo constellations. Their deployment is scheduled to begin in 2013 and if they were to be launched at the same pace as their predecessors, the system could achieve full operational capability by 2016, constituting an important commercial competitor.

Finally, as the country's satellite launch programme reaches maturity, more and more funding is routed to space science and exploration programmes, allowing for the pursuit of related missions that were placed on hold during the past decade due to the more pressing space infrastructure replenishment needs.

1.2.6 Japan

2010 and 2011 were very active years for Japanese space activities, especially in the field of scientific research and innovation. For example, the aforementioned period saw the success of the Ikaros satellite mission, with the deployment on 11 June 2010 of its experimental solar sail, made from a polyamide resin thinner than a human hair, but nevertheless measuring more than 20 m in length. The satellite was launched on 21 May, with the Venus-bound Akatsuki probe aboard an H-24 rocket from Japan's Tanegashima Space Centre. The spacecraft is positioned 770 km above the Earth. The main idea, expressed by JAXA, was to acquire navigation technology through the solar sail. With this satellite they could start measuring and observing the power generating status of the thin film solar cells, accelerate the satellite by photon pressure and verify the orbit control through acceleration.[69]

At the same time however, efforts to further commercialise the service's sector in Japan also continued. For example, the Japanese government was expected to reach a decision by the end of 2010 on whether to partially privatize Japan's Advanced Land Observing Satellite (ALOS) system, which could be of importance to U.S. and European use of ALOS data. Agreements between JAXA, NASA and ESA were signed to permit free distribution of ALOS Data, which will expire in 2020. ESA has integrated ALOS into Europe's third-party mission network and financed development of an ALOS European data Node. An ALOS-2 satellite is being planned for 2013 but Japanese authorities invited industry to submit proposals to operate the satellite as a commercial undertaking. The form of this partnership could permit a private sector operator to access the data.[70]

On an institutional level, JAXA and the Asian Development Bank signed a letter of intent to improve cooperation. Under this agreement JAXA will provide technical assistance to bank-supported projects using satellite data and the ADB will

[69] "Japan's Ikaros Spacecraft Unfurls Thin-Film Solar Sail" Space News, 14 June 2010: 3.

[70] "Partial Privatization Eyed for Japan's ALOS System." Space News, 5 July 2010: 15.

promote the use of satellite technology for disaster management, climate change mitigation and adaptation, forest monitoring and water resource management.[71]

In a separate development, Japan's Space Activities Commission (SAC) produced a 33 page report which described the technical goals for a follow up Hayabusa-2 mission, recommended for launch before the end of March 2015. The purpose is to deploy a miniature rover and collect and return a sample of a C-class asteroid that could give insight into the formation of the solar system. The report also recommended the revision of JAXA's space programme, the development of the next-generation Epsilon solid-rocket that will replace the M-V. This rocket will be used for low Earth orbit. The cost was estimated around 3.8 billion yen per launch. The budget for the Hayabusa-2 was estimated around 16.4 billion yen ($190 million), not including the cost of the H-2A rocket that will launch it.[72]

At the same time, the Japanese budgetary allocation to space activities rose to a 339.7 billion yen budget, requested by Japan's ministries and government agencies, for the financial year beginning on 1 April 2011. A supplementary request of 77.7 billion yen was also in line, in order to fund programmes and initiatives that could help stimulate Japan's economy.[73] In a separate development, Banri Kaieda was nominated as Japan's State Minister in charge of space development by the Prime Minister Naoto Kan, replacing Seiji Maehara, who was appointed as Foreign Minister.[74]

Finally, Japan's IGS-Radar 2 satellite stopped sending radar data on August 23, 2010, due to an undisclosed problem in the electrical system. This problem reduced Japan's fleet of reconnaissance satellites to two operational satellites and it will be without radar coverage until the launch of the next satellite in 2011. Japan's reconnaissance satellite programme was designed to provide coverage of North Korea and East Asia.[75]

1.2.7 China

Chinese space activities continued their full expansion in scope and volume in 2010 and 2011. In August, Chinese authorities announced the completion of the first module of the space station, Tiangong-1. It weighs 8,500 kg and was expected to be launched by the first half of 2011 on a Chinese long March 2F carrier rocket. The Shenzhu 8 vehicle was expected to launch after that. China's space programme includes the space station and, eventually, plans manned spaceships to dock with Tiagong-1 and crews to the station. Tiagong-2 is scheduled to launch in 2013, to be

[71] "JAXA Agrees to Promote Satellite Use in Asia-Pacific". Space News 2 Aug. 2010: 9.
[72] "Japanese Panel to OK Hayabusa-2 Development." Space News 9 Aug. 2010: 8.
[73] "Japan's Space Budget Request Flat for 2011". Space News 6 Sept. 2011: 3.
[74] "Japanese Cabinet Reshuffle Installs New Space Minister". Space News 20 Sept. 2011: 3.
[75] "Japan's Spy Satellite Fleet Down to Two Operational Spacecraft". Space News 6 Sept. 2011: 9.

followed by the Tiangong-3 between 2014 and 2016. Once fully assembled, the Chinese space station is expected to be operational for 3 years.[76]

At the same time, China continued the exponential growth of its satellite fleet with the launch of the Yaogan 10 Earth observation satellite from the Taiyuan Satellite Launch Centre aboard the Long March 4C rocket. Its main functions will be crop monitoring, resource mapping, disaster mitigation and support to scientific missions. Its predecessor Yaogan 9 was launched in March 2010 from the Jiquan Satellite Launch Centre.[77] Finally, China's hectic launch activity during August 2010 was concluded with the launch of the Tianhui-1 Earth mapping satellite, developed by the China Aerospace Science and Technology Corp., on a Long March 2D rocket from Jiuquan Satellite Launch Centre. The satellite was scheduled to perform scientific experiments, Earth mapping and provide assistance to land resource surveys.[78]

1.2.8 India

On 25 May 2010, India's Defence Ministry released the "Technology Perspective and Capability Roadmap" prepared by The Defence Research and Development Organisation (DRDO), which confirmed the country's plans to build ASAT weapons. Indeed, the roadmap included development of anti-satellite weapons, both for electronic interference and physical destruction of satellites, as well as steps to protect the nation's own vital in-orbit assets against such attacks. It also manifested the intention to develop satellite systems for intelligence, surveillance and reconnaissance, commencing with the design and building of a microsatellite with an electronic intelligence payload. Other projects include: the delivery of highly mobile, high data-rate satellite communications; the development of imaging satellites with sub-meter resolution and robust onboard data analysis capabilities; and the creation of precision-guided weapons compatible with India's regional satellite navigation system, as well as a secure mobile network based on the Indian National Satellite System, to eliminate the country's dependence on foreign satellites for tactical and strategic communications.

With regards to space services commercialisation, on 30 June 2010 India's telecommunications regulator confirmed its support for raising the ceiling on foreign ownership of direct-to-home television broadcasters and aligning the foreign-ownership rules with those in place for telecommunications carriers. The Telecom Regulatory Authority of India (TRAI) intends to raise the percentage of foreign ownership allowed for satellite-television broadcasters as well as for mobile-television providers, up to 74%. TRAI affirmed that no differences should

[76] "1st Piece of Chinese Station Assembled for 2011 Launch". Space News 23 Aug. 2010: 8.

[77] "Chinese Earth Observer Launched by Long March". Space News 23 Aug. 2010: 9.

[78] "Long March Rocket Lofts Chinese mapping Satellite". Space News 30 Aug. 2010: 9.

be valid between telecommunications and broadcasting. Relaxing the limits for the foreign ownership was seen as a way to obtain substantial capital investment to move from analogue to digital. Households subscribing to cable on satellite television grew by 10%, to 95 million, in 2009 and direct-to-home satellite rose from 11.1 million, in December 2009, to 21.3 million in March 2010.[79]

On the other hand, reduced hardware reliability continued to plague India's commercial space sector. For example the Insat-4B telecommunications satellite, launched in March 2007, lost half its broadcast capacity following a failure of one of its solar arrays. Antrix, ISRO's commercial arm that had developed and built the satellite, was working to restore at least some of the capacity on the satellite. ISRO confirmed that it had shut down half of the Ku-band and C-band capacity because of a power supply anomaly in one of the satellite's two solar panels. Insat-4B is based on ISRO's I-3K platform, the same platform that caused the failure of Eutelsat's W2M telecommunications satellite in 2009.[80]

India's ambitious exploration programme however, would not be deterred by such setbacks. Indeed, in August seven scientific payloads were selected for the 2013 lunar mission, which will be launched aboard India's Geosynchronous Satellite Launch Vehicle. India has finalized the instruments that the 2,650 kg Chandrayaan-2 will carry. These are improved versions of instruments used in the previous mission. The chosen payloads were recommended by a committee of experts, including three completely new and two upgraded versions of payloads flown before, aboard the Chandrayaan-1. The five selected payloads were: a Large Area Soft X-ray spectrometer and Solar X-ray Monitor an L- and S-band synthetic aperture radar, an imaging Infrared Spectrometer, a Natural Mass Spectrometer, a Terrain Mapping Camera, a Laser Induced Breakdown Spectroscope and an Alpha Particle Induced X-ray Spectroscope.[81]

In a similar development, the second test of the L110 liquid core stage was successfully completed by the Indian Space Research Organisation (ISRO) at the Mahendragiri facility. The L110 was set to a 200 s test but it stopped at 150 s due to an anomaly. The L110 was designed to lift communications satellites weighing 4,500–5,000 Kg. The test flight was scheduled for 2011.[82]

Finally, the Indian space programme improved its focus on bringing the benefits of space to the common man and applying advanced technologies to the everyday problems of man and society. The world's largest national constellation of Earth remote-sensing spacecraft is used to find the best places to fish, dig wells, plant crops and to provide specialist diagnoses and treatment for locals. All of those applications are accessible at the ISRO facilities, at 275 village Resource Centres, scattered across 16 Indian States. The system also helps central government

[79] "Indian Telco Regulator OKs Foreign Capital." Space News, 5 July 2010: 3.

[80] "Power Glitch Hobbles India's Insat-4B Sat." Space News, 12 July 2010: 3.

[81] "India Finalises Payloads for Chandrayaan-2 Mission". Space News 6 Sept. 2011: 8.

[82] "India Tests GSLV-3 Liquid Core Stage". Space News 13 Sept. 2011: 16.

planners forecast nationwide crop yields. ISRO is planning to orbit its own C-band synthetic aperture radar (SAR) on a Risat-1 spacecraft, possibly before the end of the year.[83]

1.2.9 Emerging Space Powers

The South African Maritime Safety Authority demonstrated the ability to improve security during the World Cup tournament through an ongoing trial of space-based ship detection.[84]

MEASAT, a company that owns four in-orbit satellites and has started expanding into southern and central Asia, announced in 2010 that a Malaysian billionaire wanted to purchase a portion of their satellite fleet, which was valued at around $500 million. MEASAT Global Network Systems Sdn. Bhd (MGNS) offered, at the time, 4.20 Malaysian ringgits per share, for 150 million shares. This price represented a 10% premium over the price on the Kuala Lumpur stock market. Later 2010, MEASAT agreed with the government of Azerbaijan to purchase a satellite to be launched into the orbital slot of 46° east. This deal was not presented as an alternative to the MGNS offer.[85]

Astrium Satellites signed a contract with the Vietnam Academy of Science and Technology, valued at €55.2 million, to build a medium-resolution optical Earth observation satellite. The VNREDSat-1 – Vietnam Natural Resources, Environment and Disaster Monitoring Satellite – would weigh about 150 Kg at launch and is expected to be launched in 2013. Similar programmes were established between Astrium and Thailand, South Korea, Algeria, Chile and Kazakhstan.[86]

On June 4, the Arabsat Badr-5 telecommunications satellite was placed into geostationary transfer orbit by an International Launch Services (ILS) Proton, to be co-located with the Badr-4 and 6. The Arabsat Badr-5 was built in a joint venture between Astrium Satellites and Thales Alenia Space of Europe. This satellite carries 56 Ku-and four Ka-band transponders and was designed to provide 14 kW of power to its payload. This launch was the fourth of a planned seven commercial missions that ILS was planning for 2010. Khrunichev, ILS's owner and prime contractor for the Proton rocket, was producing at a rate to assure one launch per month.[87]

Nilesat chief executive Salah Hamza said, during the World Satellite Business Week conference, that at least two different sources had intentionally jammed

---

[83] Morring, Frank Jr and Mathews, Neelam. "Indian Space Delivering On Promises To Villages." Aviation Week & Space Technology 14 June 2010: 66.

[84] Morris, Jefferson. "Satellite Tracking Of Ships Demonstrated." Aviation Week & Space Technology 23 August 2010: 22.

[85] "Measat Shareholder Bids For Remainder of Company". Space News 2 Aug. 2010: 3.

[86] "Astrium to Build Imaging Satellite for Vietnam". Space News 9 Aug. 2010: 9.

[87] "ILS Proton Rocket Launches Arabsat's Badr-5 satellite." Space News, 7 June 2010: 3.

Egypt's Nilesat satellite operator, to block entry to Iran of news broadcasts from the BBC Persian television service and also the 2010 FIFA World Cup. The jamming was thought to be caused by the Iranian government. The International Telecommunications Union (ITU), a Geneva-based United Nations affiliate that regulates satellite slots and broadcast frequencies, was alerted in order to stop Iran. The claim had no results and no evidence was found against Iran.[88]

EADS Astrium was selected to supply a new satcom terminal for the United Arab Emirates' Yahsat sitcom network. Astrium will also provide operations and maintenance services to the UAE armed forces, including training and logistics support, for 5-years. Simultaneously Astrium, in partnership with Thales Alenia Space, is building two satellites and the ground system.[89]

Finally, the President of the AEB – Brazilian Space Agency – has criticized the effort to create profit-making business out of the collection of satellite Earth observation data, defending, therefore, free access to data. These remarks were made during a roundtable session with the Canadian and German Space agencies, which support the sale of data collected. The head of DLR said that the data collected from the TerraSAR-X and TanDem-X satellite is free of charge if the final purpose is science and research. On the other hand, the Canadian space agency was likely to modify its relationship with the private sector as the government was proceeding with the next-generation Radarsat Constellation mission, however long-term data storage and retrievability was not assured in the Radarsat system. This topic was also approached by ESA's director general, who underlined the importance of storage of collected Earth observation data.[90]

1.3 Worldwide Space Budgets and Revenues

The year 2010 saw the results of a steady growth of worldwide national budgets. World government expenditure totalled $71.5 billion.[91] However, the rate of increase (Compound Annual Growth Rate (CAGR)) considerably slowed down to 2% between 2009 and 2010 whereas it amounted to 9% during the period 2004–2009. In terms of commercial revenues of space activities, the Space Report 2011 reports the total revenue of commercial satellite services to have been about $102 billion comprising telecommunications, Earth observation and positioning services (which represents an increase of 9% from 2009). The revenue of space-related commercial infrastructure including manufacturing of spacecrafts and in-space platforms, launch services as well as ground equipment is estimated

[88] "TV Broadcasts Bound for Iran Intentionally Jammed". Space News 13 Sept. 2011: 14.

[89] Morring, Frank Jr. "UAE, Canada Select Astrium Satcom Terminal." Aviation Week & Space Technology 28 Feb 2011: 18.

[90] "Brazil's Space Chief Slams Earth Science Privatization". Space News 26 July 2010: 12.

[91] Figures in this section are based on Euroconsult data.

to have reached around $87.39 billion (a decrease of 8% compared to 2009 due to a lack of launch capacity). In conclusion, the commercial space revenues of 2010 can be summed up at $189.39 billion.[92]

1.3.1 Overview of Institutional Space Budgets

The total institutional spending on space in 2010 can be estimated to be approximately $71.5 billion, a figure that shows a nominal increase of 9% compared to 2009.[93] This space spending is comprised of $37 billion in civil expenditure (52% of the total) and $34 billion in defence expenditures (48%). Consequently, the balance between civilian and military spending remained nearly the same compared to the previous year's figures. Nevertheless, overall budget levels could have even decreased for the first time in 15 years if they had not been supported by exceptional growth in SatCom expenditure (+49%, or $2.8 billion), mostly driven by demand from the U.S Department of Defence.

Out of the estimated $34 billion of defence related space expenditures worldwide, $28 billion were spent by the United States, representing a share of 82% which was a minor percentage decrease compared to the year before. These funds came from the Department of Defence (DoD), the National Reconnaissance Office (NRO), the National Geospatial-Intelligence Agency (NGA) and other government entities. It should be borne in mind that not all relevant funding is made public, resulting in a degree of uncertainty regarding the exact figures of expenditure on defence space activities. Relevant U.S. activities are therefore a major support to worldwide activity, particularly in the defence area.

State ranking with regard to expenditure on space activities related remained quite similar to 2009. Unsurprisingly, the U.S. had the largest budget including civil ($20.3 billion) and defence ($20.3 billion) expenditure (Fig. 1.1). Russia came second, and it should be noted that the assessment of the country's budget should be considered only as a low estimate of its real expenditure, its intensive military activity and regular classified launches, in a fashion similar to the U.S. Other noticeable developments in 2010 included the rise of China ($2.4 billion) that came even closer to the fourth biggest spender France ($2.5 billion) than in 2009, while India overtook Italy, fielding the seventh biggest budget and approaching the level of German expenditure. It is worth noticing that the second actor is not a state but an international organization namely the European Space Agency with a budget of ($5.32 billion[94]) managing the joint investment of 18 Member States; the five most important contributors being France 18.2%, Germany 16.7; Italy 9.9; UK 6.8;

---

[92] The Space Foundation. The Space Report 2010. The Space Foundation: Colorado Springs, 2010: 30.

[93] Figures in this section are based on Euroconsult data.

[94] Source ESA figures, exchange rate 1 EUR = 1,42158 USD.

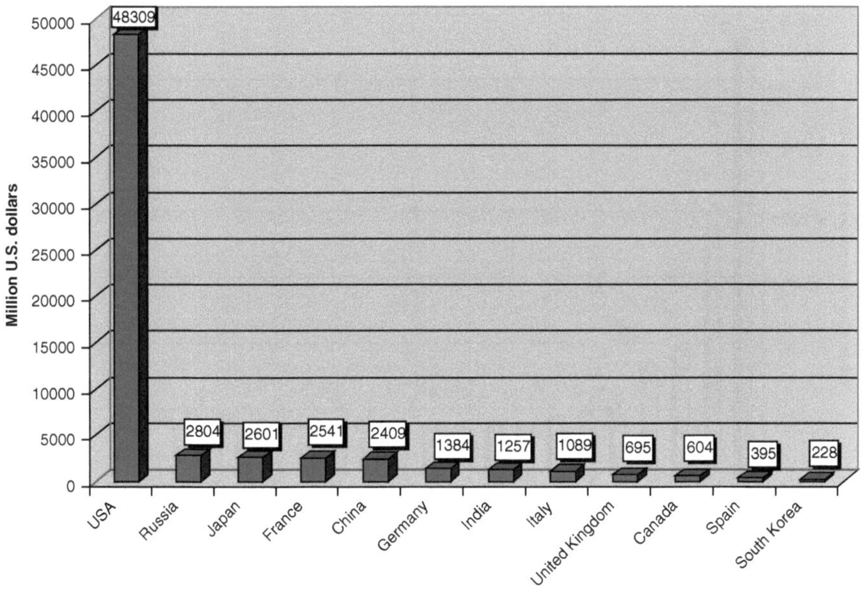

Fig. 1.1 Public space budgets of major space powers in 2010 (Based on Euroconsult data)

Spain 5.2. Japan's budget has sharply decreased compared to 2009. It is therefore noticeable that compared to 2009 the U.S., European countries, ESA and Russia spend some 82% of the world budget allocated to space activities.

In order to measure the concrete effort of a country in the space sector it is necessary to put these figures into perspective with regards to GDP[95] (Fig. 1.2). Indeed, consulting the absolute numbers alone only tells one side of the story, as comparisons between countries with different economic conditions like prices or wage levels can be misleading. The U.S.'s space budget confirms its strong engagement in the space effort but we notice a stabilisation or even a slight decrease. The information concerning Russia as previously explained must be taken with reserve, it shows a slight fall which is due to be counterbalanced by recent statements from the Russian government to increase Russia's activity in this field.[96] Following is France that stayed stable (0.1%) and India, which significantly increased its effort (0.09%). Most of the space leading countries in Europe spent approximately between 0.05 and 0.03 of their GDP on space activities.

Another perspective is furnished by the space budget per capita (Fig. 1.3) which is still largely dominated by the U.S. ($155.7), then by France $40.4. The budget per capita of Luxembourg $39.8 and Belgium $21.4 are mainly due to their strong

---

[95] The data used is the nominal GDP converted to current U.S. dollars using the official exchange rates as indicated by the International Monetary Fund.

[96] Gleb, Bryanski. "Russia targets bigger role for space program." Reuters 8 Apr. 2011 http://www.reuters.com/article/2011/04/08/us-russia-space-putin-idUSTRE7366RB20110408.

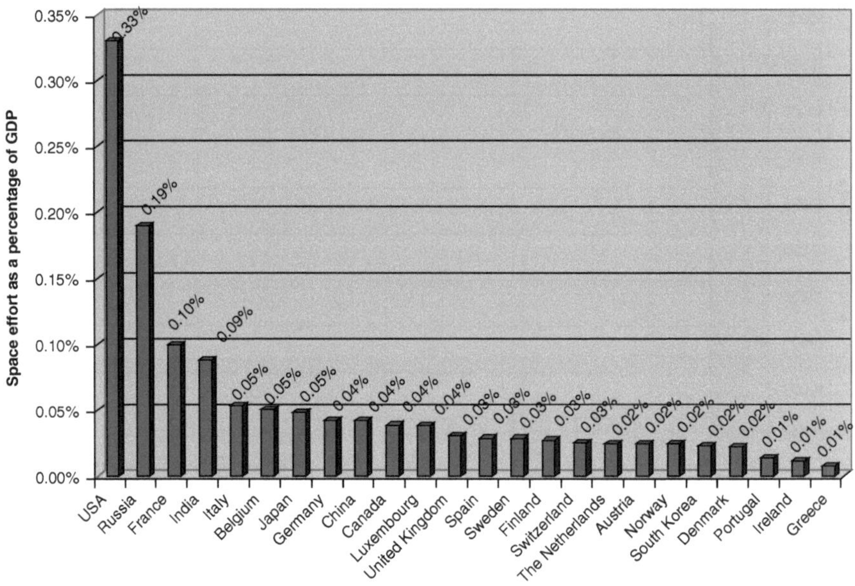

Fig. 1.2 Public space budgets (selection) as a share of nom. GDP in 2010 (Source: Euroconsult/IMF)

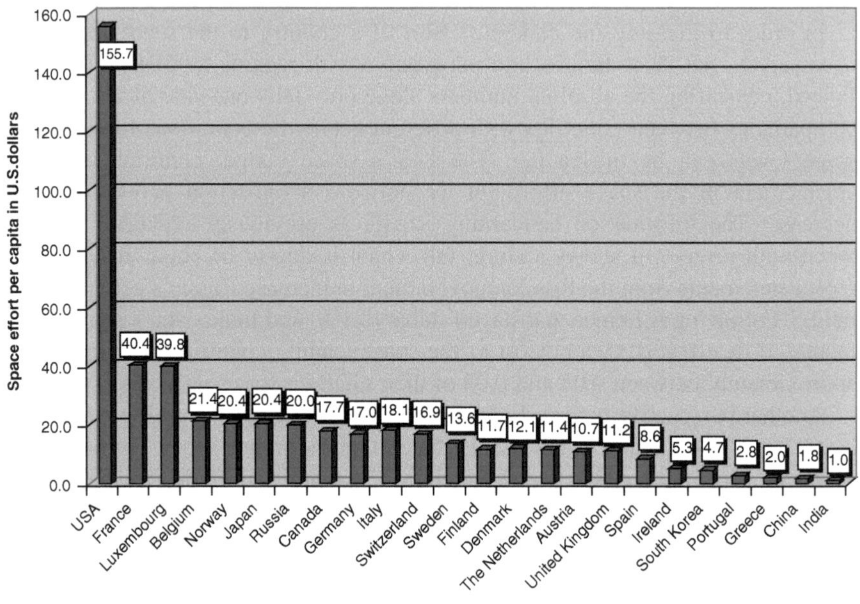

Fig. 1.3 Public space budgets per capita (selection) in 2010 (Source: Euroconsult/UN World Population Prospects)

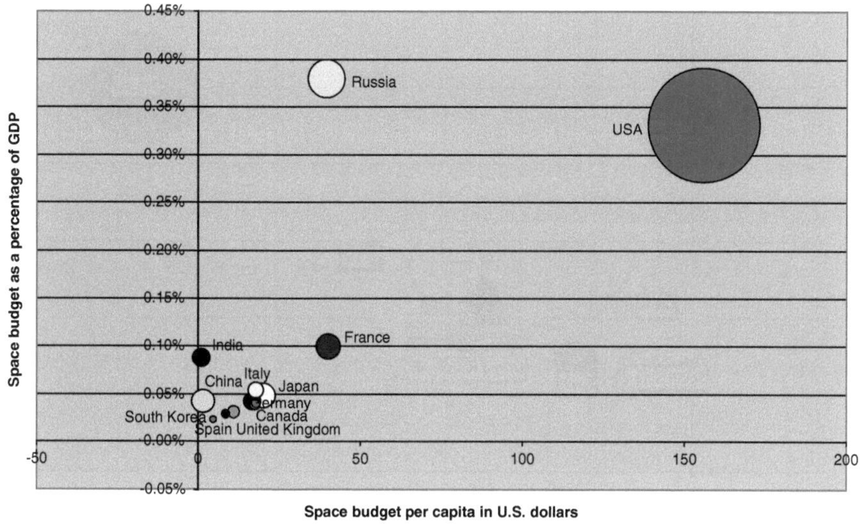

Fig. 1.4 Public space budgets as share of GDP mapped against space budgets per capita in 2010. The bubble size indicates the absolute space budget (Based on Euroconsult data)

participation in ESA (Belgium 4.3% and Luxembourg 0.3% relatively to their modest economic importance). Finally Norway and Japan have a budget per capita of $20.4. These figures are symptomatic of the milestone that 2010 represents, as an apogee for space budgets. The space budget per capita of France, Japan, Belgium and most of the countries which follow show a slight decrease compared to 2009. Growing space actors such as India or China are of course underrepresented in such chart due to their socio-economic characteristics that include a large population.

It is also possible to compare the GDP share of public space funds against the public space funds per capita. This is done in Figs. 1.4 and 1.5, with the latter excluding the United States and Russia to display the other countries more clearly. The U.S. of course dominates in both measurements. France moves to second place in a more balanced way than Russia whose position must be considered cautiously in view of the unclear data. There is however a noteworthy proportional discrepancy between per capita spending and GDP share. Japan follows, leading a group which composes the majority of second tier space powers namely Germany, Canada, Italy, UK, Spain South Korea who display comparable values in both dimensions. China and India have a rather singular position with a large part of their GDP spent on space but per capita expenditure lagging behind the other nations. This situation is mainly the consequence of their huge populations in comparison to the other nations.

1.3.2 Overview of Commercial Space Markets

In 2010 and 2011 the satellite services industry showed remarkable resilience to the adverse global financial conditions. It was able to maintain its upward trend mostly

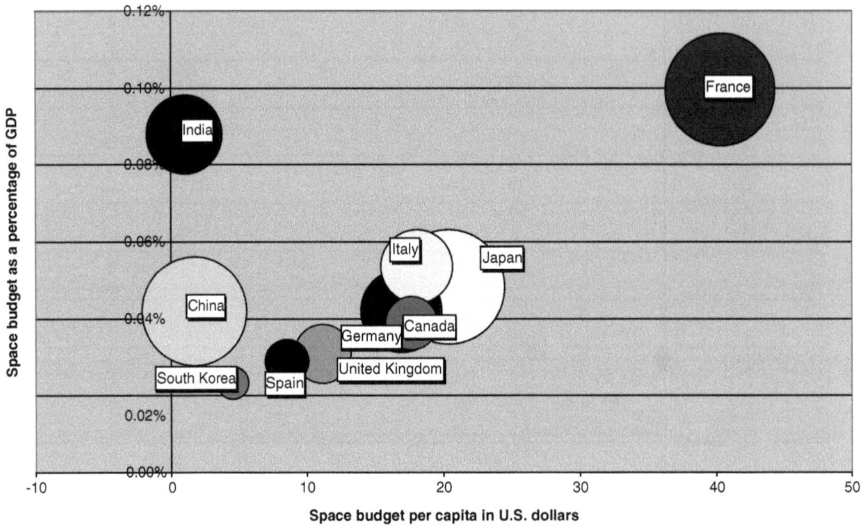

Fig. 1.5 Public space budgets as a share of GDP mapped against space budgets per capita in 2010. The bubble size indicates the absolute space budget (Based on Euroconsult data)

thanks to its inherently global nature, which allowed it to profit from the quick economic recovery of emerging markets, especially in the SE Asia and South America regions. Booming demand in developing regions allowed for a sustained expansion of satellite capacity and corporate revenues. Worldwide satellite capacity rose above 7,000 36 MHZ transponder equivalents in 2010, while revenues are expected to have exceeded the $160 billion plus figure of 2009.[97] Below is a breakdown of the industry's key developments and trends, according to the nature of the services provided.

1.3.2.1 Direct Broadcast Services

Direct Broadcast Services (DBS) refer to direct-to-home satellite television and radio broadcasts. This section of the industry showed considerable development in 2010 and 2011, fuelled by the quantitative expansion in emerging markets and the qualitative increase in new technologies and services in developed ones. Demand in the latter manifested signs of recovery compared to the relatively flat revenues of 2009. In 2009 DBS revenues increased from $64.9 to $71.8 billion, or over 10%. This trend was expected to continue into 2010 and 2011 as the number of HDTV

[97] "2010 Futron Forecast of Global Satellite Services Demand: Executive Summary". 3 Nov. 2010. Futron 21 Feb. 2011 http://www.futron.com/upload/wysiwyg/Resources/Briefs/2010_Futron_Forecast_of_Global_Satellite_Services_Demand_Exe_Summary.pdf.

channels has been growing exponentially (e.g. by 82% between 2008 and 2009).[98] At the same time, according to current projections the annual growth rate of DBS is expected to exceed that of the entire satellite services' sector as DBS will replaces more traditional services such as video distribution.[99]

All major satellite operators have increased their investments in new technologies and products, especially towards developed markets. Such an example was 3DTV, interest in which spread quickly after the World Cup football. Satellite operators increased their support to this broadcasting platform: Eutelsat, which had been running a 3DTV demo channel, provided capacity for the Paris Open tennis broadcasts; Intelsat announced it would carry ESPN's 3DTV channel and Russian Satellite Communications Co. is planning the allocation of 25% capacity to HD and 3DTV applications on its Express AT1 and 2 satellites. Despite current technical issues, the prediction is that by 2015 twenty million homes worldwide will be watching 3DTV. Concurrently, even non-commercial actors have demonstrated their interest, including the European Space Agency, which aims to demonstrate a complete 3DTV service and is also studying the end-to-end 3DTV chain and testing viable TV products, content and reception sites.[100]

1.3.2.2 Fixed Satellite Services

Fixed Satellite Services (FSS) refers to the use of spacecraft that utilise land terminals in fixed positions to broadcast. They include broadband internet, communications and network television and radio broadcasts. In 2010 and 2011 the fixed satellite service (FSS) outlook remained buoyant, as operators continued to profit from previous investments in new capacity, as well as from the sustained demand for TV and broadband services.

Although some reports anticipated a drop in new satellite investment towards 2012–2013, the effect of the current boom in FSS is expected to continue throughout the decade. Industry-wide FSS revenues climbed 5.3% in 2009, to $10.3 billion and the prediction is that, between now and the end of the decade, they will grow by 50%, to $14.8 billion. The growth is explained by the insatiable demand for video and broadband. To meet this demand, 30 new FSS spacecrafts were ordered in 2009. Eutelsat revenues are a clear example of this upward trend. Revenues climbed to €1.05 billion and EBITDA to €827.8 million. In 2010–2011, revenues were expected to reach €1.12 billion and EBITDA to exceed €875 million. Hispasat also

---

[98] "State of the Satellite Industry Report". Aug. 2010. Satellite Industry Association and Futron Corp 5 Mar. 2011 http://www.sia.org/wp-content/uploads/2011/06/2010-State-of-the-Satellite-Industry-Report-Presentation-August-Employment-Update-2010_08.pdf.

"Satellite Industry Revenues Toped $160B Globally in 09" Space News, 14 June 2010: 8.

[99] "2010 Futron Forecast of Global Satellite Services Demand...": 3.

[100] Taverna, Michael A. "Satellite Operators See New TV Standard as Next Holy Grail." Aviation Week & Space Technology 7 June 2010: 36.

reported a solid performance in 2009–2010. Revenues rose 9.5% to €151 million, while EBITDA climbed 5% to €116.7 million.[101]

As a result of increasing demand, most commercial operators have invested in technological upgrades that will allow them to absorb it in the most profitable way. For that purpose, a number of Ka-band satellite projects for broadband connectivity were announced in 2010 and 2011. For Example, Inmarsat confirmed the creation of a Ka-band broadband system named Global Xpress to complement its existing L-band satellites. The Inmarsat- 5 spacecraft, scheduled to orbit by 2014, will provide 10 times more speed than the Broadband Global Area Network (BGAN) service and will have major advantages in price per bit, antenna size and downsizing costs. Inmarsat predicts that the new system will be generating $500 million in revenue within 5 years of launch and estimates the system's full cost at $1.2 billion.[102]

At the same time, the Global Xpress project, based on a deal between Inmarsat and Boeing Satellite Systems (BSS), marked a step forward in the latter's goal to return to the commercial satellite business. With the deployment of all three spacecraft envisaged by 2014, the Global Xpress satellites will give Boeing a foothold in the market for high-throughput commercial Ka-band satellites and offer a quick response to the demand for broadband Internet access.[103] Finally, Inmarsat was evaluating air-interface technologies and ground segment suppliers for the Global Xpress broadband network. By the end of 2011, it was scheduled to have its channel management plan completed, so that marketing to government and military customers could start.[104]

At the same time, its competitor Eutelsat launched its first KA-SAT satellite in December 2010, bringing back to life the International Launch Service's Proton M. The KA-SAT spacecraft was the first of a new generation of high-throughput spacecraft that are expected to improve the competitiveness of satellite communications compared to terrestrial cable networks. The second high-throughput satellite – ViaSat-1 – is scheduled for launch in 2012.[105]

Apart from investing in innovative technologies and services, FSS providers have moved to consolidate their market positions through better satellite and bandwidth management. For example SES announced that in the next 5 years it would be reconfiguring its orbital assets so that by 2015 it would have 30% more capacity spread over a fleet that would be reduced from 43 to 38 satellites in service today. Furthermore, SES expected to spend between €430 and €495 million per

[101] Taverna, Michael A. "Satellite Sector Revenues Expected To Double By 2020." Aviation Week & Space Technology 9 Aug. 2010: 26.

[102] Taverna, Michael A. "Inmarsat Set To Provide High-Speed Internet Service." Aviation Week & Space Technology 16 Aug. 2010: 27.

[103] Taverna, Michael A. "Sale Provides Opening In Fast-Growing Ka-Band Segment." Aviation Week & Space Technology 16 Aug. 2010: 28.

[104] Morring, Frank Jr. "Inmarsat Advances On Global Xpress." Aviation Week & Space Technology 6 Dec. 2010: 22.

[105] Taverna, Michael A. "Eutelsat Spacecraft Is First With High-Throughput Internet Access." Aviation Week & Space Technology 3 Jan. 2011: 29.

year to replace capacity. The plan for 2011 was to spend €790 million on satellites and other projects. This investment should drop to below €390 million by 2014 and remain stable for some years. The company currently has 16 satellites under construction, including its share in the Yahsat satellite for the United Arab Emirates. Significantly, only four of these will occupy new orbital slots and will be directed to new markets. The remaining 12 will be satellites large enough to replace existing spacecraft while still offering expansion capacity.[106]

Smaller satellite operators on the other hand, have opted for more traditional commercial strategies. One such case is Asia Broadcast Satellite (ABS), which has pursued the low-risk expansion strategy of acquiring aging spacecraft from other operators. For example, it purchased the Koreasat-3 spacecraft from Korea Telecom Corp. of S. Korea, renamed ABS-7, in order to provide Ku – and Ka-band VSAT cellular backhaul, broadband and government services to the Middle East. In July 2009, it acquired KT Corp's Koreasat-2, also to serve the Middle East market, and in November 2009 it acquired Mabuhay Sat Corp. of Manila, along with its Agila-2 satellite and extensive ground facilities, renamed ABS-5. This more conservative approach is intended to provide market growth and cash flow until the company's first all-new satellite – ABS-2, expected to be launched in 2012.[107]

1.3.2.3 Remote Sensing

Commercial remote sensing refers to commercial companies that provide optical and radar images to the open market, mostly to government entities that have been increasingly outsourcing such capabilities over the past few years. After an impressive 37% increase in 2009, commercial remote sensing revenues continued to grow in 2010 and crossed the $1 billion benchmark. Although the share of the sector's private clients has been on the rise, government (and particularly military) demand was again instrumental in boosting the revenue of remote sensing services providers.

For example, in an effort to boost the U.S. industry's global competitiveness, the U.S. government undertook an important initiative to deliver more work to remote sensing companies. On August 2010, the National Geospatial-Intelligence Agency (NGA) awarded 10-year contracts totalling $7.3 billion to two publicly traded satellite operators, in order to supply imagery for the U.S. intelligence community and Defence Department. DigitalGlobe Inc. received a $3.5 billion contract, while GeoEye Inc. a $3.8 billion contract.[108] Both deals were unprecedented in scale and

[106] De Selding, Peter B. "SES Expansion Program to Raise Capacity 30%." Space News, 7 June 2010: 6.

[107] Taverna, Michael A. "Deal Is Hong Kong Operator's Fourth In Past Year." Aviation Week & Space Technology 5 July 2010: 41.

[108] Anselmo, Joseph C. "Big Government Contracts Bolster Satellite Imagery Companies." Aviation Week & Space Technology 16 Aug. 2010: 10.

contract duration providing an important cash flow base for their respective contractors.

Such government contracts effectively secure the industry's investment projects, allowing them to expand their activities. RapidEye for example was searching for a new investor to help sustain it until the business model for its system could be shown to be viable. The company's five satellite fleet provides imagery in five spectral bands and the project is mainly funded by the private sector. The company is seeking $200 million in equity to fund new products and applications, as well as to design a follow-on constellation to be launched by 2014–2015. In this case also, the U.S. government stepped in with $337 million to defray the cost of building and launching GeoEye2, which is estimated at a total of $800 million.[109]

1.3.2.4 Mobile Satellite Services

To prevent geostationary commercial communications satellites from colliding or causing signal interference, a new cooperative data tool, the Space Data Centre (SDC), was developed. This new Space Data Centre automatically plots conjunctions in the orbits of satellites owned by participating operators and alerts their control centres to the problem. The organization that runs the system held discussions with the U.S. Strategic Command about contributing relevant data to an integrated near space situational awareness picture.[110]

At the same time, several tests were planned to determine the potential for interference and identify ways to prevent jamming of GPS signals. The result of those tests will have an effect on a Federal Communications Commission (FCC) decision to grant LightSquared conditional approval to deploy 40,000 terrestrial base stations to augment its mobile satellite service capacity. Lightsquared closed a $586 million additional debt deal to help fund the $7 billion project.[111]

RascomStar-QAF, an African Operator, in partnership with ViaSat, is introducing a low-power portable GSM cellular uplink station. This new cellular system intends to reduce satellite price disparities with respect to terrestrial networks and will permit lower costs for serving cell-phone customers in rural areas. It will mainly benefit Africa and other areas with insufficient terrestrial communications infrastructure.[112]

[109] Taverna, Michael A. "EO Sector Needs More Public Support To Stand On Its Own." Aviation Week & Space Technology 4 Oct. 2010: 77.

[110] Morring, Frank Jr. "Satellite Operators Now Tracking Each Other's Spacecraft." Aviation Week & Space Technology 9 Aug. 2010: 50.

[111] Warwick, Graham and Taverna, Michael A. "Tests Planned On Cell-Phone Jamming Of GPS." Aviation Week & Space Technology 28 Feb. 2011: 48.

[112] Taverna, Michael A. "New Cellular System Will Debut In Africa." Aviation Week & Space Technology 28 Feb. 2011: 49.

In a separate development, in June 2010 Iridium Communications awarded Thales Alenia Space of France a $2.1 billion contract to build Iridium's next-generation constellation of low-orbiting voice and data communications satellites. In June, the required credit guarantee by French export credit organisation Coface was secured, allowing all involved parties to proceed with their venture, after an initial $53 million contract between Iridium and Thales Alenia Space. Coface's loan covered 95% of a $1.8 billion credit. This new constellation will feature inter-satellite links, a feature that should permit Iridium to be less dependent on ground stations to relay signals.[113] Three serious challenges were identified on Iridium's path, as it started to build its second-generation constellation of 72 satellites. These were: securing additional funding outside the scope of the Coface-guaranteed part of the budget; meeting the ambitious satellite delivery schedule and in fact accurately assessing the true commercial potential of this market.

1.3.2.5 Satellite Manufacturing

Satellite manufacturing revenues in 2010 saw an increase compared to 2009. The total revenues of satellite manufacturers that have built satellites both for governmental and commercial launches are estimated to have reached $14.5 billion in 2010[114] which is a rise of 7% from the $13.5 billion in 2009.[115] This augmentation marks a substantial increase after the abrupt end in 2009 to the trend of slightly decreasing revenues from 2006 on. The actual predictions forecast a regular increase of revenue to reach $15 billion in 2013.[116] It is expected that the demand by region will be stable to 2019 with a distribution hardly different from 2009 except for the growing market in South Asia.

1.3.2.6 Commercial Space Launch

In 2010 there was a total of 23 commercial launches, carrying 44 payloads into orbit, out of which 31 were for commercial operators. Russian companies had again the lion's share with a total of 13, or 57%, followed their European (26%) and U.S. (17%) competitors. All 23 launches accounted for approximately $2.45 billion in revenues, only a slight increase of $43 million on 2009. European revenues were again the highest reaching $1.32 billion, followed by Russia ($826 million) and the

[113] De Selding, Peter B. "Iridium Selects Thales Alenia To Build Iridium Next Constellation." Space News, 7 June 2010: 5.

[114] "2010 Futron Forecast of Global Satellite Services Demand: Executive Summary". 3 Nov. 2010. Futron 21 Feb. 2011.

[115] "State of the Satellite Industry Report.", SIA June 2010 http://www.sia.org/news_events/2010_State_of_Satellite_Industry_Report.pdf.

[116] "2010 Futron Forecast of Global Satellite Services Demand: Executive Summary…"

U.S. ($307 million). In spite of the marginal increase in the industry's revenues, all launch service providers increased their turnover thanks to the absence of Chinese or multinational commercial launches in 2010. However, since these figures take into account the value of the past year's activity they should only be considered as indicative, since contracts are typically prepaid 1–2 years prior to launch.[117]

All six European launches were carried out by Arianespace onboard Ariane 5 rockets. However, in the course of 2010 concerns started to be raised regarding Arianespace's own financial survival. In October, the company warned the members of its supply chain and the European governments of the necessity to bear more of the operating costs, in order to build capacity for a launcher system expansion to deal with tough competition. The attention to competitiveness has led Arianespace to convene the manufacturers of the Ariane 5 and two new systems that are to enter service at its Kourou launch complex next year, to discuss measures to ensure it. To ensure competitiveness, several different measures were analysed to make Arianespace more efficient against new arrivals, such as SpaceX, Sea Launch and ILS.[118] NASA is more quickly implementing its new program to stimulate the private sector to develop commercial space transport services. With the first successful flight in December 2010, SpaceX was pushing to dock with the ISS on the next launch in mid-2011. A new successful flight could mean the start of a commercial cargo run to the ISS by year-end.[119]

1.3.2.7 Ground Equipment

Ground equipment revenue includes elements of infrastructure, such as mobile terminals, gateways and control stations, and consumer equipment, such as very small aperture terminals (VSAT), ultra small aperture terminals (USAT), DTH broadcast dishes, satellite phones and digital audio radio satellite (DARS) equipment. Portable Navigation Devices (PND) form one of the sub-segments of end-user electronics incorporating GPS chip sets. The PND market continued its regular increase but with a notable slow down in comparison to 2008. The sharp rise of 34% previously experienced from 2007 to 2008 fell to 8% between 2008 and 2009, nonetheless constituting a global turnover of $49.9 billion.[120] In 2009 ground equipment represented 31% of the world space business revenue[121] showing an increase of 3% from the global distribution of the previous year.

---

[117] Federal Aviation Administration. Commercial Space transportation: 2010 Year in Review. Washington DC: FAA, Jan. 2011: 3.

[118] Taverna, Michael A. "Space Arianespace Says Its Long-Term Viability Is At Stake" Aviation Week & Space Technology 18 Oct. 2010: 31.

[119] Warwick, Graham. "Commercial Space Transport Rockets On." Aviation Week & Space Technology 24 Jan. 2011: 149.

[120] "State of Satellite_Industry_Report". SIA. June 2010 http://www.sia.org/news_events/2010_State_of_Satellite_Industry_Report.pdf.

[121] Ibid.

Optimistic expectations concerning 2010 revenues for the two companies leading the market were partially met. After the fall in revenue in 2009, the news was quite good for TomTom which announced total revenue of $1.52 billion in 2010,[122] an increase of 2.5%. The situation was less favourable for Garmin, which experienced a drop of revenue from $2.95 billion in 2009 to 2.69 billion in 2010,[123] a noticeable drop of 9.6%.

In order to better understand these figures, they will be broken down into their respective type of product and sale areas. Garmin which had a particularly flat 2010, experienced a decrease in automotive products from $2.054,127 billion in 2009 to $1.668,939 billion in 2010, whereas its segments related to marine, aviation and outdoor fitness improved significantly despite being unable to mitigate the overall loss in revenue caused by the drop in automotive products revenue. Meanwhile, TomTom significantly increased its sales in automotive systems by 55.9% during the same period with global revenue of $179 million in this market. Geographically speaking, Garmin reinforced its sales in Asia with a revenue increase of 47% (an increase of $70 million) but lost 16.52% in the U.S. falling from $1.97 billion to $1.65, while remaining stable in Europe. TomTom was not significantly more successful than its counterpart in North America with a decrease of 7.54% of its sales in this area – representing however a less severe loss of $31 million. TomTom increased its European revenues by 6.26% in its core area, and 12.9% in the rest of the world. Europe and Asia seem to be currently particularly favourable for the PND market.

1.3.2.8 Insurance Sector

As the space industry has continued to demonstrate increased hardware reliability, low accident rates and booming growth in recent years, insurance costs have been decreasing.[124] A safer market has brought about more competition between space insurance suppliers and a change of behaviour from customers. The trend is now to take out insurance 2–3 years before the launch thus reducing the price. The rates of space insurance have been progressively reducing throughout the 2000s leading to the withdrawal of some actors from this field. A number of companies such as Swiss Re in 2010 did not deploy their entire insurance capacity. Given that commercial space launches are expected to grow in the coming years and the technologies involved have proven their worth in practise, one can expect this trend to continue. To summarize, insurance rates for launch and in-orbit coverage have been

[122] "TomTom Annual Report and Accounts 2010." 23 May 2011. http://ar2010.tomtom.com/home.html.

[123] "Garmin Anuual Report 2010." 23 May 2011. http://www8.garmin.com/aboutGarmin/invRelations/reports/10K_2010.pdf.

[124] Peter B., de Selding. "Insurance Premiums Stay Flat Despite W3B Satellite Failure." Space News, 20 May 2011 http://www.spacenews.com/satellite_telecom/110520-insurance-premiums-flat.html.

exceptionally low for the 2010–2011 reporting period.[125] This situation is mainly due to the few claims and substantial profits made by the underwriters. Indeed, the most important accident was the $343 million loss of Eutelsat W3B, insured by Space Consortium in London, part of Brit Insurance Holdings N.V, but this notably was isolated. The failure of a Proton-M launch carrying three Russian global orbiting navigation satellite system satellites in December was only covered for $3 million. The insurance rates were therefore expected to fall in 2011.[126]

1.3.3 Developments in the Space Industry

1.3.3.1 Industrial Developments in Europe

As mentioned above, a financial guarantee from Coface, a French export credit agency, syndicated through French and international banks, permitted Iridium to continue with the biggest commercial space deal ever, a new low-Earth-orbit satellite system. Coface covered 95% of a $1.8-billion facility that ensured most of the financing for the project. The new constellation, dubbed Iridium Next, will replace Iridium's existing 66-satellite LEO constellation. The costs to build and launch were estimated at around $2.9 billion. Paris-based Thales Alenia was selected to supply the $2.1 billion space segment, which includes 66 operational satellites, six in-orbit spares and nine ground spares. In this project two U.S. satellite makers have a role, Ball Aerospace and Boeing's Intelligence and Security Systems Mission Operations group. The first one will handle the satellite assembly, integration and testing and the second one will deliver the ground system hardware. The first new satellites will be in orbit by the first quarter of 2015, and the process will be concluded by 2017.[127]

At the same time, SES announced that it would be reconfiguring its orbital assets so that by 2015 it would have 30% more capacity spread over a fleet that nonetheless would decline to 38 satellites from the 43 in presently in service.

In addition to the above, SES announced the sale of its ND Satcom subsidiary. Through this, SES was expecting to be able to show investors an immediate improvement in gross-profit margins. A broader strategy was to put limits on the vertical expansion of satellite fleet operations and to turn to services markets to

[125] De Selding, Peter B. "Satellite Insurers on Track for Highly Profitable 2010." 3 Sept. 2010. Space News 3 Sept. 2010 http://www.spacenews.com/satellite_telecom/100903-satellite-insurers-profitable.html.

[126] Collins, Stuart. "Space insurance rates expected to fall in 2011." 13 Feb. 2011. Business Insurance 24 Feb. 2011 http://www.businessinsurance.com/apps/pbcs.dll/article?AID=/20110213/ISSUE03/302139999.

[127] Taverna, Michael A. "French Government Credit Guarantee Assures Iridium Next Go-Ahead." Aviation Week & Space Technology 7 June 2010: 34.

refocus on selling satellite bandwidth. In 2009, ND Satcom accounted for 25% of what SES's calls its "services" revenue, which totalled $443 million.[128]

In a separate development, on 19 June 2010 Europe's Arianespace Launch Consortium and Russia's Roscosmos signed a contract to build 10 additional Soyuz rockets scheduled for delivery in 2014 and 2015. At that time, Russian and European authorities were reviewing possible dates for the inaugural launch of Russia's Soyuz from Europe's Guiana Space Centre in French Guiana and set a mid-December date for that, but depending on various factors that date could change and the launch be pushed to 2011, said the Arianespace Chief Executive. Arianespace continued to accumulate orders for Soyuz missions.

On 7 July, Astrium Space Transportation announced a €10 million contract for 15 months with ESA to produce designs and cost estimates for a next-generation rocket. This project had secured €1.5 million financing by the company itself, while the remaining amount would be provided by the 18-nation ESA. ESA governments scheduled a decision on the development of a successor to the Ariane 5 rocket within 2012 that would allow them to maintain Europe's independent access to space in the most efficient and affordable manner. In the meantime, design work on the new rocket would focus on creating a modular vehicle that would be able to lift different type and mass category payloads with the use of strapped boosters. This new rocket was not expected to fly before 2025, replacing at that time both the Ariane 5 and medium-lift Soyuz rockets.[129]

Hispasat agreed with Space Systems/Loral to build the Amazonas 3 telecommunications satellite, to replace Amazonas 1. It is scheduled to be launched in late 2012, into Hispasat's 61° west orbital slot for broadcasts in the Americas and Europe and it was designed to deliver 14 Kw of power to its payload. Amazonas 1, launched in 2004 and built by Astrium Satellites of Europe, had a projected 15-year service life, but this was reduced because of a defect in its on-board fuel system. Madrid-based Hispasat contracted with Palo Alto, Calif. – based Space Systems/Loral for the Hispasat 1E satellite, which launched on 28 December 2010. This project is seen as a way of maximising the value of the satellite investment and a way to get new orders from the same customer.[130]

Norway has an annual space budget of around €100 million per year, yet its space sector generates revenue of about €650 million. Norway's Kongsberg Group has developed an Automatic Identification system – AIS – terminal for space-based maritime traffic monitoring in Norwegian waters, unlike commercial AIS systems that were developed in Canada and USA. Norway's AISSat-1 spacecraft cost €4.4 million and was one of the passengers on the Indian Polar satellite launch vehicle

[128] De Selding, Peter B. "SES Eyes Improved Gross-Profit Margins Following Sale of ND Satcom Unit." Space News, 7 June 2010: 6.

[129] "Astrium to Lead Studies of Successor to Ariane 5." Space News 12 July 2010: 8.

[130] "Hispasat Orders Another Satellite From Loral." Space News 12 July 2010: 9.

rocket. A possible broadening of the AIS effort was discussed by Norway and ESA, integrated in a European programme.[131]

Italy's e-Geos Earth observation satellite services, a company that is 20% owned by the Italian Space Agency and 80% by Telespazio of Rome, reported €90 million revenue in 2009, 13% return on sale and growth to 300 employees. The company's principal product is radar imagery from Italy's Cosmo-SkyMed satellite constellation, placed into orbit in 2007. In addition, e-Geos has struck image-distribution agreements giving it rights to distribute imagery from satellites operated by GeoEye and DigitalGlobe, MDA Corp. of Canada and ImageSat of Israel.[132]

Hispasat reported a 50.3% increase in net profit for 2009. In 2009 the company reported revenue of €150.8 million, an increase of 9.7% over 2008. EBITDA was 77.4% of revenue, down from 80.1% a year earlier and net profit was €70.6 million. In 2009 Hispasat distributed a record €12.2 million in dividends. The Hispasat President saw in this growth proof that the expansion into Americas with the Amazonas satellite was justified. Amazonas 2 was launched in October 2009 and the Amazonas 3 will be launched in 2013. The 30° west orbital position, that serves Europe and North America, accounted for 2/3 of revenue in 2009; the remaining is due to the 61° west position, which serves the Americas. Hispasat AG-1 satellite is under construction by OHB Technology of Germany, and scheduled for launch in 2012.[133]

Telenor Satellite Broadcasting of Norway reported a 6.1% increase in revenue for the 3 months ending June 30, 2010. Oslo-based Telenor was expecting UPC Direct, one of its major customers, to complete the re-pointing of the rooftop antennas of UPC's more than 300,000 subscribers. Telenor reported revenue of $43.7 million for the same period and EBITDA was 69% of that revenue. To offset the decline in subscribers, Telenor moved its aging Thor3 satellite to a new slot; which can work for another 6–10 years.[134]

Europe's Astrium space hardware and services provider reported higher pre-tax profit on revenue, as well as the Astrium services division, which has multiyear contracts with NATO, British and German defence forces. Astrium reported revenue of €2.11 billion for the first semester of 2010, dropping 4% from the previous year. Pre-tax profit was 5% of revenue.

Telecommunications satellites for Nilesat of Egypt and Rascom of Mauritius were launched successfully by Arianespace's Ariane 5 ECA rocket. France-based Arianespace won contracts to launch two satellites – the first was Intelsat's IS-20, designed to replace both Intelsat's IS-10 and the IS-7 satellites. The other satellite is the Gsat-10 built by the Indian Space Research Organisation (ISRO), which will carry 12 Ku-, 12-C- and 12 extended-C- band transponders.

---

[131] "Norway Considers Second Ship-tracking Satellite." Space News 12 July 2010: 12.

[132] "Italy's e-Geos Company Turned Profit last Year." Space News 12 July 2010: 12.

[133] "Hispasat: Growth Validates Expansion." Space News 19 July 2010: 3.

[134] "Telenor Revenue Boosted by Nordic HDTV Demand." Space News 26 July 2010: 8.

RapidEye AG was seeking to raise €20 million in equity to expand into new product areas, to expand the market reach of its five-satellite constellations, add capacity to handle new business and set a solid base to manage the growth of the RapidEye's second generation system. The company targeted €25 million in revenue for 2010[135]; however, its revenue fell far short of that mark, earning nearly €14 million.[136]

GeoEye contracted with Lockheed Martin Commercial Launch Services of Denver to launch the GeoEye-2 imaging satellite aboard an Atlas 5 rocket. Another division of Lockheed Martin was contracted in March by Dulles, Va.-based GeoEye to build GeoEye-2, which has an estimated cost of between $750 and $800 million. The launch is expected in 2013. GeoEye-2 was the only commercial Atlas 5 launch. Delta-4 launched the Intelsat-14 hosting an experimental U.S. military communications payload.

Thales Alenia Space started working on Turkey's Gokturk high-resolution optical Earth imaging satellite system, the first satellite sold for export with such high resolution. It is scheduled for launch in 2013, carrying an optical imager similar to France's two Pleiades imaging satellites. Furthermore, the contract included the creation of a satellite integration and test centre in Turkey and Gokturk ground infrastructure.[137]

Space Exploration Technologies (SpaceX) Falcon 1 rocket has been marketed by Europe's Astrium space Hardware and Services Company to European governments. This agreement gives the opportunity to Astrium and SpaceX to secure a launch for small government satellites, before the debut of Europe's Soyuz and Vega Rockets. The European governments had expressed the will to launch satellites aboard European rockets, as much as possible. Falcon 1 has a $10.9 million cost per launch and will be able to place into circular orbit satellites weighting up to 1,010 kg.[138]

Two contracts were signed between European governments and industry to start the production of the first Vega small-satellite launch vehicle and to cover the delivery of the five subsequent Vega rockets. The first contract covers the first Vega Launch in 2011 from Europe's Guiana Space Centre. It was signed between Arianespace launch consortium and ESA. The second contract between Evry France-based Arianespace and Vega prime contractor ELV of Colleferro, Italy, established the delivery terms of the five Vega vehicles. Under contracts with Arianespace, ELV will build Vega and 2/3 of its construction will be done in Italy.[139]

---

[135] "RapidEye in Hunt for New Equity Financing." Space News 13 Sept. 2010: 3.

[136] De Selding, Peter B. "Earth Imagery Firm RapidEye Seeking Bankruptcy Protection." Space News 6 June 2011: 6.

[137] "Thales Alenia Begins Work on Turkish Sat." Space News 13 Sept. 2010: 9.

[138] "Astrium to Market Space X Falcon 1 Launches in Europe." Space News 13 Sept. 2010: 14.

[139] "Arianespace Inks Vega Rocket Contracts." Space News 13 Sept. 2010: 15.

Under an extension of the 2007 agreement, International Launch Services (ILS) will loft six satellites for fleet operator SES aboard Russian Proton rockets. ILS launched the SES-1 in April 2010, as part of the MultiLaunch Agreement, and launched SES-3 in July 2011; SES-4 and SES-5 would launch in 2012. The two other ILS missions are planned within the 2012–2014 timeframe.[140]

Indra – a Spanish information and defence technology provider – purchased 49% of IndraEspacio for €39.2 million. This has permitted Indra to strengthen is position in the space market and has encouraged the company to look forward for new applications in navigation and Earth observation. IndraEspacio is specialized in satellite ground systems and the deal gave an implied value of €80 million to the company, which reported €70 million revenue in 2009. However the financial crises can interfere with space spending for the coming years.[141]

Government budget cuts in 2010 were of concern to Astrium, because these could compromise its activities. The EADS space unit saw sales rise 4% last year, to around €5 billion, and orders leap 43% to €6 billion. Although Astrium had various disappointments during 2010, for example, those failed attempts to win the first batch of Galileo navigation satellites and the space segment of Europe's Meteosat Third Generation weather satellite system, the company is investing large sums to permit the service business to continue to drive future growth. However, Astrium's position as the world's top provider of geospatial intelligence system (GIS) products, secure telecommunications satellite capacity, are at risk.[142]

1.3.3.2 Industrial Developments in the United States

Space Exploration Technologies (SpaceX) Corp.'s Falcon 9 rocket was launched on 4 June 2010 after months of delay, from Cape Canaveral Air Force Station, Fla., at 2:45 p.m. EDT. The Falcon 9 is a two-stage liquid oxygen-and- kerosene-fuelled rocket. This new rocket was considered a boost for NASA's controversial commercial spaceflight initiatives. Notwithstanding all the criticisms about reaching orbit with the Falcon 9, there was also much praise. NASA Deputy Administrator, Lori Garver, William Pomerantz the senior Director of Space Projects at the Xprize Foundation and Bett Alexander, President of the Commercial Space Flight were pleased with the launch and the day was considered "historic", establishing the historical track record for new vehicles. However the enthusiasm was not universally shared. U.S. Sen. Kay Bailey Hutchison (R. Texas) called the launch "a belated sign that efforts to develop modest commercial capabilities are showing some promising signs".[143]

---

[140] "ILS and SES Extend Launch Services Deal." Space News 13 Sept. 2010: 16.

[141] "Indra Buys Remaining 49% of IndraEspacio." Space News 20 Sept. 2010: 9.

[142] Taverna, Michael A. "Waning Support May Undercut Satcom Growth." Aviation Week & Space Technology 17 Jan. 2011: 30.

[143] Klamper, Amy. "Falcon 9 Reaches Orbit in Debut." Space News 7 June 2010: 1.

On June 2, during a visit to NASA's Kennedy Space Centre (KSC) in Florida, Hilda Solis, the U.S. Secretary of Labour, announced a $15 million federal grant to Florida's Brevard County for assistance to 3,200 area workers who were in danger of losing their jobs when NASA retired its space shuttle fleet in the end of 2010. One press article released at that time said that the grant money was intended to help workers from major KSC contractors, including the United Space Alliance, ASRC Aerospace Corp. and Boeing. Up to 20,000 contractor and subcontractor employees were expected to be laid off at the end of the shuttle programme.

In February, the White House unveiled a new direction for the U.S. spaceflight programme and marked Constellation and Orion for termination. However in April President Obama revised the plan, and affirmed that the Orion would be back as the space station crew lifeboat. U.S. lawmakers demanded documentation to support the decision to terminate Constellation and insisted that the information provided by NASA should include analysis of the executability of the proposed plan through 2025.[144]

The U.S. Justice Department announced on 23 June 2010 that Northrop Grumman had agreed to pay U.S. $12.5 million to settle allegations that it knowingly submitted false claims to multiple government agencies, when it supplied electronic parts for navigation systems for military transportation. Allegedly Northrop's Woodland Hills based Navigation systems Division failed to test commercial grade electronic components to ensure they would function in the extreme temperatures required for military and space use. This misconduct affected the U.S. Navy, Army, Air Force, Defence Logistics Agency, Coast Guard, and Forest Service.[145]

The industry is planning to offer a significant number of suborbital flights, with a cost of between $50,000 and $100,000 per person. One idea that was discussed was the possibility for the suborbital market to offer global, high-speed, point-to-point transportation. The manager of the Commercial Reusable Suborbital Research Programme (CRuSR) did not estimate how many people would travel on suborbital flights but said that thousands of payloads could be expected. In February 2010, CRuSR offered microgravity flights for technology development and research.[146]

In order to cover the London-based company's three-satellite contract with Boeing Space and Intelligence Systems, Inmarsat sought a $500 million financial package from the U.S. Export–import Bank, however Inmarsat had sufficient capital for the $1.2 billion Global Xpress project. The negotiation was expected to finish at the end of 2010. Each satellite provides 12 kW of power at the end of its 15 year life, has 89 fixed spot beams and weighs about 6,000 kg. Inmarsat was negotiating with 4 companies to provide the Global Xpress ground network, expecting to select 1. Boeing has agreed to purchase 10% of the Global Xpress

[144] Klamper, Amy. "Lockheed With Less NASA Oversight, Orion Lifeboat Could Cost $4.5 Billion." Space News 21 June 2010: 1.

[145] "Northrop to Pay $12.5 M In False Claims Case." Space News 28 June 2010: 8.

[146] "Suborbital Price Drop Foreseen by 2014." Space News 26 July 2010: 3.

satellite capacity for the 5 years of the system's operations. Boeing Satellite Systems International created a new division to sell Inmarsat Global Xpress capacity. By the end of the fifth year, it is expected to generate $500 million in revenue.[147]

On 12 August 2010, SpaceX completed its first high-altitude drop test to validate the parachute deployment systems and recovery operations of the Dragon space capsule. The test was held with success, in advance of the launch atop the SpaceX Falcon 9 rocket.[148]

On 18 August, 2010, the U.S. Department of Agriculture's Rural Utilities Services announced, that the U.S. government had selected four satellite-broadband providers to provide links to rural areas. Hughes Network Systems of Germantown received $58.8 million; Carlsbad, Calif. – based ViaSat Inc.'s Wild Blue Communications, $19.5 million; EchoStar XI Operating LLC: $14.2 million and Spacenet of McLean Va. $7.5 million.[149]

1.3.3.3 Industrial Developments in Russia

The Soyuz TMA-17 Russian spacecraft landed on June 2 at about 9:25 a.m. in Kazakhstan, to return the cosmonaut Oleg Kotov, the USA astronaut Timothy Creamer and the Japanese Soichi Noguchi. They were half of the space station's full six-person crew and had lived in the orbiting laboratory since December. The crew hosted three visiting NASA space shuttle missions, during their 163 days in orbit. NASA's shuttle, Atlantis, visited in May and delivered a $200 million a Russian Research module called Rassvet. On May 31, Kotov handed control of the space station over to its new Expedition 24 Commander, Skvortsov. On June 15, the Soyuz TMA-19 spacecraft launched a 6 month mission, with Fyodor Yurchikhin, Douglas Wheelock and Shanon Walker.[150]

On June 17, at 6:21 p.m., the Soyuz TMA-19 linked up with the ISS, carrying two American astronauts and one Russian cosmonaut. This marked the return to full strength of the space station crew and a milestone for women in space: two female astronauts, Shannon Walker and Tracy Caldwell Dyson, were part of the long-term spaceflight.

A Proton rocket was launched on September 2, at 4:53 a.m. from the Baikonour Cosmodrome in Kazakhstan, carrying three Glonass-M Navigation satellites. These were the second trio of Glonass satellites launched in 2010. Roscosmos affirmed that the satellites reached their intended orbit.[151]

[147] "Inmarsat Seeks U.S. Ex-Im Backing for Satellite Deal." Space News 9 Aug. 2010: 4.

[148] "Space-X Conducts First Drop Test of Dragon." Space News 23 Aug. 2010: 3.

[149] "Federal Broadband Grants Go to 4 U.S. Satellite Firms." Space News 23 Aug. 2010: 3.

[150] "Three Spaceflyers Return to Earth in Russian Soyuz." Space News, 7 June 2010: 8.

[151] "Russian Proton Launches Trio of Glonass Satellites." Space News 6 Sept. 2010: 8.

Sea Launch Co. was purchased by Energia, a large Russian space hardware manufacturer, after the approval of the U.S. government. This purchased was a way for Energia to enter the global commercial launch business. Chicago based Boeing Co. remained a Sea Launch supplier but ended its management role (it was the general contractor and principal shareholder). Sea Launch was emerging debt-free from Chap. 11 and could survive on three or four launches per year.[152]

1.3.3.4 Industrial Developments in Japan

Japan's Servis-2 satellite, built by Mitsubishi Electric Corp was launched on June 2 by the Russian-German Eurockot Launch Services Company (a joint venture between Astrium of Europe and Moscow-based Khrunichev), from Russia's Plesetsk Cosmodrome. The 736 Kg Servis-2 will test commercial off-the-shelf components to use on future space missions and was placed into a 1,200 km sun-synchronous orbit.[153]

1.3.3.5 Industrial Developments in China

China Satcom, which was merged in 2009 into China Aerospace Corp. (CASC), reported a 15% increase in revenue in the same year, from its five in-orbit satellites. The revenue totalled 930 million Yuan ($136 million) which placed the company in the 14th place for 2009 in the Space News Top Fixed Satellite Service Operators ranking.[154]

On 31 July, China successfully launched the fifth spacecraft to join its Beidou-Compass satellite navigation and positioning network. The satellite lifted off at 5:30 a.m. from the Xichuang Satellite Launch Centre, in Sichuan Province. The network is expected to provide navigation, timing and short message services in Asia and Pacific region by 2012 and global navigation services by 2020.[155]

AsiaSat of Hong Kong reported a 27% increase in revenue on the 6 months ending 30 June 2010. The whole revenue was 689.8 million Hong Kong dollars. Speedcasts, an AsiaSat subsidiary, increased its revenue by 34% during the same period and accounted for 13% of AsiaSat's revenue. The company was encouraged by its joint venture with Echostar of Englewood, Colo. to provide 36 enhanced standard-definition and high-definition television channels in Taiwan. In addition to the four satellites already operated by AsiaSat, the AsiaSat 7 was launched during

[152] "Sea Launch Acquisition Receives U.S. Approval." Space News 13 Sept. 2011: 3.
[153] "Japanese Test Satellite Launched Atop Rockot." Space News, 7 June 2010: 3.
[154] "China Satcom'09 Revue Rose by Nearly 16%." Space News 2 Aug. 2010: 8.
[155] "Long March Rocket Lofts Chinese Navigation Satellite." Space News 9 Aug. 2010: 9.

2011. At the beginning of 2011, AsiaSat had received 760 million Hong Kong dollars in new orders.[156]

The Chinese Sinosat-6, which carries 24 C-band, 8 Ku-band and 1 S-band, replaced the Sinosat-w satellite at 125° east, but had a leak in its helium-pressurization system that could result in reduced operational life, from 15 to 10 years. The system is unable to deliver full helium-provided pressure to the satellite's fuel tank. The entire satellite was insured for $200 million and for this problem $60 million was claimed. The Sinosat-6 was the first claim for launch and satellite failures in 2010. The glitch also affected the reputation of the DFH-a platform that Sinosat-6 uses.[157]

Bremen-based OHB Technology signed an agreement with China to develop a small constellation to measure atmospheric levels of methane and carbon dioxide. This partnership was seen by the chief executive of OHB Systems as a serious commitment by both parties to the implementation of the post-Kyoto protocol. The constellation could be known as the Disaster Monitoring Constellation, constituted by Earth imaging satellites. The main idea is to provide a globally accepted source of climatic data.[158]

1.3.3.6 Industrial Developments in India

Private investment is flowing into India's aerospace industry. There are many reasons for the growth of the industry in India. On thc onc hand, India has a low cost industry and, on the other is the increasing availability of defence work for private companies and the opportunity to gain business from foreign suppliers compelled to offset their sales into India with domestic purchases.

India is still compared with China, unavoidably, but big differences appear between them. India's industry is smaller. The dominant company, Hindustan Aeronautics Ltd., had revenue of $2.6 billion in 2009 and is encourage by private companies. China's Avic had about 10 times larger revenue and is dominated by state giants. General Electric is one notable foreign participant in India's industry and their business is growing 10–15% a year.[159]

---

[156] "AsiaSat Sales and Earnings Post Double-digit Gains." Space News 23 Aug. 2010: 3.

[157] "China's Sinosat-6 hampered by Leak." Space News 13 Sept. 2011: 3.

[158] "OHB, China Sign Pact For CarbonSat Mission." Space News 20 Sept. 2011: 8.

[159] Perret, Bradley. "Private Investment Leading The way in Indian Aerospace." Aviation Week & Space Technology 14 June 2010: 51.

Table 1.1 Worldwide launches in 2010 by country and commercial status (Source: FAA)

Launchers	Commercial	Non-commercial	Total number of launches
Russia	13	18	31
USA	4	11	15
China	0	15	15
Europe	6	0	6
India	0	3	3
Japan	0	2	2
Israel	0	1	1
South Korea	0	1	1
Total	**23**	**51**	**74**

1.3.4 Industrial Overview

1.3.4.1 Launch Sector

The launch sector is particularly crucial for the satellite industry. It is a key element of each independent space policy. However, the revenues it generates are far less important than those originating from the satellite manufacturing and satellite services business.

The year 2010 was slightly less active for the launch sector than 2009, with a total of 74 launches (23 of which commercial) conducted by launch providers from Russia, the United States, Europe, China, India, Japan, South Korea and Israel. There were some important events that year, such as the two non-commercial launch failures, a Proton-M launch carrying three Russian global orbiting navigation satellite system satellites in December,[160] and a second south Korean's KSLV I which was caused by the Russian stage according to the South Korean official investigation.[161] The failure of the Eutelsat satellite was mainly due to technical problems after the launch.[162]

When looking into specific countries (Table 1.1), Russia was again the world leader in the number of launches, representing approximately 42% of the total number. It was followed by the United States (app. 20.3% of the total), Europe (app. 8.1%), China (app. 20.3%), Japan (app. 2.7%), India (app. 4.1%), South Korea and Israel which launched one vehicle each, or approximately 1% of the total launch figure.[163]

[160] Ibid.

[161] K.S, Jayaraman. "ISRO Team Says Cable Rupture Caused Rocket Failure." 31 Dec. 2010. Space News 31 Dec. 2010 http://www.spacenews.com/launch/101231isro-team-says-cable-rupture-caused-rocket-failure.html.

[162] Peter B, de Selding, "Eutelsat W3B Declared Total Loss.", Space News 29 Oct. 2010 http://www.spacenews.com/launch/101029-eutelsat-w3b-declared-total-loss.html.

[163] Federal Aviation Administration. Commercial Space Transportation: 2010 Year in Review…

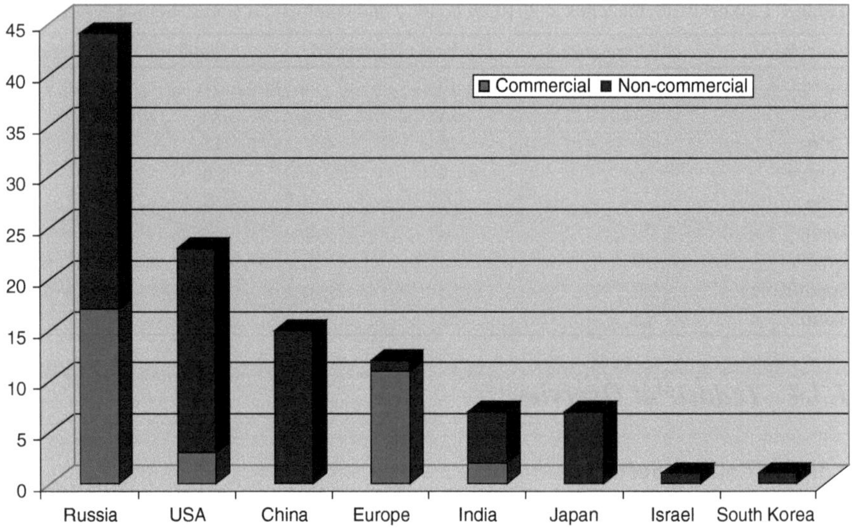

Fig. 1.6 Payloads launched in 2010 by country and commercial status (Source: FAA)

Russia launched 31 vehicles using 11 different launch system configurations. Concerning the U.S., 8 different launcher configurations were used for a total of 18 launches, and China, with the same amount of launches, used 6 configurations. India used a different launch configuration for each of its three launches. The exception is Europe which only employed the Ariane 5 launcher for six launches. Japan with two launches, Israel and South Korea with one each had recourse to one. The total number of different configurations of launchers is 32 for 2010, an increase of 3 compared to 2009.

The commercial launches market is shared between three actors, namely Russia 56.5%, Europe 26.1% and U.S. 17.4%. However these figures must be put into perspective due to the European launcher Ariane V carrying two payloads.

The activity of the two leaders in commercial launches (Europe and Russia) is thus nearly equivalent. Concerning only non-commercial launches the dominance of Russia is also not so obvious. Indeed, other countries carried out intensive national programmes such as the U.S. with 21.6% of non-commercial activity and especially China with 29.4%, not far from Russia (35.3%).

The distribution between commercial and non-commercial payloads launched shows no significant difference from the previous year (Fig. 1.6). 2010 confirms the concentration of Europe on commercial activities with 11 commercial payloads. On the other hand, China and the USA focused on non–commercial activity. This is particularly true for China with its 15 non-commercial launches, a country that is currently intensively developing its national programmes in remote sensing and especially the Beidou navigation system with 5 launches dedicated to this sole purpose in 2010. In the same manner but to a lesser extent, many of Russia's 44 launches concerned non-commercial activity with 27 payloads destined to programmes such as the achievement of the Glonass system. U.S. commercial payloads are still marginal with 3 compared to the 20 non-commercial. This observation is valid for South Korea, Israel and Japan as well because their payloads were solely non-commercial.

With regard to the global share of payloads launched in 2010, Russia was once again the leader, increasing its advantage by 2.2% compared to 2009, with 44 payloads launched representing 40% of total launches. The U.S. share slightly decreased with 23 payloads and 20.9%, while China increased its share and was third with 15 payloads reaching 13%. Europe was next with 12 payloads and 10.9%. Japan and India each had seven payloads. They thus remained the same as in 2009 with 6.4%. Finally South Korea and Israel sent only one payload into orbit. The global number of payloads launched in 2010 was stable with 2009 and the hierarchy among space powers also mirrored the previous year.

In 2010 the market for launchers for GEO satellites seen the number of actors considerably narrowed. Indeed, there were only three protagonists compared to five in 2009 with China Aerospace Corporation and Mitsubishi Heavy Industries dropping out. As in 2009, Arianespace took the lion's share. Its Ariane V ECA dominated the market with about 57% of market share for mainly telecommunications satellites such as Hispasat, Rascom, Astra 3B and so on.[164]

Vega, a smaller European launcher more suitable for low orbit, was expected to be operational during 2011 in spite of repeated delays due to export controls concerns with France.[165] Arianespace's revenue for 2010 dropped by about 10% compared to 2009 forcing it to request some subsidies from governments and the ESA. This raised crucial questions about future European strategies concerning launchers. The situation of the market could change with the arrival and the improvement of the U.S. launcher Falcon 9 from SpaceX which in 2010 carried out one successful launch transporting to NGSO light charges such as Dragon COTS 1, SMDC ONE, QbX-1 and 2, Perseus 0, 1, 2 and 3, Mayflower & CAERUS in one launch.

Revenues from the 23 commercial launch events in 2010 amounted to an estimated $2.45 billion, an increase of $43 million compared to 2009. The increase is therefore less impressive than the previous year but indicates nonetheless that the sector did rather well. Europe confirmed its position with revenue of $1.32 billion, while Russia and U.S. respectively made $826 and $307 million.[166]

---

[164] Ibid.

[165] Peter B., de Selding. "Delays Continue To Affect Vega, European Soyuz Programs." 30 Apr. 2010. Space News 4 Mar. 2010 http://www.spacenews.com/launch/100430-delays-vega-european-soyuz.html.

[166] Federal Aviation Administration. Commercial Space Transportation: 2010 Year in Review...

1.3.4.2 Satellite Manufacturing Sector

In 2010, 110 payloads were launched. 40% of the launched payloads were manufactured by Russia, 21% by the U.S. and 13.6% by China. Europe accounted for only 11% of the payloads launched, India and Japan 6.4% and Israel and South Korea around 1%.[167] Beyond these figures it is important to distinguish between commercial and non-commercial activity, because only 30% of the payloads were commercial, slightly more than in 2009, when they represented 23%. Europe with 33.3% and Russia with 51.5% were the leaders in commercial payloads. The U.S. depended heavily on government programmes because only 9.1% were commercial (Fig. 1.7).

61.5% of the satellites launched were GEO satellites.[168] In this field, 40% of the satellites manufactured came from the U.S. (five by SS/L, two Boeing and three Orbital) while 28.6% from Europe (three by Thales Alenia Space and seven by EADS), 20% China (CAST) and only 3% for Russia which focused on low orbit. In contrast, concerning the non GEO orbit, Europe (nine Thales Alenia Space and three EADS Astrium), Russia and Asian manufacturers took the lead with respectively 28.6% of satellite launches followed by a far measure by the U.S. with 8.9%. National programmes had a big influence, for instance the Glonass constellation completion in MEO or the Beidou navigation system in GEO for Chinese figures.

2009 was a prodigious but unique year in terms of GEO satellites with 40 orders. Unfortunately 2010 was not as successful. The figures were mainly the continuation of 2008 (21 orders) with 26 contracts awarded in 2010. There were also only 8 companies represented in this market whereas 2009 counted 11 different manufacturers. The U.S. companies Boeing and SS/L were each awarded five GEO satellite contracts while their direct followers Thales Alenia and EADS Astrium respectively got four and three. EADS Astrium won a contract of $750 million for the construction of four satellites for SES Luxembourg.[169] The success of Boeing is also due to contracts from Inmarsat of London to build three Ka-band satellites to cover the world's oceans and surrounding regions as part of a $1.2 billion investment in Inmarsat's Global Xpress system.[170]

ISS-Reshetnev was awarded three satellite orders (Express-AM8, Express-AT1 and Express-AT2) but in association with Thales Alenia for the payload[171] while ISS Reshetnev will provide the satellite platforms. This association is going to be extended by Russia's Gazprom Space Systems, RSCC's principal Russian

---

[167] Ibid; Payloads are assigned to the nation that commissioned them, not according to the nationality of the manufacturer.

[168] Ibid.

[169] Space News Staff. "2010: The Year in Review" 13 Dec. 2010. Space News 13 Dec. 2010 http://www.spacenews.com/civil/2010-year-review.html.

[170] Ibid.

[171] Peter B, de Selding. "RSCC Expansion Continues with Three-satellite Order." 22 Sept. 2010. Space News 22 Sept. 2010 http://www.spacenews.com/satellite_telecom/100922-rscc-expansion-satellite-order.html.

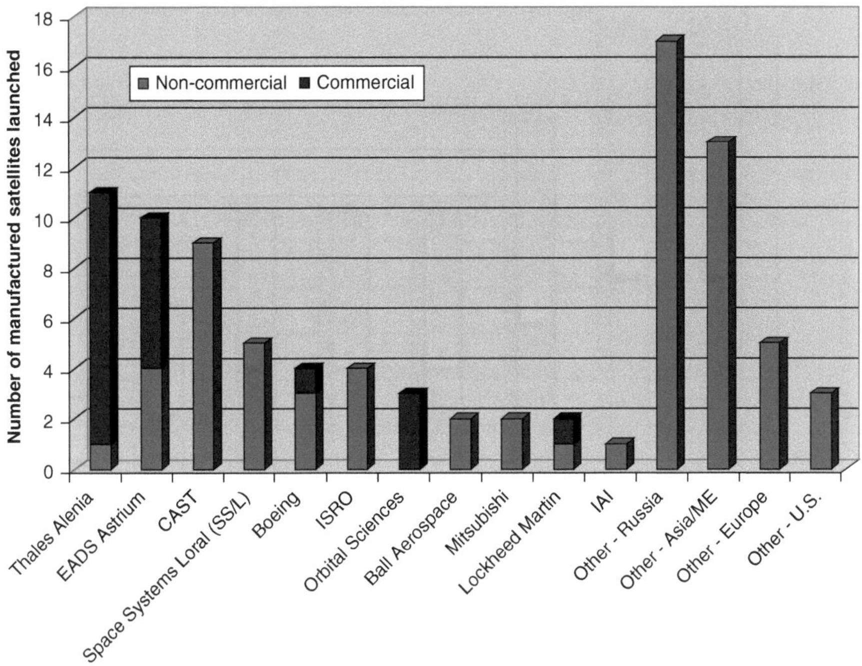

Fig. 1.7 Satellites launched in 2010 by manufacturer and commercial status (Source: Futron)

competitor that has ordered two large satellites from Thales Alenia Space, one of which includes major work by ISS Reshetnev.[172] ISRO confirmed its entrance into the market with two satellites ordered whereas it had just one in 2009 and will not be penalized by its recent GEO launch failure.[173] The core of the competition is thus between U.S. companies with 53.8% of the contracts awarded against 26.9% for European ones. Most of these contracts are made with private entities and are not due to governmental orders, which is often important to take into account especially with U.S. figures. In 2010 several contracts were made between the U.S. Airforce and satellite manufacturers such as Boeing for the seventh global Satcom satellite.[174] The main buyers in 2010 with three satellites ordered were Eutelsat with models W6A, W5A, W3D, Inmarsat, and Mexsat. SES, contrary to the busier previous year, ordered just one.

Of the 98 satellites launched, 26 were commercial. The majority of commercial satellites launched were European or U.S. built: 7 of them were European,

---

[172] Ibid.

[173] K.S, Jayaraman. "ISRO Team Says Cable Rupture Caused Rocket Failure." 31 Dec. 2010. Space News 31 Dec. 2010 http://www.spacenews.com/launch/101231isro-team-says-cable-rupture-caused-rocket-failure.html.

[174] Turner, Brinton. "U.S. Talks with Allies About buying into WGS." 10 Sept. 2010. Space News 10 Sept. 2010 http://www.spacenews.com/military/talks-with-allies-about-wgs.html.

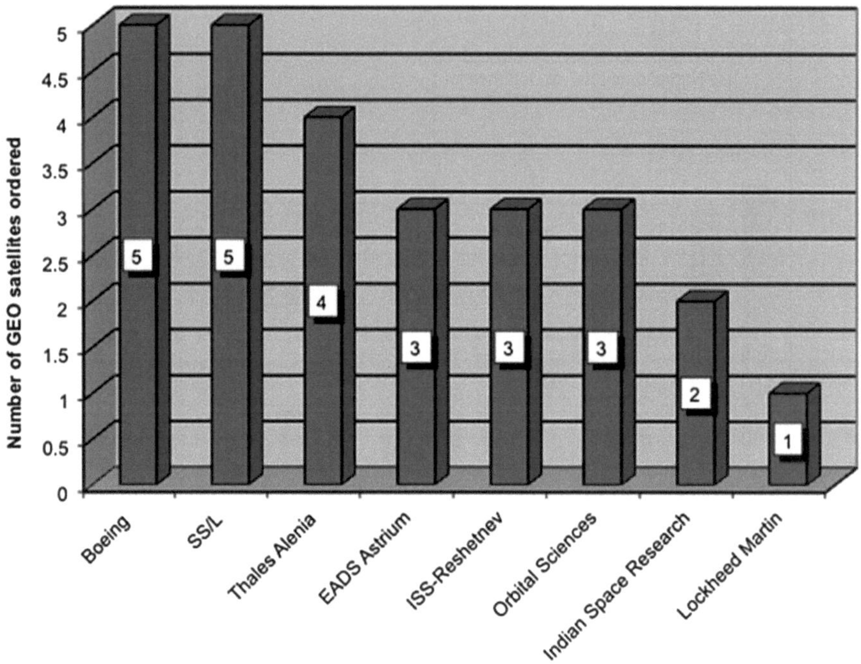

Fig. 1.8 GEO satellite orders in 2010 by manufacturer

representing only 27% of the total number of commercial satellites launched, whereas 17 were manufactured in the U.S., accounting for 65% of the total. Twenty-eight satellites were launched to GEO and 70 into other orbits. When looking at GEO satellites, Europe lost the lead it held in 2009: 25% of the GEO satellites launched in 2009 were European (three made by EADS Astrium and four by Thales Alenia). In contrast, 61% of the GEO satellites launched were U.S. built, whereas Russia accounted for 11% of the total figure (Fig. 1.8).[175]

1.3.4.3 Satellite Operators Sector

As was stated above, satellite operators managed to preserve in 2010 and 2011 their forecasted business growth rates despite the economic turmoil. However, all major operators were acutely aware that this development was at least partly attributable to the industry's long investment cycles. Consequently, they were bracing for the difficult period that might very well lie ahead. In preparation for such an eventuality, all major operators proceeded with restructuring their fleets, hoping to reduce their operating costs in time, while simultaneously expanding their market

[175] Ibid.

portfolios. For example, Luxembourg-based SES planned to spend around $1 billion in 2010 for the replacement of midsize satellites with larger ones and will expand its in-orbit fleet in 30% by 2014. SES also announced that it would cut expenditure in the short term. The company predicted that up to 2014 it would spend 50% less per year, before its next satellite replacement cycle reaches maturity. Intelsat, on the other hand, also expected to decrease spending and it was preparing an initial stock offering to enable its private-equity owners to receive returns from their investment. As a demand driver in the coming years, Intelsat planed to focus on the governmental market. The company's own capital expenditures were estimated at between $825 and $900 million in 2010.[176]

1.4 The Security Dimension

1.4.1 The Global Space Military Context

Space-related military spending remained stable in 2010, reaching a total of $34 billion, or 3% more than in 2009. As in the previous year, the U.S. military budget represented the bulk of global spending, with $27 billion, or over 80% of the total.[177] However, this figure only takes into account the U.S. Department of Defence budget. If the total sum of U.S. spending for national security purposes (including intelligence gathering and mapping services) is taken into account, total U.S. expenditures rise to $46 billion and their global share close to 90%. Furthermore, even this latter figure should be considered conservative as it does not take into account U.S. government programmes of a dual use nature, such as the use of NOAA weather satellites' data by the military.

It is believed that Russia and China followed the U.S. in military space spending in second and third place respectively, but exact figures are difficult to calculate since they are entirely classified. According to some sources, the Russian budget could be comparable to the country's civil space budget of $2.8 billion and China's could reach as much as $1 billion.[178] Other reports calculate the total figure of worldwide military space expenditures outside the U.S. at less than $3 billion, suggesting much more conservative spending levels for both countries.[179] Furthermore, fluctuating currency exchange rates, as well as different purchasing power levels and employment costs in these countries would make a direct comparison of their budgets a very relative exercise.

[176] De Selding, Peter B. "Fleet Recapitalization Cycle Likely to End in 2-3Years." Space News, 5 July 2010: 12.

[177] Based on Euroconsult data.

[178] Ibid.

[179] Space Foundation. The Space Report 2011. Colorado Springs 2011.

Concerning developed regions outside the U.S., in 2010 and 2011 European budgets were the most significant, with a total estimated value of $770 million, excluding dual use systems. This would imply that the downward trend in European military space observed in 2009 and 2010 has continued, but it should be considered as a natural result of the European programmes' investment cycles, rather than specific policy decisions to limit them. Indeed, during the period in question a number of programmes concluded their operational development, while their successors had not yet reached their full development phase. For the same reason, this downwards trend is expected to be reversed in the next 2 years. At the same time, the development of the GMES Earth observation constellation is expected to increase the part that dual use systems play in the European security space architecture, consequently increasing their budget share compared to the purely military spending of member states. It should also be kept in mind that spending is not always clearly allocable, because some budget positions can be assigned to various categories. Finally, in 2010 and 2011 Japan maintained its position as the country with the fifth largest military space budget, mainly related to the development of its intelligence gathering satellite constellation. Its budget remained stable at approximately $650 million, mainly driven by the Japanese government's continued commitment to improving its intelligence apparatus in the face of the continuously unstable geopolitical environment in the region.

1.4.2 Europe

Military space programmes in Europe continue to belong to the competence of individual member states. They are carried out by all major European space faring nations, with an even larger number of European countries participating on the basis of bilateral or multilateral agreements and arrangements. Although the associated European budgets dropped significantly in 2010 and 2011, to a great extent this can be attributed to a curve in the investment cycle in a number of projects, and budgets are expected to return to their average figures in the medium term. However, the trend of outsourcing services and reducing public budget costs through PPPs has appeared in the field of security space applications too. Together with the increasing number of cooperation agreements, it demonstrates the willingness of European countries to maintain key operational capabilities through cooperation and innovative funding schemes. The key security space mission areas for European countries include Earth observation and dedicated military satcom services.

France has the highest investment in national military space programmes, believed to amount to approximately $264 million in 2010. However, the exact amount is difficult to calculate, since almost all of the development work is conducted under the country's civil space agency CNES. This figure is considerably lower when compared to 2009, mainly due to the conclusion of the development of key programme, such as the Syracuse-3 satcom spacecraft. Earth observation satellite development also remained at historical lows, as the Helios programme completed its development and the upcoming Pleiades system has not entered its full development phase yet. Nevertheless, related spending is expected to increase in the next 2 years, as the programme will mature further. In fact, a first sign of this development appeared already in 2010, since the French defence Earth observation budget more than doubled to $86 million, from $30 million in 2009.[180]

Germany reported the third highest spending in military space programmes in 2010, mostly thanks to the deployment of its dedicated SatCom BW military communications satellites, with the two spacecraft constellation completed in 2010. The United Kingdom, which was second with roughly $215 million, also dedicated the bulk of its budget to pay the Skynet-5 constellation operator for its use. Although the system performed as expected, increased communications needs that were the result of the UK's forces multiple engagements all over the world, and especially in Afghanistan, almost saturated the network. The Italian defence budget also decreased, for the same reasons as in the case of France, especially as far as Earth observation missions were concerned, with the launch of the fourth and last Cosmo-Skymed satellite in 2010 and its next generation satellites not scheduled for launch before 2016.

1.4.3 The United States

In 2010 and 2011 the U.S. Department of Defence remained the biggest spender by far in the field of military space activities, accounting for at least 80% of expenditures worldwide. This figure increases to almost 90% if one considers the overall security related space programmes and services, which include the related budgets of the National Reconnaissance Office, the national Geospatial Intelligence Agency, or Homeland Security; dual purpose assets, such as NOAA's weather satellites could also be included, insofar as they provide service to military as well as civil authorities. However, for the purposes of this report only DoD programmes will be examined, principally for reasons of comparability with the equivalent expenditures of other space faring nations.

In 2010 and 2011 U.S. military programmes continued their trend of annual increase that has been uninterrupted for over a decade. However, the increase in 2010 was limited to approximately $800 million, or only half of the annual average reported in the previous 10 years. This relative stabilisation might be the first sign of a longer period of reduction in military space spending in the U.S. Although this trend will have to be confirmed in the following 10 years, there are certain factors that may contribute to its development. These include the Obama administration's decision to cut military budgets in the medium term, the general trend in U.S. military procurement to outsource services to the private sector (especially in the

[180] Based on Euroconsult data.

fields of EO and satcom), as well as to reduce operating costs by replacing non critical space assets with hosted payloads in commercial spacecrafts. Furthermore, the fact that a number of DoD programmes that have been under development in recent years will enter their deployment phase is expected to reduce the current level of investments.

For the time being however, the U.S. Department of Defence continues the development of a number of dedicated military systems, with special emphasis on the mission areas of navigation, observation, communications and space situational awareness. Regarding satellite navigation programmes, efforts to develop the new generation of GPS satellites continued in 2010 and 2011. Lockheed Martin, which had been contracted in 2008 to build the 12 new spacecraft for a total budget of $3 billion, continued its development work, but concerns continue to be raised on whether the new satellites will start their deployment in 2014 as planned, to replace the ageing second generation spacecraft. If these concerns are confirmed, a gap in the system's performance could arise in the middle of the decade. On the other hand, the key trend of outsourcing services to the private sector can be observed in the field of Earth observation, where DoD expenditures in 2010 fell further to $432 million. This was primarily the result of the NPOESS cancellation, as well as of the entry into effect of the Enhanced View commercial EO services purchase contract. Military satcom was also an area of activity in 2010 and 2011, with the entire military satcom fleet under upgrading.

1.4.4 Russia

Russia has a long tradition of space military activities, and a great number of its launches in 2010 were classified.[181] Rather than investing in new projects, 2010 expenditure was dedicated to completing previous projects, such as the Glonass GNSS system. Russia has thus invested $925 million[182] in its navigation system overtaking, for the first time, the U.S in navigation systems investments. The failure of a rocket carrying three Glonass satellites[183] prevented Russia in 2010 from successfully achieving its constellation by the beginning of 2011 as previously expected. Although in principle a dual use system, Glonass is also funded by the military and provides a standard precision signal for civil purposes, and a high precision signal for military ones. It is currently composed of 26 satellites and could become a crucial strategic asset for Russia, if it attracts international customers. Negotiations are under way with countries such as Ukraine, Egypt, Brazil and former CIS states.

[181] Federal Aviation Administration. Commercial Space Transportation: 2010 Year in Review...

[182] According to Euroconsult data.

[183] See above.

Another priority of the Russian national security space programme is the need to replace aging military communication satellites. The Russian army operates three types of spacecraft that are due to be modernised: the Raduga and Meridian constellations are likely to be progressively replaced by one launch per year, while the Rodnick system will have three per year. The latter is already under modernisation, providing an enhanced service according to officials. In the same way, Russia is increasing its efforts to renew its earth observation fleet with a dedicated budget of $294 million. Since its content is classified, it is particularly difficult to accurately determine Russian policy in this field. However, it is known that two of four radar satellites are approved to be launched as part of the Kondor series. Russia has also planned to launch several low-cost Kobalt new generation reconnaissance spacecraft in the coming years.

As often with Russia, it is very difficult to gauge its military space effort. The country has had significant experience in developing ELINT systems and it is likely that it focuses its effort in systems such as early warning series (Kosmos/Tselina and Prognoz), which is a priority for Russia that already owns three operational early warning satellites. A new generation of such spacecraft has been recently reported, such as Lotos-S launched in 2009, and Oko which will be integrated in the SPRN constellation. The network, which also includes dedicated ground control, is designed to detect and track global ballistic missile launches and the spacecraft in the system uses telescopes to detect infra-red radiation emitted by the exhaust of the rocket engines. Its budget is estimated at $350 million, representing a 3% increase compared to 2009. The country also envisages a programme concerning space debris mitigation that would focus on sweeping satellite debris, but little information is known about it.

1.4.5 Japan

Officially, the Japanese space programme does not have a military component and is entirely oriented to the peaceful use of space. However, the uncertain geopolitical environment, especially with respect to developments in the Korean peninsula, has raised concerns over the past 10 years about Japan‘s capacity to reconcile its peaceful space programmes with its own security requirements. The latter have gradually become a key aspect of its national space policy. This trend is mainly reflected by the constant $500–600 million spent each year, mainly directed to intelligence gathering capabilities. Although the Japanese budget has slightly decreased compared to 2009, it can be considered stable. However, a further slowdown cannot be excluded due to the particularly harsh impact of the economic crisis on the Japanese economy, as well as recent natural catastrophes which will presumably consume a great deal of public spending that under different conditions would be spent elsewhere.

The core of the Japanese space security programme is Earth observation. It includes three areas, namely intelligence gathering for national security purposes and for assisting peacekeeping operations; continuous observation and surveillance for rapid response, presumably focused on the North Korean missile development programme; and information sharing and communication in support of Japan's obligations under its alliance with the U.S. Moreover, Earth observation has the advantage of being used for both military and civil purposes. The base of the Japanese security Earth observation effort is constituted by the Information Gathering Satellites (IGS) devoted to providing information to the MoD and the Japanese Defence Agency (JDA). The system is made up of optical satellite whose development started in 1998 and is so far regularly upgraded by newer satellites manufactured by Mitsubishi Electric. The overall cost of the MoD IGS programme is around $700 million, accounting for the total of Japanese security related expenditures in space. Currently, the IGS system consists of two optical and two radar imaging satellites. The next generation of its satellites will be designed to be more compact, and it is expected that an optical and a radar imaging spacecraft will be launched in 2011 and 2012. Further satellites currently under development are already scheduled to fly towards the mid of the decade. They are expected to benefit from greatly improved operational performance that will include higher resolution (0.4 m instead of 0.6). The first spacecraft of the series will be IGS-Optic 5 (scheduled for launch in 2014).

Satellite navigation is also an important part of the Japanese effort and, like EO, its utilisation could serve both military and civil policy objectives. Jaxa continues the deployment of its navigation constellation named "Quasi-Zenith Satellite System" (QZSS), which will be composed of three satellites and will be devoted to regional navigation over Japan and the Asian Pacific area. This system should constitute a strategic asset in the Asian-pacific region for Japan, since both South Korea and Australia, Japan's closest allies in the region, are expected to benefit from the system. Like its European counterpart, there is a sense of uncertainty as to how to finance the project, especially given its failure to attract private funds. The system can only be operational with three satellites in orbit, which could cost as much as $2.2 billion to complete.

In spite of the fact that Japanese authorities continue to adhere to the use of dual use systems, the officially announced satellite development programmes bring the country's space applications closer and closer to traditional security and defence mission types. This was mainly noticeable from preliminary work on the development of a new ELINT/SIGINT satellite, as well as from the country's involvement in the U.S. led joint ballistic missile defence system that has inspired the development of an infrared early warning satellite to monitor regional missile launches by 2015. Consequently, space surveillance and infrared missile warning sensor technologies were also included in the country's medium term security space research objectives. Although the deployment of such satellites can hardly be described as of dual use nature, Japanese authorities continue to be discrete on the subject, whereas their respective budgets continue to be extremely reserved.

1.4.6 China

Estimating the Chinese effort in space military programmes is a difficult task. Beyond the official statements, it is not always easy to discern the strict reality. China has unquestionably made a great endeavour to catch up in military space capabilities and establish itself as a major actor. The evidence of this ambition is the steady augmentation of its space budget dedicated to military activities, to approximately \$1.108 billion in 2010,[184] which demonstrates its commitment to independent capabilities in the most visible and crucial space activities that are represented by launchers, a space station and manned flights.

The manned flight budget is estimated to have tripled over the past 10 years to \$624 million[185] representing about 24% of the total Chinese budget, and in 2003 China became the third country to acquire manned spaceflight capabilities with the Shenzhou-5 spacecraft, followed by significant successes through the Shenzhou programme in 2007 and 2008. It is important to recall that these programmes and the development of launchers are managed by the People's Liberation Army and must be therefore be considered as defence expenditure. Another facet of this phenomenon is the strong commitment for China to establish its own space station by 2022 outside the ISS. However, that does not prevent China from developing cooperation with NASA and other agencies so long as that does not entail any form of technological or strategic dependence. China's perspective concerning the future "Tiangong station" is scheduled in this way: firstly improve the technique of "rendezvous" in space and secondly launch in orbit by 2012–2013 a small laboratory where astronauts could live for short periods. The next step is to improve life support systems and logistics by 2013–2014, to have finally by 2016 a station of 13 t and a capacity of up to 5.5 t able to dock two manned Shenzhou vessels and one cargo spacecraft. The ultimate objective is to have a station of 30 t by 2022. It is a necessary for China to develop efficient launchers to cope with its high objectives, this is the reason the launcher programme is the second largest budget in its global space effort with \$484 million in 2010. The success of the development of Long March 5 and 6 will be crucial in the near future to achieve its ambitions.

Anther crucial area is navigation and communication satellites which once more illustrate the will of China to acquire its own constellation in this critical field related to both the military and public domains. Concerning navigation systems, we can easily see the strong commitment of China, which in 2010–2011, launched numerous satellites destined for its Beidou constellation.[186] Its investment in the Satnav project has progressively increased from \$100 million in 2005 to \$300

[184] Based on Euroconsult data.

[185] Ibid.

[186] Federal Aviation Administration. Commercial Space Transportation: 2010 Year in Review…

million in 2010.[187] The system has the advantage of combining quick positioning, low data rate mobile telecommunication and short messages. The generation Beidou II should count 5 satellites GEO, 5 IGSO and 4 MEO to provide comprehensive coverage of the Asia Pacific region, and will be achieved by 2020 with the third generation comprising 5 GEO, 3 IGSO and 27 MEO to be a constellation of 35 satellites.

In the same way the Satcom policy aims to reduce dependence on foreign technologies and use the platform developed to penetrate the international market. That has been accomplished so far through the DongFangHong-4 (DFH-4) satellite platform. Actually, China has seen a slow down in this kind of activity, preferring to develop low cost satellites for emerging countries in a less competitive market than those already dominated by Europe. Significant augmentation of the governmental Satcom capacity is foreseen, with one launch by year end, in order to improve and introduce new services in S-, L- and Ka-Band. An S-Band geostationary satellite will be also launched around 2015 to provide maritime communication especially for the Chinese Navy to protect the commercial fleet against pirate attack on the strategic route near Somalia.

Earth Observation (EO) is also a critical part of military programmes and China is not an exception. In this specific area the budget of China is not disclosed but it is estimated to be up to $467 million with a significant increase from 2009 of $63 million. The main strategies envisaged by Chinese authorities are aimed at developing independent technologies and fostering an industry capable of responding to civil and military needs. The latest Yaogan satellites (YG-6, YG-7 and YG-8), categorized generically as remote sensing satellites for science, land applications and disaster monitoring purposes, as well as two ZY-2 series, could also be used for reconnaissance purposes. Moreover Fanhui Shi Wixing is a new series of reconnaissance satellites directly controlled by the People's Liberation Army. China's EO capacity merges civil and military proposes.

China is thus making a big effort to catch up in all space sectors, being determined to keep its firm independence and take the lead in certain domains. There is a special interaction between the military and civil purposes specific to Chinese society, which has chosen an offensive strategy in space.

1.4.7 India

India does not have dedicated military space programmes and ISRO's activities concern mainly civil preoccupations. This situation could change in the next decade due to the evolution of the geostrategic situation in this area. Pakistan and China provoke growing concern in India especially since the Chinese anti-satellite test in 2007. In 2008, Gen. Deepak Kapoor said at a conference in New Delhi on using

[187] Based on Euroconsult data.

space for military purposes that India urgently needs to "optimize space applications for military purposes.)[188] He noted that "the Chinese space program is expanding at an exponentially rapid pace in both offensive and defensive content." It is now expected that India will accord increasing importance to military purposes, especially concerning the opportunity to develop its own ASAT capability[189] to counterbalance and thwart the Chinese threat and consequently the Pakistani one as well. This will be done within an overall effort to improve the security of the country especially by developing a working missile defence system by 2012. In this atmosphere it is unlikely that India will neglect the considerable asset that space represents in military supremacy. The country intended to launch its first satellite totally and officially dedicated to military activities to support the Indian navy in the beginning of 2011,[190] however, the launch date has subsequently been pushed to the end of 2012 due to repeated failures.[191]

The direct military investments of India are difficult to asses. However, dual use technologies are more discernable and can provide an insight into India's strategy. The Indian National Satellite System (INSAT) is the largest domestic satellite communication system in Asia with 11 operational satellites and a total of 186 transponders. The 11th National Space Plan (2007–2012) anticipated the launch of 13 satellites over the same period and expects to reach a total capacity of 500 transponders in 2012. This endeavour has been hampered by the successive failures of the GSLV in 2006 and 2007 but still India has a very efficient constellation concerning communication. The government is strongly committed to breaking the isolation of remote areas particularly prone to be influenced by various local insurgencies.

In the same way, the Indian regional navigation satellite system (IRNSS) consisting of seven satellites with three GEO and four near-GSO satellites is due to be operational in 2012. IRNSS will be a crucial military asset as it will include positioning and missile targeting. Funds for this programme reached $57 million in 2010, accounting for a modest 5% of the agency's budget, exceeded by far by the launcher programme estimated around $531 million. The 11th plan is particularly devoted to national security issues and could represent a cornerstone in the way India considers the use of space. A payload containing GSAT-4 was lost in April 2010 in a launch failure. Two additional dual-frequency payloads will be on GSAT-8 and GSAT-10 set to be launched in 2012 to complete the fleet.

---

[188] Rabinowitz, Gavin. "Indian army wants military space program." 17 June 2008 NBC News 1 Sept. 2010 http://www.msnbc.msn.com/id/25216230/ns/technology_and_science-space/t/indian-army-wants-military-space-program/.

[189] Samson, Victoria. "India's missile defense/anti-satellite nexus." 10 May 2010. The Space Review 1 Sept. 2010 http://www.thespacereview.com/article/1621/1.

[190] Defense News Admin. "India Making Strides in Satellite Technology." 10 Nov. 2010. Defense News 10 Nov. 2010 http://www.defencenews.in/defence-news-internal.asp?get=old&id=239.

[191] Pandit, Rajat. "Military satellite delayed again by a year." 13 Oct. 2011. The Times of India 4 Aug. 2012 http://articles.timesofindia.indiatimes.com/2011-10-13/india/30274617_1_gsat-7-satellite-indian-space-research-organization.

Another fundamental dual use segment is represented by earth observation. India has shown a strong desire for independence concerning EO data by developing rather ambitions programs with civil objectives such as oceanography or to help governmental public policy in territory management. Numerous satellites are due to be launched for civil purposes, such as DMSAR-1 radar mission and GISAT, representing a budget of approximately $77 million in 2010–2011, and it is probable that they will be used for military intelligence as well. It is also expected that India will continue its series of Cartosat satellites in the next few years, which are equally appropriate for civil and military purposes.

Thus, India is step by step revealing its strategy in relation to the threatening environment in which it lives. There is a noticeable imbalance between civil and military assets in India, and it is probable that in the future the military sector will not be neglected any more given that the challenges the country has to deal with. The first necessity for India is still to develop a reliable rocket to support its military efforts which mostly depend on costly satellites.

Chapter 2
Developments in Space Policies, Programmes and Technologies Throughout the World and in Europe

Spyros Pagkratis

2.1 Space Policies and Programmes

All major space policy developments worldwide were presented in the previous section of Part I, in an attempt to clarify the principal space faring nations' strategies in 2010 and 2011. In the section bellow, there will be a brief discussion on developments in technology related areas, including access to space technologies and policies. The aim of this section is to clarify how the strategies already presented above interact with and influence specific space programmes and related research and development projects.

2.2 Space Transportation

2.2.1 Europe

European launcher development programmes are funded almost exclusively by ESA. These have been progressively decreasing since 2000, and there is considerable uncertainty about the way forward for European launcher technology.

The commercial operations of Arianespace led to significant losses in 2010 and a need for aid, said to be necessary to offset the adverse effect of currency movements.[1]

[1] De Selding, Peter B. "Arianespace Needs aid to avoid loss in 2010." 4 Jan. 2011. Space News 4 Jan. 2011 http://www.spacenews.com/civil/110104-arianespace-needs-aid.html

S. Pagkratis (✉)
ESPI, Vienna, Austria
e-mail: spyros.pagkratis@espi.or.at

P. Hulsroj et al. (eds.), *Yearbook on Space Policy 2010/2011*,
Yearbook on Space Policy, DOI 10.1007/978-3-7091-1363-9_2,

2.2.2 United States

The new push of the U.S. government towards cooperation in space was made evident in the public discourse of NASA officials immediately after the new space policy's announcement. For example, Lori Garver, NASA's Deputy Administrator said, during the Berlin Air Show that NASA would see its budget reduced by 50 % in non-human spaceflight areas if the Obama's Administration had not cancelled the Moon-focused Constellation programme. On the contrary, she praised the new focus on international partnerships, which was very well received by its German audience and could be seen as intending to invite collaboration in space exploration in areas once reserved for U.S. technologies, including possible European contributions to a U.S. led space exploration programme. Garver affirmed that the policy to keep non-U.S. contributions off the critical path has not been fully put into practice and one example was the fact that the U.S. will rely on Russia to ferry U.S. astronauts to the ISS. As she said, any definition of "critical path" would certainly include crew-transport systems. The Deputy Administrator also qualified the ISS's life extension as a nod to the concerns of the NASA's space station partners. European and Japanese laboratories were among the last major station components to be launched to the orbital complex by the U.S. space shuttle and if the ISS life is not extended, they will have little time to operate facilities.[2]

At the same time, through its "Broad Area Announcement" (BAA) NASA issued a call for "affordable" heavy-lift launch vehicle concepts that could be used by multiple entities, such as the Department of Defence, commercial corporations and international space agencies. Proposals should be presented by 29 June 2010. The total funding for the project was around $8 million and no single contract was expected to exceed $625,000. The final selection was expected in 2015.[3]

In a separate development, on August 5, the U.S. Senate passed a NASA Authorization Bill that added a space shuttle flight to 2011 as well as $1.3 billion for the proposed commercial crew initiative over 3 years. This bill required NASA to start working on a heavy-lift rocket capable of supporting manned missions beyond low Earth orbit. Sen. Jay Rockefeller, Chairman of the Senate Commerce, Science and Transportation Committee considered this bipartisan bill to help in refocusing and reinvigorating NASA. The House Science and Technology committee approved a $3.3 billion investment in a commercial crew transportation system over the next 3 years. This House Bill would also permit the continuation of much of the work being done under NASA's constellation programme, an effort to build new rockets and spacecrafts optimized for lunar missions.[4]

On the other hand, however, efforts to increase space transportation commercialisation increased. For example, Armadillo Aerospace of Rockwell, Texas and Masten Space Systems of Mojave, Calif., were awarded a contract through

[2] "Garver: New NASA Direction Opens Door to Cooperation." Space News 14 June 2010: 13.

[3] "NASA Seeks Affordable Heavy-lift Rocket Ideas." Space News 5 July 2010: 8.

[4] "NASA Authorization Bill Passes Senate." Space News 9 Aug. 2010: 3.

NASA's commercial Reusable Suborbital Research Programme (CRuSR) (for a total of $475,000) to launch seven test flights.[5] NASA officials considered this contract as the beginning of an innovative teaming relationship with private industry, in order to provide affordable access to space. Armadillo is currently building three unmanned launches of its Super-Mod vehicle from Spaceport America in New Mexico. Masten Space System's Xaero vehicle made three unmanned test flights during the winter of 2010.[6] Super-Mod and Xaero vehicles would use the global navigation satellite systems to determine their position. They are also able to broadcast position data to ground stations. In order to advance their technology and business, Masten signed a deal with XCOR Aerospace to develop unmanned launchers that could be used for Moon, Mars and asteroids missions. On the other hand, Armadillo struck a deal with Vienna, Va.-based Space Adventures which booked space tourism flights to the ISS; planning to offer seats on suborbital rocket ships that Armadillo started developing. Each trip would cost $102.000.[7]

2.2.3 *Russia*

Improving its space transportation capabilities was a key plank of general Russian space policy in 2010 and 2011. Efforts to modernise relevant space and ground infrastructure continued, while use of the aging but reliable Soyuz rocket continued. The programme's budget has experienced a tenfold increase over the past decade, bringing it to $617 million in 2010, or over 20 % of the total space budget. Furthermore, the programme's total expenditures are believed to be even higher, as it also receives funding from the military. The core of the programme consists of the simultaneous development of the new Angara launcher and the construction of a spaceport certified for human space launch at Vostochny, in the Russian Far East.[8] Significantly, the $800 million long-planned launch facility in Vostochny started its construction already in 2011, ahead of schedule, and was expected to be completed by 2015. Vostochny will provide an alternative to Baikonur Cosmodrome, allowing Russia to launch from its own territory.[9] On the other hand, after a long development period the Angara launcher is expected to fly for the first time in 2013. The rocket will incorporate significant improvements compared to its predecessors that will bring it closer to European and U.S. standards, such as a modular design easily

[5] Space.com Staff. "NASA Funds Test Flights for Private Experimental Rockets." 31 Aug. 2010. Space.com 5 Aug. 2012 http://www.space.com/9042-nasa-funds-test-flights-private-experimental-rockets.html

[6] Werner, Debra. "Suborbital Spaceflight Gets a Boost from NASA, Congress." 18 Oct. 2010 Space News 5 Aug. 2012 http://spacenews.com/venture_space/101018-suborbital-spaceflight-boost-nasa-congress.html

[7] "NASA Grants Aim To Foster Manned Suborbital Flights." Space News 6 Sept. 2011: 8.

[8] Based on Euroconsult estimates.

[9] "Russia to Break Ground In 2011 for Spaceport." Space News 30 Aug. 2010: 3.

adaptable to different mission and payload profiles, as well as integration of the latter on the launch pad. These improvements are expected to allow Russia to extend its position in both government and commercial launch activities worldwide, by increasing its launching capabilities' performance and affordability.

2.2.4 Japan

Regarding Japan's launching capabilities, a significant development arose when JAXA announced that it would be now able to use its Tanegashima Space Centre all year around, lifting longstanding restrictions that limited its activities there. This was the result of an agreement between the agency and local fishing unions, which were concerned about the spaceport's impact on their local fishing grounds. JAXA policy called for a re-examination of those restrictions, blaming them for Japan's inability to be competitive in the global commercial launch market. However, a theoretical ceiling of 17 launches per year would be maintained, according to JAXA.[10]

From Tanegashima Space Centre, Japan launched the Quasi-Zenith satellite on 11 September 2010. The satellite, built by Mitsubishi Electric Corp., was designed to rebroadcast enhanced GPS navigation signals to hard-to-reach areas in Japan. It was launched aboard a Mitsubishi Heavy Industries- built H2A rocket.[11]

2.2.5 China

During the reporting period, Chinese space launch activities continued their medium-term growth. For example, on 15 June, a Chinese Long March 2D rocket launched the Shijian 12 scientific research satellite, from Jinquan Satellite Launch Centre in China's Gansu Province. This satellite was developed by the Shanghai Academy of Spaceflight technology.[12] In addition to this, on 22 September 2010, at 10:42 a.m., China launched the Yagon11 remote sensing satellite from Jinquan Satellite Launch Centre, the fifth launch in 7 weeks. The satellite was placed into orbit atop a Long March 2D rocket and is used to conduct scientific experiments and help with natural disaster response.[13]

Great Wall Industry Corp. recovered from its 2009 underperformance during the launch of Indonesia's Palapa-D telecommunications satellites. During 2011, China launched three satellites for non-Chinese customers: the W3C satellite for Eutelsat;

[10] "Deal Permits Year-round Launches from Japan." Space News 9 Aug. 2010: 9.

[11] "Japan's H2A Rocket Lofts Quasi-Zenith Sat." Space News 20 Sept. 2011: 8.

[12] "Long March 2D Rocket Lofts Chinese Research Satellite." Space News, 21 June 2010: 9.

[13] "China Launches 5th Satellite in 7 Weeks". Space News 27 Sept. 2011: 3.

Paksat 1R for Pakistan and Nigcomsat 1R for Nigeria. China Great Wall affirmed that the anomaly in the helium-propulsion on China's Sinosat 6 satellite could delay the launch of the Pakistani and Nigerian spacecraft. The Chinese launcher has proceed with upgrades that have increased the Long March 3B rocket's payload-carrying ability from 5,200 to 5,500 kg; they also reduced the amount of time needed for launching to 25 days, permitting the vehicle to launch 10 times per year. China Great Wall has enough business to conduct 20 launches per year in the next 5 years, counting only domestic launches.[14]

2.2.6 India

India's space launch programme is based around the development and exploitation of two rockets, the PSLV and the GSLV, capable of carrying payloads to Low and Geosynchronous Earth Orbits respectively. In 2010 and 2011 PSLV marked two successful launches. First, on 12 July 2010 it placed in orbit Cartosat 2B, an Indian Earth observation satellite of the IRS series, together with a cluster of experimental small satellites for scientific research purposes that included Algeria's ALSAT 2A, Canada's AISSAT 1 and Switzerland's TISAT 1. Second, on 20 April 2011 it launched RESOURCESAT 2, as well as two research satellites: YOUTHSAT 1, a joint project by Indian and Russian universities, and X-SAT, a technology demonstration small EO satellite from Singapore that was that country's first spacecraft.[15]

Contrary to the PSLV's successful and reliable record, ISRO has met with some technical difficulties in the development of the country's heavier launcher GSLV. In its present form, the rocket weights over 400 t and is capable of lifting up to 2.5 t of payloads into GTO, using two stages equipped with liquid fuel engines and a third stage with a solid fuel one. Its principal mission is to orbit India's new and heavier communication satellites of the GSAT series. Its inaugural flight took place in 2001, but progress with its development has been slow and uneven, a fact that is also demonstrated by its moderate success rate of three out of seven unsuccessful flights, with the last two consecutive. During the reporting period GSLV faced one such catastrophic failure on 25 December 2010, when it exploded 60 s into ascent, destroying the GSAT-5P spacecraft it was carrying. Despite the GSLV's mixed success rate, ISRO kept up the pace of development of its even heavier rocket GSLV Mk3, scheduled for flight for the first time in 2012. This new launcher would in fact be an almost entirely new rocket, and not a simple upgrade of the GSLV series, with which it would only share a certain number of components. With a lift-off weight of 630 t and a lift capacity of 5 t into GTO, the system would be the heaviest Indian rocket yet and would offer the country full autonomy in the launch

[14] "China Great Wall Shakes Off Satellite's Botched Launch". Space News 27 Sept. 2011: 8.

[15] PSLV. Indian Space Research Organization http://www.isro.org/Launchvehicles/PSLV/pslv.aspx

of its new generation INSAT communication spacecraft. Finally, it is noteworthy that GSLV Mk3 would field for the first time a second stage equipped with a re-startable liquid fuel engine that would greatly improve the system's operational flexibility and commercial attractiveness.[16]

2.2.7 *Emerging Actors*

On 10 June 2010, South Korea suffered a second space launch failure with the vehicle consisting of a modified Angara first stage manufactured by *Khrunichev*, and a South Korean solid-fuelled upper stage, launching from the Naro Space Center. After 136 s into the flight, the telemetry data downlink was terminated which demonstrates that an unexpected event occurred. A senior government official said that the vehicle was believed to have exploded. The cause of this incident was investigated by Korean and Russian experts. The Korea Space Launch Vehicle (KSLV)-1 featured a Korean-developed second stage and payload fairing.[17]

2.3 Space Science and Exploration

Space science goals include understanding the origin, evolution and future of the Universe, galaxies, our Solar System, stars, exoplanets, comets and asteroids. Exploration is understood to be human spaceflight, or robotic missions that prepare for it. These aspects of spaceflight often command the most prestige, visibility and funding. Therefore, they are dominated by publicly-funded projects carried out by the traditional governmental space agencies. However, the past year has also seen continued progress in commercial exploration, and emerging space powers have begun to demonstrate the technology needed to carry out such missions.

2.3.1 *Human Spaceflight Activities*

The focus of human spaceflight activities continued to be the International Space Station (ISS): with its use formally extended until at least 2020, it provides the opportunity for continued human presence in Low Earth Orbit (LEO) until that time. NASA continued to fly the Space Shuttle's last few missions, and authorised a

[16] GSLV MARK III. Indian Space Research Organisation http://www.isro.org/Launchvehicles/GSLVMARKIII/mark3.aspx; and Bergin, Chris. "FAILURE: Indian GSLV fails during launch with GSAT-5P satellite." 25 Dec. 2010. NASAspaceflight.com 25 Dec. 2010 http://www.nasaspaceflight.com/2010/12/indian-gslv-launch-with-gsat-5p-satellite/

[17] "South Korean Rocket Fails For the Second Time." Space News 14 June 2010: 3.

final mission, STS-135 launched on 8 July 2011. Roscosmos maintained its regular supply of cargo and crew to the ISS using Progress and Soyuz, with Europe's Automated Transfer Vehicle (ATV) and Japan's H-II Transfer Vehicle (HTV) complementing this.

ESA's six new astronauts, Samantha Cristoforetti, Alexander Gerst, Andreas Mogensen, Luca Parmitano, Timothy Peake and Thomas Pesquet, graduated from their basic training at the European Astronaut Centre (EAC) in Cologne, Germany on 22 November. Luca Parmitano was the first to be assigned to a flight, ISS Expeditions 36/37, scheduled for 2013.[18]

Two members of the ESA Astronaut Corps flew. Paolo Nespoli was launched on Soyuz TMA-120 on 15 December on the MagISStra mission. He carried out an intensive programme of experiments, ranging from radiation monitoring to measurements that could improve oil recovery in petroleum reservoirs.[19] Roberto Vittori was launched on STS-134 on 16 May on a flight opportunity provided by the Italian Space Agency (ASI), in agreement with NASA. His DArk MAtter (DAMA) mission had the primary objective of installing the Alpha Magnetic Spectrometer (AMS) on the ISS, a particle physics experiment designed to search for dark matter and antimatter by studying cosmic rays. During his 16-day mission Roberto worked as a mission specialist with the Space Shuttle and its robotic arm to complete six ASI-sponsored experiments, and conducted crew performance tests.[20]

The second Automated Transfer Vehicle (ATV-2), Johannes Kepler, was launched from Kourou on 16 February, and docked with the ISS 8 days later.[21] Whilst attached to the station, ATV-2 performed six releases of oxygen, six altitude control sequences, seven ISS reboosts, one debris avoidance manoeuvre, and delivered 850 kg of fuel and 1,600 kg of cargo to the orbital outpost.[22] After undocking on June 20, it was commanded to burn up in the atmosphere. The Advanced Re-entry Vehicle (ARV), a development of the ATV with the capability to return cargo to Earth and possibly eventually carry humans, underwent its Phase A preliminary requirements review during the summer.[23] However, the future of the vehicle remains unclear.

---

[18] "ESA – Human Spaceflight and Exploration – Astronauts – Graduation of Europe's new astronauts." 22 Nov. 2010. European Space Agency 25 Aug. 2011 http://www.esa.int/esaHS/SEMRFLIRPGG_ astronauts_0.html

[19] "ESA – MagISStra – About the MagISStra mission." 20 May 2011. European Space Agency 18 Aug. 2011 http://www.esa.int/esaMI/magisstra/SEM7E6QOHEG_0.html

[20] "ESA – DAMA mission– A DAMA for Roberto." European Space Agency 18 Aug. 2011 http://www.esa.int/SPECIALS/DAMA_mission/SEM4UE4SZLG_0.html

[21] "ESA – ATV – ATV-2: Johannes Kepler." 2 Mar. 2011. European Space Agency 18 Aug. 2011 http://www.esa.int/esaMI/ATV/SEM8HX6K56G_0.html

[22] Beskow, C. "ATV-2 is no more... | Edoardo Amaldi#more-3188#more-3188." 22 June 2011. European Space Agency 18 Aug. 2011 http://blogs.esa.int/atv/2011/06/22/atv-2-is-no-more%e2%80%a6/#more-3188

[23] "Annual Report 2010 of the International Space Exploration Coordination Group." International Space Exploration Coordination Group (ISECG) 25 Aug. 2011 http://www.globalspaceexploration.org/c/document_library/get_file?uuid=927d699a-2126-4205-b182-2fc8409eff88&groupId=10812

Within ESA, an Inter-Directorate Exploration Scenarios Working Group began defining a vision for space exploration, and awarded Exploration Scenario Studies contracts to various European companies. The end goal is a Strategic Plan for Human Spaceflight and Exploration by mid-2012.[19]

Besides ESA activities, national space agencies within Europe progressed in their exploration and space science efforts. Together with CNES, DLR developed a joint Report on Exploration, which was presented to the members of the ESA council meeting in Paris on 15 December 2010. Both agencies agreed that a European strategy considering both human and robotic missions should be proposed. In October 2010, the French government and CNES signed the "Contract between the State and CNES for the Period 2011–2015", which asked CNES to "make proposals to promote an international exploration programme of the Solar System in renewed governance." The French minister for Higher Education and Research proposed the creation of an international political forum on space exploration. The COmmittee for SPAce Research (COSPAR) meeting in Bremen from 18 to 25 July created a new panel which prepared a comprehensive report on robotic and human exploration of the Moon, Mars & near-Earth asteroids.[23]

NASA launched two space shuttles on ISS assembly flights. OV-103 Discovery was launched on the STS-133/ISS ULF 5 mission on 24 February 2011. The crew were Alvin Drew, Nicole Stott, Eric Boe, Steven Lindsey, Michael Barratt and Steve Bowen. The main payloads were the Permanent Multipurpose Module (PMM) and Express Logistics Carrier 4 (ELC4).[24] OV-105 Endeavour was launched on the STS-134/ISS ULF 6 mission on 16 May. The crew were Mark Kelly, Gregory Johnson, Michael Fincke, Greg Chamitoff, Andrew Feustel and Roberto Vittori. The payloads included the Alpha-Magnetic Spectrometer (AMS-02) and Express Logistics Carrier 3 (ELC3).[25]

In November, 10 years of permanent human presence on the ISS was celebrated. Over 600 science and technology experiments have been carried out in that time, leading to advances in medicine, environmental systems and scientific understanding of the Universe. The shift from assembly to utilisation also means that the ISS will be used as a testbed for research, technology and operations demonstrations related to exploration beyond LEO. As part of this, NASA has been directed, through the 2010 Authorisation Act, to select a research organisation to run a part of the U.S. segment as a national laboratory.[26]

JAXA launched its H-II Transfer Vehicle (HTV) Kounotori 2 on 22 January 2011, docking with the ISS on 27 January and undocking on March 28.[27] Russia

---

[24] "NASA – STS-133." NASA 25 Aug. 2011 http://www.nasa.gov/mission_pages/shuttle/shuttlemissions/sts133/main/index.html

[25] "NASA – STS-134." NASA 25 Aug. 2011 http://www.nasa.gov/mission_pages/shuttle/shuttlemissions/sts134/main/index.html

[26] ISECG...

[27] "H-II Transfer Vehicle "KOUNOTORI2" (HTV2)." 24 Aug. 2011 http://www.jaxa.jp/projects/rockets/htv/index_e.html

launched four Soyuz spacecraft: TMA-19 on 15 June with the Expedition 24/25 crew of Fyodor Yurchikhin, Shannon Walker and Douglas Wheelock, TMA-01M on 7 October with the Expedition 25/26 crew of Aleksandr Kaleri, Oleg Skripochka and Scott Kelly, TMA-20 on 15 December with the Expedition 26/27 crew of Dmitri Kondratyev, Catherine Coleman and Paolo Nespoli, and TMA-21 on 4 April with the Expedition 27/28 crew of Aleksandr Samokutyayev, Andrei Borisenko and Ronald Garan.[28] Russia also continued its regular resupply of the ISS with the launches of five Progress cargo transfer vehicles on 30 June 2010, 10 September 2010, 27 October 2010, 28 January 2011 and 27 April 2011,[29] While China did not launch any human missions this year, it continued to prepare for the launch of its first space station, Tiangong 1.

Looking to the future use of the ISS to prepare for human exploration beyond LEO, a "Call for Ideas: ISS for Exploration" was released on 15 October 2009 by ESA, attracting 181 proposals from 19 countries and ESA staff which covered a broad spectrum of areas of interest such as crew assistants, countermeasures for long-duration spaceflight, the monitoring of astronauts' health, robotics, maintenance, failure management and on-orbit repair, tele-operations as well as other topics. On 17 March 2011 ESA formally committed to extending Europe's participation in the ISS until 2020.

Analogue campaigns continued to be a feature of preparation for human exploration. The Mars500 520-day isolation study began on 3 June at the Institute of Biomedical Problems in Moscow. The crew was made up of two Europeans, one Chinese and three Russians. The six crewmembers, Romain Charles, Sukhrob Kamolov, Alexey Sitev, Alexandr Smoleevskiy, Diego Urbina and Yue Wang conducted simulated Martian surface operations beginning on 14 February 2011.[30]

ESA conducted field testing of the Eurobot Ground Prototype (EGP) and Long-Term Medical Survey system (LTMS) in association with the Austrian Space Forum (ÖWF) in Rio Tinto, Spain during April.[31]

NASA ran their Desert Research and Technology Studies (RATS) campaign from 31 August to 15 September in the Utah desert. Space Exploration Vehicles (SEVs), sample collection and communications tools, navigation systems, and the Habitat Demonstration Unit (HDU) were among the technologies tested.[32] The Haughton-Mars Project on Devon Island, Nunavut, Canada, concluded its 2010 field season on 10 August. The Mars Society continued its activities at the Mars

---

[28] "Starsem headlines." Starsem: The Soyuz Company 24 Aug. 2011 http://www.starsem.com/news/headlines.html

[29] "Spaceflight Now: Tracking Station: Launch Log." 1 Aug. 2011 http://spaceflightnow.com/tracking/launchlog.html

[30] "ESA – Mars500" European Space Agency European Space Agency 25 Aug. 2011 http://www.esa.int/esaMI/Mars500/

[31] "ESA Portal – Exploring Rio Tinto Eurobotically." 2 May 2011. European Space Agency 25 Aug. 2011 http://www.esa.int/esaCP/SEMT1DZGRMG_index_0.html

[32] "NASA – Desert Research and Technology Studies (RATS)" NASA 25 Aug. 2011 http://www.nasa.gov/exploration/analogs/desertrats/index.html.

Desert Research Station (MDRS) in Utah.[33] Technologies to support exploration, including Micro-Ecological Life Support System Alternative (MELiSSA) continued to progress, with study also continuing on Advanced Closed-Loop System (ACLS) and In-Situ Resource Utilisation (ISRU) technologies.

The Canadian Space Agency (CSA) continued to pursue its exploration efforts. Its robotic arm, Dextre, was fully certified on the ISS. The Exploration Core programme made significant progress due to the release of additional funds from the country's economic action plan. This programme funds the development of exploration technologies. The elaboration of a new Canadian Space Exploration Strategic Plan that will have a 25-year vision and a 10-year plan[34] has started.

2.3.2 *Lunar Exploration*

As the celestial body closest to Earth, science and exploration missions to the Moon remained active over the past year. Europe, along with India, China, and Japan progressed toward their first lunar landings. NASA continued its strong portfolio of lunar missions. Russia moved forward with plans to resume lunar exploration after a decades-long hiatus, mirroring the resumption of its robotic Mars exploration programme.

The ESA Lunar Lander programme took a significant step forward. The Phase B1 study contract was signed with EADS-Astrium in Berlin, Germany on 16 September. The contract will culminate in a Preliminary System Requirements Review in 2012, which will provide the basis for the final design of the mission and lander. The mission will land autonomously near the previously unexplored south pole of the Moon and is described as a precursor for future human exploration. The south pole is a region of interest due to the near-continuous illumination of the surface and potential access to water.[35]

The Lunar Reconnaissance Orbiter (NASA), launched in June 2009, was tasked with preparing for future lunar exploration, including finding landing sites, locating resources such as water ice and hydrogen, and investigating the long-term effects of the lunar environment. The mission brought an impressive list of achievements to a close, ending its exploration phase on 16 September. These included creating the most precise and complete topographic maps of the moon yet, finding the coldest spot ever discovered in the Solar System (−248 °C), and determining areas of the moon that are in near-continuous sunlight. Such areas could be valuable to power hardware in support of a robotic or human mission. New data was also announced

[33] "The Mars Society: Mars Desert Research Station." The Mars Society 24 Aug 2011 http://www.mdrs2011.com/

[34] ISECG...

[35] "ESA Portal – Next step for ESA's first Moon lander." 16 Sep. 2010. European Space Agency 19 Aug. 2011 http://www.esa.int/esaCP/SEMUV2KOXDG_index_0.html

from this mission, in collaboration with the Lunar CRater Observation and Sensing Satellite (LCROSS), which found that the soil in the shadowed parts of the Moon's craters is rich in water ice, and that the Moon is chemically active and has a water cycle.[36]

NASA re-tasked two spacecraft from the Time History of Events and Macroscale Interactions during Substorms (THEMIS) Earth-orbiting mission to take part in the Acceleration, Reconnection, Turbulence and Electrodynamics of Moon's Interaction with the Sun (ARTEMIS) mission. The spacecraft moved to two different Lagrange points, gravitationally semi-stable points near the Moon. The new mission focuses on the interactions between the solar wind and the Moon's surface.[37]

JAXA discussed the strategy for lunar exploration and issued a final report, "Lunar Exploration Strategy of Japan – World-Leading Robotic Lunar Exploration and Establishment of Technology Base towards Manned Space Activity" in July. This has the stated goal of assembling an exploration base at the South Pole of the moon. As a first step, Japan aims to demonstrate a soft landing by 2015.

China took major steps forward in its lunar exploration programme with the launch on 1 October of Chang'e 2, its second lunar orbiter. Flying in a lower orbit than its predecessor, it provided higher resolution images of the Moon's surface. The spacecraft completed its primary objectives within 6 months, including a low pass of 15 km altitude to image Sinus Iridum, or the Bay of Rainbows, the proposed landing site for future Chang'e missions.[38]

India announced the final payloads for its second lunar mission, Chandrayaan 2. It will consist of an orbiter and a rover, delivered by a Russian-supplied lander. Due to failures of the Geosynchronous Satellite Launch Vehicle (GSLV), however, the launch date has slipped to 2014.

Russia continued work on its Luna-Glob and Luna-Grunt series of missions, scheduled to begin in 2012. These missions include a lunar orbiter and surface penetrators, and will contribute to knowledge about the moon's formation.

The Google Lunar X PRIZE is a competition with a $30 million prize for the first privately-funded organisation to land a rover on the moon. To win, the rover must travel at least 500 m and send high-definition video and images back to the Earth. To provide additional impetus, the prize will reduce in value after a government-funded mission explores the lunar surface. Twenty-nine teams had registered by the deadline of 31 December. In press releases on 15 October and 20 December, NASA announced that it would purchase data and contract with some of the teams to

---

[36] "NASA – Lunar Reconnaissance Orbiter (LRO)." NASA 25 Aug. 2011 http://www.nasa.gov/mission_pages/LRO/main/index.html

[37] "NASA – ARTEMIS Mission Site." NASA 25 Aug. 2011 http://www.nasa.gov/mission_pages/artemis/

[38] "China announces success of Chang'e-2 lunar probe mission." 8 Nov. 2010. Xinhua News 25 Aug. 2011 http://news.xinhuanet.com/english2010/china/2010-11/08/c_13596211.htm

demonstrate technology in high technical risk areas associated with low-cost lunar missions.[39]

Upcoming missions of note include the Lunar Atmosphere and Dust Environment Explorer (LADEE), scheduled to launch in 2013, which will characterise the atmosphere and dust environment on the Moon,[40] and the Gravity Recovery And Interior Laboratory (GRAIL) mission was launched on 10 September 2011, with aims to determine the structure of the lunar interior.[41] Further in the future, an International Lunar Network (ILN) is proposed for 2018, that would involve operating all upcoming lunar landing missions as nodes in a geophysical network.[42]

2.3.3 Mars Exploration

The focus for Mars science remains the investigation of the planet's habitability, in particular the presence of water. A number of reports during the year suggested that Mars was once partially covered by large oceans, and that life would have been possible on many locations on its surface.

ESA's Mars Express mission, launched in June 2003, arrived at Mars in December of that year. It has the objectives of imaging the entire surface of the planet at high resolution, including maps of the mineral composition and atmosphere, and determining the structure of the sub-surface to a depth of a few kilometres, the effect of the atmosphere on the surface, and the interaction of the atmosphere with the solar wind. This year, the orbiter produced a unique video of an entire orbit, returned results that suggest that liquid water was once widespread over the surface of the whole planet, found evidence of debris from landslides, conducted a 100 km altitude flyby of the Martian moon Phobos, and returned high-resolution imagery of numerous craters, volcanoes and fractures on the surface of the planet.[43]

The joint NASA-ESA ExoMars mission, envisaged as an orbiter to be launched in 2016 with a rover following 2 years later, completed its payload selection on 2 August.[44] The orbiter features a 1000-fold increase in sensitivity over previous Mars orbiters, and will study the chemical composition of the planet. The System

[39] "Google Lunar X PRIZE." Google Lunar X Prize 25 Aug. 2011 http://www.googlelunarxprize.org/

[40] "Missions – LADEE – NASA Science." NASA 25 Aug. 2011 http://science.nasa.gov/missions/ladee/

[41] "Missions – GRAIL – NASA Science." NASA 25 Aug. 2011 http://science.nasa.gov/missions/grail/

[42] "Missions – ILN – NASA Science" NASA 25 Aug. 2011 http://science.nasa.gov/missions/iln/

[43] "ESA – Mars Express." European Space Agency. 25 Aug. 2011 http://www.esa.int/esaMI/Mars_Express/index.html

[44] "NASA – NASA and ESA's First Joint Mission To Mars Selects Instruments." 2 Aug. 2010. NASA 11 Aug. 2011 html://www.nasa.gov/home/hqnews/2010/aug/HQ_10-181_Joint_Mars_Mission.html

Preliminary Design Review for the orbiter and the rover was completed in December.[45] However, on 29 March, NASA advised ESA that it could not fund its rover, leading ESA to issue stop-work orders for all ExoMars contracts. Contract payments were expected to resume in July.[46]

The Mars Odyssey mission, launched in April 2001, continued. During its unprecedented length, it has returned the first global Martian maps of the amount and distribution of several chemical elements and minerals, provided evidence leading to the discovery of subsurface water ice, and recorded the radiation environment. The spacecraft also acts as a communications relay for the two Mars Exploration Rovers. During the last year, the orbiter expanded educational outreach programmes, with U.S. middle school students finding a lava tube using imagery from the orbiter.[47] The Mars Exploration Rover (MER) Opportunity continued to drive towards the Endeavour crater, stopping to examine various interesting scientific targets on the way. However, NASA stopped attempts to communicate with its twin rover, Spirit. The last communication was received from Spirit on 22 March 2010.[48]

The Mars Reconnaissance Orbiter (MRO) was launched in 2005 with the goals of determining whether or not life has existed on Mars, characterising the climate and geology, and preparing for future human exploration. During a very productive year, it returned detailed images of numerous different features; explained how wind changes the configuration of gullies at the polar ice caps; developed high-resolution 3D maps of the surface in collaboration with Microsoft; conducted a study of the atmosphere to prepare for the arrival of the Mars Science Laboratory (MSL) in August 2012; found hydrated silica deposits that provide the best evidence yet of prior hydrothermal environments such as a hot springs; provided mineral maps of areas near to the Mars Exploration Rover Opportunity, helping it to decide where to explore on the surface; observed changing sand dunes; provided new insights into buried carbonates, which may help to explain why Mars' atmosphere is much thinner that it used to be; and discovered that the total amount of atmosphere changes significantly as the tilt of Mars' axis changes.[49]

Development of the NASA Mars Science Laboratory (MSL) and its Curiosity rover continued, with a 26 November 2011 launch date. The rover will be the biggest ever landed on Mars, and will study Mars' habitability. Hardware assembly and testing continued, which was monitored by a public webcam watched by over

[45] ISECG...

[46] "ESA Expects by July To Restart Work on 2016 Mars Mission." Space News 30 May 2011: 6.

[47] "Mars Odyssey." NASA Jet Propulsion Laboratory, California Institute of Technology 25 Aug. 2011 http://mars.jpl.nasa.gov/odyssey/

[48] "Mars Exploration Rover Mission: Home." NASA Jet Propulsion Laboratory, California Institute of Technology 25 Aug. 2011 http://marsrover.nasa.gov/home/index.html

[49] "Mars Reconnaissance Orbiter." NASA Jet Propulsion Laboratory, California Institute of Technology 25 Aug. 2011 http://marsprogram.jpl.nasa.gov/mro/.

one million individuals.[50] Looking ahead to other future Mars missions, NASA awarded the launch services contract for the 2013 Mars Atmosphere and Volatile Evolution Mission (MAVEN) that aims to explore the planet's upper atmosphere, ionosphere and interactions with the Sun and solar wind.[51]

The Phobos-Grunt joint Russian-Chinese sample return mission to the Martian moon Phobos continued to progress towards a 8 November 2011 launch window. Its objective is to land on Phobos, collect 200 g of the Martian moon, and return them to Earth in 2014. The mission will also carry a Chinese sub-probe called Yinghuo-1, which will travel to Mars on the Russian spacecraft before separating and entering Mars orbit.[52]

2.3.4 Saturn Exploration

The Cassini-Huygens mission, a joint mission of NASA, ESA and ASI, was launched in 1997. After reaching Saturn in 2004, Cassini dropped the Huygens probe onto the Saturn moon Titan, where it made the first landing in the outer Solar System. The nominal Cassini mission ended in June 2008, however, the spacecraft continued to operate well on its extended Solstice mission and provided new insights, particularly into the plumes being emitted from the moon Enceladus. The scientific highlight of the year, reported on 22 June, was a flythrough of the plumes executed by Cassini. The scientific results provided strong evidence for the existence of large-scale saltwater reservoirs beneath the moon's icy crust.[53]

Other recent findings about Enceladus included the observation of distinctive coloured bands and patches on the inner, mid-size moons thought to be partially caused by material emitted from Enceladus; fissures, nicknamed 'tiger stripes', with warm edges and a heat output an order of magnitude higher than expected; and an electrical connection via magnetic field lines between Saturn and the moon, allowing electrons from the moon to create an auroral 'footprint' on the planet.[49] Cassini also provided new insights into Titan, including the detection of hydrogen molecules flowing down through the atmosphere and disappearing at the surface; a lack of acetylene, which could be the best candidate energy source for hypothetical methane-based life; the possible existence of cryovolcanism; and the observation of the seasonal methane precipitation.[49]

---

[50] "Mars Science Laboratory." NASA Jet Propulsion Laboratory, California Institute of Technology 25 Aug. 2011 http://marsprogram.jpl.nasa.gov/msl/

[51] "MAVEN." University of Colorado at Boulder, Laboratory for Atmospheric and Space Physics 25 Aug. 2011 http://lasp.colorado.edu/home/maven/

[52] Friedman, Louis D. "Timeline for the Phobos Sample Return Mission (Phobos Grunt) – The Planetary Society Blog | The Planetary Society." 27 Oct. 2010. The Planetary Society 25 Aug. 2011 http://www.planetary.org/blog/article/00002738/

[53] "Cassini Solstice Mission." NASA Jet Propulsion Laboratory, California Institute of Technology 25 Aug. 2011 http://saturn.jpl.nasa.gov/index.cfm

2.3.5 Venus Exploration

Venus Express, an ESA mission, was launched in 1995. It reached Venus in 2006 with the objective of studying the atmosphere and clouds with unprecedented detail and accuracy. Amongst its discoveries, it detected the escape of ionic hydrogen and oxygen in the ratio of two to one. From this, it was inferred that solar ultraviolet radiation streams into the atmosphere and breaks up the water molecules into atoms. The results of several low polar passes, with the aim of measuring the density of the upper polar atmosphere, have shown that the atmosphere high above the poles is 60 % thinner than predicted. This could indicate that unanticipated natural processes are at work in the atmosphere, and may present a difficulty in using atmospheric drag to circularise Venus Express' elliptical orbit. This would be necessary to prevent fuel exhaustion due to manoeuvres needed to compensate for orbital perturbations caused by the Sun's gravity.[54]

The mystery of high-altitude sulphuric acid clouds was solved, with findings that may have implications for artificial mitigation of global warming on Earth. Venusian sulphuric acid clouds form at altitudes of 50–70 km, when sulphur dioxide from volcanoes combines with water vapour to make sulphuric acid droplets. Above 70 km, sulphur dioxide should be destroyed rapidly by intense solar radiation. However, another layer of sulphur dioxide was detected between altitudes of 90 and 110 km in 2008; the explanation being that some sulphuric acid droplets evaporate at high altitude, creating gaseous sulphuric acid that can then be broken apart by sunlight to release sulphur dioxide gas.

Based upon the global cooling of Earth caused by the ejection of sulphur dioxide into the Earth's atmosphere after the 1991 eruption of Mt. Pinatubo in the Philippines, it had been suggested that the artificial injection of sulphur dioxide, which would form sulphuric acid droplets, into the atmosphere may help to mitigate the effects of global warming by reflecting sunlight. However, this new evidence from Venus Express suggests that these droplets may be rapidly changed back into gaseous sulphur dioxide, which is transparent to sunlight, reducing their effectiveness in cooling the Earth.[50]

The JAXA Venus Climate Orbiter AKATSUKI (PLANET-C) was launched on 21 May 2010. On 7 December, it failed to inject into its planned 300 by 80,000 km elliptical orbit. JAXA maintained communication with the orbiter, and currently plans to reattempt orbit insertion during the next available opportunity, 6 years after the initial attempt. Results of the investigation into the failure have determined that it was most likely caused by damage to the thruster nozzle of the orbital manoeuvring engine. The mission was intended to make comprehensive observations of the planet's atmosphere and surface, in particular the particles escaping from the atmosphere, using five cameras operating at different wavelengths. In addition, the mission would have taken close-up photographs to

[54] "ESA – Venus Express." European Space Agency 25 Aug. 2011 http://www.esa.int/esaMI/Venus_Express/

observe the 'super-rotating' winds that blow on the surface, which can reach speeds of up to 100 m per second, faster than the planet itself rotates. It had also aimed to confirm the presence of active volcanoes and thunder.[55]

Included in the Russian Federal Space Programme is the Venera-D mission, its first Venus exploration mission in three decades, scheduled for a launch date in 2017. The mission has been simplified from a complex mixture of orbiter, two balloons at different altitudes, several microprobes delivered from the balloons, and a lander, to just a lander, orbiter and subsatellite. The lander will study the formation and evolution of Venus, in particular the elemental and mineralogical composition of the surface, geology, iron-containing phases and the distribution of iron oxidation states. During its descent, the lander will make meteorological measurements, record the isotopic composition of the atmosphere, measure the structure, chemistry and microphysics of clouds, and monitor electromagnetic radiation. The orbiter will be in a daily polar orbit and will study the atmosphere from the surface to an altitude of 160 km, using spectrometers from the ultraviolet to millimetre ranges. The subsatellite will allow the simultaneous measurements of plasma and magnetic fields. Europe and China have been invited to participate in the project.[56] The next Russian mission to Venus after Venera-D is called Venus Globe. It is targeted for a launch date of 2021. The major difference will be a long-lived lander.[52]

2.3.6 Mercury Exploration

BepiColombo will be Europe's first mission to Mercury, and it is a joint ESA-JAXA mission. Targeted for a launch in 2014 with arrival at Mercury in 2020 for a 1-year mission, it is currently in the implementation phase. The mission is made up of two spacecraft: the Mercury Planetary Orbiter (MPO), a three-axis stabilized spacecraft provided by ESA that will study the planet's geology, composition, inner structure, and exosphere, and the Mercury Magnetospheric Orbiter (MMO), a spin-stabilised spacecraft provided by JAXA that will study the planet's magnetic field, atmosphere, magnetosphere and inner interplanetary space. The MMO underwent testing in the European Space Technology and Research Centre (ESTEC)'s Large Space Simulator (LSS) in January.[57, 58]

---

[55] "Venus Climate Orbiter mission of Japan." JAXA 25 Aug. 2011 http://www.stp.isas.jaxa.jp/venus/top_english.html

[56] "VENERA-D: ВЕНЕРА: Изучение продолжается." 3 Oct. 2011. Roscosmos 25 Aug. 2011 http://venera-d.cosmos.ru/index.php?id=692&tx_ttnews[tt_news]=1288&cHash=f9bfd2c6e7616171412b316d206d73a4

[57] "ESA Science & Technology: BepiColombo." European Space Agency 25 Aug. 2011 http://sci.esa.int/science-e/www/area/index.cfm?fareaid=30

[58] "ESA Science & Technology: ESA's Mercury mapper feels the heat." 18 Jan. 2011. European Space Agency 25 Aug. 2011 http://sci.esa.int/science-e/www/object/index.cfm?fobjectid=48254

MErcury Surface, Space ENvironment, GEochemistry and Ranging (MESSENGER), one of NASA's Discovery-class missions, was launched in August 2004. On 18 March, it became the first spacecraft to orbit the planet, following three flybys. Orbital data collection is scheduled to continue until March 2012. The mission is designed to address six broad scientific questions: why Mercury is so dense, the planet's geologic history, the nature of its magnetic field, the structure of its core, the nature of the unusual materials at the poles, and what volatiles are important at Mercury. MESSENGER is designed and built by the Johns Hopkins University Applied Physics Laboratory (APL).[59]

2.3.7 *Jupiter Exploration*

The Europa Jupiter System Mission (EJSM), a proposed joint NASA-ESA mission also known as Laplace, was placed under review due to the combined effect of not being ranked as a top priority in the U.S. National Research Council (NRC)'s decadal survey of astronomy and planetary science, and the overall U.S. budget outlook. It is now unlikely that EJSM/Laplace will be implemented as a NASA-ESA mission as originally planned.

Juno, a NASA New Frontiers mission to Jupiter, launched on 5 August 2011. The mission's objectives are to determine how much water is in Jupiter's atmosphere, measure the atmospheric composition, temperature, cloud motion and other properties, map the magnetic and gravitational fields, and explore the magnetosphere near the poles, especially the planet's aurorae.[60]

2.3.8 *Solar Observation*

Observations of the Sun continued to focus on improving understanding of the star's interior, corona and solar wind, as well as predicting 'space weather'. This included the monitoring of large Coronal Mass Ejections (CMEs), events in which the Sun emits larger than usual amounts of high-energy charged particles that travel towards Earth. Such CMEs often have negative effects on telecommunications and electrical infrastructure.

ESA's PRoject for OnBoard Autonomy (PROBA)-2 mission, part of the Agency's small, low-cost In-orbit Technology Demonstration Programme launched in November 2009, continued its solar observation. Proba-2 tracked CMEs,

---

[59] "MESSENGER: Mercury Surface, Space Environment, Geochemistry, and Ranging: Mercury Orbit Insertion." Press kit. NASA http://www.nasa.gov/pdf/525164main_Mercury MOI_PK.pdf

[60] "NASA – Juno." NASA 25 Aug. 2011 http://www.nasa.gov/mission_pages/juno/main/index.html

observed a partial solar eclipse, and an alignment of the Sun, Earth and Moon. As of 9 November, it had returned more than 180,000 images of the Sun in addition to solar and space weather information and demonstrated a range of technologies. Calibrated data was made available from several instruments, including a small solar ultraviolet radiometer that observes solar irradiance at least 20 times per second. Such rapid sampling enables the detailed analysis of how energy is released during solar events.[61]

Fifteenth of June saw the launch of the CNES solar metrology mission Picard. Its objectives are to improve our knowledge of how the Sun functions and the influence of solar activity on the Earth's climate. Picard will accomplish this by measuring absolute total and spectral solar irradiance, solar diameter and shape, and by probing the interior of the Sun using the helioseismology method. It is anticipated that the lifetime of the satellite will be between 2 and 3 years. The first image of the Sun was captured on 22 July, with subsequent images taken once per minute. In-flight commissioning ended on 8 October, and the system was declared ready for use. On 4 January, Picard observed a partial eclipse of the Sun by the Moon.[62]

The Solar Dynamics Observatory (SDO) is a NASA mission that is part of the Living With a Star (LWS) programme. It has the objectives of determining how the Sun's magnetic field is generated and structured, and how this stored magnetic energy is released in the form of the solar wind, energetic particles and variations in the solar irradiance. Launched in February 2010, the spacecraft comprises three scientific experiments: the Atmospheric Imaging Assembly (AIA), EUV Variability Experiment (EVE) and the Helioseismic and Magnetic Imager (HMI).[63]

The SOlar and Heliospheric Observatory (SOHO) continued to operate during the year. It was launched in December 1995, and has the scientific objectives of investigating the solar interior, explaining the high heating of the solar corona, and the mechanism by which the solar wind is produced and accelerated. Some of its key results include discovering new dynamic solar phenomena such as coronal waves and solar tornadoes, vastly improving our ability to forecast space weather, by giving up to 3 days' notice of adverse space weather, and monitoring the total solar irradiance, which is important in understanding the impact of solar variability on the Earth's climate. Having had the mission extended five times, SOHO has exceeded its expected lifetime of 2 years and is currently approved until the end of 2012.[64]

Along with SOHO, the Solar TErrestrial RElations Observatory (STEREO) spacecraft of NASA continued to operate. STEREO is made up of two spacecraft.

---

[61] "Proba-2's Lyra Solar Intensity Measurements Available." 11 Mar. 2011. ESA 11 Mar. 2011 http://www.esa.int/esaMI/Proba/SEMN91YTVKG_0.html

[62] "PICARD." CNES 25 Aug. 2011 http://smsc.cnes.fr/PICARD/

[63] "SDO | Solar Dynamics Observatory." NASA Goddard Space Flight Center 25 Aug. 2011 http://sdo.gsfc.nasa.gov/

[64] "Solar and Heliospheric Observatory Homepage." NASA 25 Aug. 2011 http://sohowww.nascom.nasa.gov/

They have provided new insights into CMEs, including detecting a CME heading for Earth on 2 August, and completing the processing of data that is intended to enable the tracking of CMEs all the way from the Sun to the Earth. STEREO also contributed to crowd-sourced data analysis, using data analysed by the public to make a prediction of a solar storm that reached Earth on 13 December, captured the first-ever images of the entire surface of the Sun, and has been used to discover more than 122 new eclipsing binary stars and hundreds more variable stars.[65] The Deep Space Climate ObserVatoRy (DISCOVR), originally built to conduct observations of the Earth's climate and never launched, may be re-purposed as a space weather and solar storm warning satellite.[66]

The Hinode (Solar-B) probe, led by JAXA in collaboration with NASA, the Science and Technology Facilities Council (STFC, U.K.) and ESA, was launched in September 2006, with the mission of studying the solar magnetic field; the project explores the solar magnetic fields of the Sun to better understand the mechanisms that power the solar atmosphere and drive solar eruptions.[67] The Advanced Composition Explorer (ACE), a NASA mission launched in August 1997, with its primary science objective of measuring the composition of the solar corona, wind, interplanetary particles, the interstellar medium and galactic matter, continued to collect data to improve forecasts and warnings of solar storms, and is expected to maintain its orbit until 2024.[68] The Reuven Ramaty High Energy Solar Spectroscopic Imager (RHESSI), a NASA SMall EXplorer (SMEX) mission with the objective of exploring the basic physics of particle acceleration and energy release in solar flares has also continued to operate. Data from the mission has led to several hundred publications, the first gamma-ray images of a solar flare, and the first measurements of terrestrial gamma-ray flashes.[69]

2.3.9 Outer Solar System Exploration and Observation

With near-Earth objects becoming a more widely-discussed target for human exploration, there was renewed interest in studying these asteroids and comets. This section also aims to cover some of the significant discoveries made about

---

[65] "NASA – STEREO." NASA 25 Aug. 2011 http://www.nasa.gov/mission_pages/stereo/main/index.html

[66] Clark, Stephen. "Spaceflight Now | Breaking News | NOAA taps DSCOVR satellite for space weather mission." 2 Feb. 2011. Spaceflight Now 25 Aug. 2011 http://spaceflightnow.com/news/n1102/21dscovr/

[67] "Hinode (Solar-B)." NASA 25 Aug. 2011 http://solarb.msfc.nasa.gov/

[68] "ACE Mission." California Institute of Technology 25 Aug. 2011 http://www.srl.caltech.edu/ACE/ace_mission.html

[69] "RHESSI Homa Page." NASA Goddard Space Flight Center 25 Aug. 2011 http://hesperia.gsfc.nasa.gov/hessi/index.html

celestial bodies outside the Solar System, including those in galactic astronomy and cosmology.

Rosetta completed a successful flyby of the asteroid Lutetia on 10 July, and also discovered that an object previously thought to be a comet is in fact a pair of colliding asteroids. In 2014, Rosetta will release a lander that will make the first landing on a cometary nucleus, that of 67P/Churyumov-Gerasimenko.[70] The Herschel infrared telescope made several significant discoveries. It confirmed that ultraviolet starlight is a key ingredient for making water in space, captured the most detailed infrared image of the Andromeda galaxy yet taken, revealed a population of galaxies shrouded by dust that do not need as much dark matter as previously thought to trigger bursts of star formation, found a link between star formation and shockwaves in interstellar clouds, and observed molecular gas streams emerging from galaxies at speeds of up to 1,000 km per second.[71] Herschel's counterpart Planck, a microwave observatory, was launched in May 2009 to study the Cosmic Microwave Background (CMB) – the ancient radiative 'fingerprint' of the Big Bang. Planck provided its first all-sky image, which gave scientists new insight into star and galaxy formation, and the formation of the infant universe soon after the Big Bang.[72]

The COnvection, ROtation and planetary Transits (COROT) space telescope operated by CNES is an astronomy mission that was launched in December 2006. It focused on probing the inner structure of stars using stellar seismology and detecting extrasolar planets. In June, the discovery of six new planets was announced.[73]

On 4 November, the Extrasolar Planet Observations and characterisation/deep impact eXtended Investigation (EPOXI) flew past comet Hartley 2. An extension of the Deep Impact mission, the spacecraft returned images of the comet that provided new information on the role comets may have in planetary formation.[74] NASA also announced its first asteroid sample return mission, the Origins Spectral Interpretation Resource Identification Security Regolith Explorer (OSIRIS-REx).[75] The JAXA Hayabusa mission successfully returned samples from the asteroid Itokawa on 13 May. About 1,500 particles were identified, most of them of extraterrestrial origin.[76]

---

[70] "ESA – Rosetta." European Space Agency 25 Aug. 2011 http://www.esa.int/esaMI/Rosetta/

[71] "ESA – Herschel." European Space Agency 25 Aug. 2011 http://www.esa.int/SPECIALS/Herschel/

[72] "ESA – Planck." European Space Agency 25 Aug. 2011 http://www.esa.int/SPECIALS/Planck/index.html

[73] "Corot." CNES 25 Aug. 2011 http://smsc.cnes.fr/COROT/

[74] "NASA – EPOXI." NASA 25 Aug. 2011 http://www.nasa.gov/mission_pages/epoxi/index.html

[75] "NASA – NASA Selects OSIRIS-Rex as Next New Frontiers Mission." 25 May 2011. NASA 11 Aug. 2011 http://www.nasa.gov/centers/goddard/news/releases/2011/11-037.html

[76] "JAXA: Identification of origin of particles brought back by Hayabusa." 16 Nov. 2010. JAXA 11 Aug. 2011 http://www.jaxa.jp/press/2010/11/20101116_hayabusa_e.html

Exoplanets are planets orbiting stars other than the Sun. Until a few years ago their existence was hypothetical, but recent work has shown that they are quite common. Kepler is a NASA space telescope with the specific aim of finding exoplanets, particularly Earth-sized planets in the habitable zone of solar-like stars, where liquid water could exist on their surfaces. During the year several significant discoveries were made. On 26 August, Kepler discovered two planets transiting the same star for the first time, and followed up with the discovery of the first rocky exoplanet Kepler 10-b on 11 January, which is approximately 1.4 times the size of Earth and the smallest planet ever discovered outside of our Solar System. It also found candidate planets in the habitable zone of their stars.[77]

NASA's NEO Wide-field Infrared Survey Explorer (WISE) mission completed its survey of the sky for small bodies, asteroids and comets. Its list of discoveries includes 20 comets, more than 33,000 main belt asteroids, and 134 Near-Earth Objects (NEOs).[78]

The Spitzer infrared space telescope discovered 14 of the coldest stars known in our universe, leading to speculation that one of these brown dwarf stars could be a companion to the Sun, and contributed to observations of a dust cloud trailing Earth that may help astronomers to find Earth-like planets around other stars.[79]

The James Webb Space Telescope (JWST), the successor to the Hubble Space Telescope (HST), continued to progress towards a tentative 2018 launch date. It will investigate the formation of the first galaxies, planetary systems, and stars. Segments of the large primary mirror underwent cryogenic testing throughout the year.[80] However, the project is in political trouble, with an independent review finding, on 10 November, that its cost overrun was already $1.5 billion.[81]

2.3.10 International Cooperation in Space Exploration

It is almost universally acknowledged on a political level that future space exploration efforts will require international cooperation. However, the concept of 'autonomous' access to space remains a strong one. This has led, for example, to the

[77] "Kepler: Home Page." NASA Ames Research Center 25 Aug. 2011 http://kepler.nasa.gov/

[78] "NASA – NASA's NEOWISE Completes Scan for Asteroids and Comets." 1 Feb. 2010. NASA 25 Aug. 2011 http://www.nasa.gov/mission_pages/WISE/news/wise20110201.html

[79] "NASA Spitzer Space Telescope." NASA 25 Aug. 2011 http://www.spitzer.caltech.edu/

[80] "The James Webb Space Telescope." NASA 25 Aug. 2011 http://www.jwst.nasa.gov/

[81] Klamper, Amy. "JWST's Latest $1.5B Cost Overrun Imperils Other High-priority Projects." 12 Nov. 2010. Space News 25 Aug. 2011 http://www.spacenews.com/civil/101112-jwst-cost-imperils-priority-projects.html

U.S., Europe, Russia and Japan all developing their own methods of transferring cargo to the ISS.

At the second International Conference on Space Exploration, co-organised by ESA, attendees agreed that action was needed to make sure than Europe plays a significant role in future space exploration. EU Ministers and ESA Member States adopted the resolution "Global challenges: taking full benefit of European space systems" with section five dedicated to the European Vision on Space Exploration.[82]

The International Space Exploration Coordination Group (ISECG) continued to discuss architecture options for future human space exploration. On 23 June, senior managers from Canada, China, Europe, Japan, South Korea, Russia, Ukraine and the U.S. met to discuss human and robotic exploration, including the development of a Global Exploration Roadmap, that would constitute an international exploration architecture. Also released was the ISECG Reference Architecture for Human Lunar Exploration, which detailed a human Moon mission using the assets of many agencies.[83]

2.4 Satellite Applications

2.4.1 Space-Based Communications

Satellite communications operators continued to grow in 2010 and 2011, both qualitatively and quantitatively. With the help of improved financial results, the industry has undertaken significant space infrastructure modernisation efforts, especially in the field of new broadband technologies with the manufacturing of sizeable Ka-band spacecraft. Further investments were made in consolidating capabilities through the ordering or launch of larger spacecrafts with enhanced signal power and transponder capacity. The way the industry has reacted to the challenge of the 2008 financial crisis clearly demonstrates an acute rate of responsiveness to the changing conditions, as well as an increased sense of extrovert corporate behaviour and confidence in the sector's future prospects. Achieving the right mixture of investing in innovating technologies and new services on the one hand, while consolidating current operations on the other has boosted the industry's revenues for one more consecutive year.

---

[82] ISECG…

[83] "ISECG – Homepage." ISECG 25 Aug. 2011 http://www.globalspaceexploration.org/

2.4.2 *Space-Based Positioning, Navigation and Timing Systems*

During the reporting period, the development of GNSS systems continued, with all relevant actors increasing their efforts to complete their full satellite constellations as soon as possible.

In Europe, the European Commission continued the development of its Galileo GNSS constellation with the completion of four-in-orbit validation (IOV) satellites, constructed by a consortium led by Astrium Satellites and Thales Alenia Space. Construction of the IOV satellites were scheduled for launch well in advance of the initial 14 fully operational spacecraft that were already under order with OHB Technology of Bremen. However, in June 2010 IOV's contractors announced an additional 7-month delay to the programme, created by the need to replace Chinese-built search and rescue payloads that were already integrated to the nearly complete satellites. This departure form the spacecrafts' originally foreseen architecture was requested by the European Commission that reversed a previous decision and insisted that no non-European payloads would be allowed on the Galileo spacecraft. At that time, the IOV satellites were scheduled for a late 2010 launch, onboard two separate Soyuz flights.[84]

As mentioned above, the restoration of Russia's Glonass GNSS constellation to its full operational capacity remained a top priority in 2010 and 2011. The development of the system's new generation satellites was already underway in 2010. Known as the Glonass K series, the new spacecraft would most likely incorporate significant technical improvements, including a new more accurate timing device and a non-pressurized structure, bringing its operational performance close to U.S. and European standards. As mentioned above, their deployment was at that time scheduled to begin in 2013 and could be concluded as soon as 2016.

2.5 Technology Developments

New developments in space-related technologies merit discussion, as they indicate current trends in space technology and reveal the focus of relevant policy decisions. The following chapter provides an overview of recent activities, spanning across all major space faring powers and institutions, both established and emerging ones.

2.5.1 *Propulsion*

Significant advancements have occurred within Europe, the US, and China, in the field of propulsion. Improved rockets and other methods of propulsion are being explored with favourable results.

[84] "Galileo Validation Satellites Facing Another Launch Delay." Space News 14 June 2010: 3.

In Europe, Snecma's Vinci M4 engine, Europe's next generation upper-stage rocket, has reached the halfway milestone of its development. This cryogenic, liquid oxygen/hydrogen-fuelled expander cycle engine is designed to produce three times more thrust and increase payload capacity by 20 % more than the current upper stage rocket used by the Ariane 5. It can also be restarted up to five times while in flight, providing greater flexibility in meeting various orbit requirements.[85] Upon its completion, it is planned to be used in ESA's Ariane 5ME (Midlife Evolution) programme.

In the US, NASA is preparing to install a prototype 200-kw variable specific impulse magnetoplasma rocket (VASIMR) engine, produced by the Ad Astra Rocket Company, onto the International Space Station in 2014.[86] Using Argon gas as propellant, the VASIMR will convert electrical current (supplied by either a solar or a nuclear power source) into radio-frequency waves that ionize the argon gas to produce plasma; superconducting magnets then constrain and channel the super-heated plasma into thrust. While this experimental engine is designed to perform on 200-kw, the Ad Astra is working toward developing 100-MW larger plasma propulsion system that would have the capability of flying cargo and astronauts to Mars in less than 40 days.[87]

In another development, Pratt & Whitney Rocketdyne (PWR) has assembled the first complete J-2X upper-stage rocket engine for use by NASA's Orion spacecraft, designed to lift cargo, equipment, and science experiments into orbit.[88] Building on previous J-2 engines used on the Saturn V launch vehicle, the J-2X engine was originally planned for use on the decommissioned Ares I crew launch vehicle. Ares I was part of the Constellation programme that was terminated following the Obama administration's Fiscal 2011 NASA budget request. However, the J-2X continued to undergo tests while its future remained unclear, and PWR has already concluded that the engine can run on methane as an alternative to hydrogen fuel. Even if the J-2X was not selected for a heavy-lift launcher, PWR's $900 million investment could be salvaged by using the engine as a test-bed for turbine blade materials, avionics, and engine controls.[89]

Another promising propulsion technology development programme was the U.S. Navy's plan to launch a pair of small satellites in 2011 that would be able to use the Earth's magnetic field for propulsion.[90] Still in development, this

[85] Norris, Guy. "Validating Vinci – Next generation upper-stage rocket development effort nears halfway point." Aviation Week & Space Technology 28 Mar. 2011: 32.

[86] Klotz, Irene. "Flight-Test Engine – NASA preparing to demonstrate advanced plasma engine on ISS." Aviation Week & Space Technology 7 Feb. 2011: 56.

[87] Ibid; See also "VF-200." Ad Astra Rocket Company 16 Nov. 2011 http://www.adastrarocket.com/aarc/VF200

[88] Morring, Frank Jr. "Down To The Wire – J-2X engine ready for testing as funding comes to an end." Aviation Week & Space Technology 15 Nov. 2010: 49.

[89] Ibid.

[90] Matthews, William. "Experiment Designed To Harness Magnetic Field for Propulsion." Space News 12 July 2010: 14.

experiment involves tethering two small satellites together, where an electron collector from one end gathers electrons from the Earth's plasma (a low density cloud of electrons that exists in space) while an electron emitter on the other end of the tether discharges the electrons back into the plasma. This propulsion method would work best between the altitudes of 500–1,000 km. While this electrodynamic motor's capacity is limited to changes in altitude of about 5 km per day, it would operate significantly longer than standard liquid-fuelled thrusters, thus providing greater autonomy and longer operational life span to the spacecrafts that will use it.

At the same time, the U.S. Army is also looking into innovative and affordable access to space technologies for small payloads. One approach currently under development involves launching swarms of nanosatellites into orbit through the use of affordable nanomissiles that are designed to put 10 kg payloads into space.[91] Known as the 'Multipurpose Nanomissile System', the booster is 3.6 m tall and 60 cm in diameter, and would produce 3,000 lbs of thrust.[92] Upon successful development, this system will be available commercially for suborbital and orbital missions. However, as this nanomissile is part of Space Missile and Defence Command's (SMDC) Nanosatellite Technology Programme, its other objective is to produce battlefield-relevant satellites.[93] If produced in quantity, these nanomissiles could have a price tag as low as $150,000 per unit.

Across the Pacific, China is also advancing its propulsion technology with the development of a staged-combustion kerosene and liquid oxygen engine that is likely to offer greater performance for its Long March 6 and 7 launchers.[94] This engine provides 18 metric tons of thrust and will offer a high specific impulse, allowing for greater payloads to be launched into orbit. The Long March 6 may be operational before Long March 5, and the light launcher will be able to lift 1,000 kg to an orbit of 600 km.[95] The staged-combustion engine will also be used in the second stage of China's medium-heavy Long March 7 launcher.

---

[91] Brinton, Turner. "Nanomissile Being Designed To Launch the Smallest Satellites Affordably." Space News 9 Aug. 2010: 11.

[92] "2011 U.S. Commercial Space Transportation Developments and Concepts: Vehicles, Technologies, and Spaceports." Jan. 2011. Federal Aviation Administration 30 Nov. 2011: 20 http://www.faa.gov/about/office_org/headquarters_offices/ast/media/2011%20devcon%20report.pdf

[93] Ibid.

[94] Perrett, Bradley. "Next Stage – An advanced propulsion technology is readied for new Long Marches." Aviation Week & Space Technology 10 Jan. 2011: 35.

[95] Ibid.

2.5.2 Information Technology

In Europe, Eutelsat began providing commercial broadband service with its Ka-Sat satellite on May 31.[96] This satellite, the first of a fleet of high throughput satellites under development by Astrium, has a throughput of about 70 gigabits per second. Approximately 44,000 Surfbeam 2 terminals have already arrived in the Netherlands in anticipation of the demand for broadband capacity. Ka-Sat distributes its bandwidth through 82 spot beams, each carrying 900 megabits per second of capacity. These beams are evenly distributed to provide services in Europe and Africa. Yet despite the ability to shift bandwidth between nearby spotbeams, the satellite lacks the onboard processing capacity needed to maximize bandwidth and location flexibility. The population coverage of the satellite is about 800 million, nearly three times that of its U.S. counterpart, ViaSat's WildBlue consumer broadband service. It is expected that European demand for Ka-Sat might be higher in suburbs rather than rural areas, which are often the subject of government broadband stimulus programmes.[97]

2.5.3 Spacecraft Operations and Design

In Europe, UK developer Surrey Satellite Technology Ltd. (SSTL) is upgrading its constellation to improve its disaster monitoring capabilities. Its DCM3 will be optimized for commercial use, and will work along side SSTL's future medium-resolution constellation, Earthmapper, to attract additional non-disaster users. In addition to disaster monitoring, this system aims at natural resource and environmental monitoring, deforestation and forest fire applications, and land use mapping.[98] Earthmapper will get an entirely new bus, SSTL-100 v.3.0, allowing it to support higher resolution instruments and provide emergency response users with the ability to zoom in on disaster areas, while also meeting the needs of other users. The DCM3, expected to be operational in 2014, will be based on the SSTL-300S1, a new highly-agile submetric-resolution bus, which carries a new camera that provides a spatial resolution of 75 cm (the spatial resolution on Nigeriasat-2 is 2.5 m), and supports a high-speed downlink and 45° off-pointing slew rate.[99]

In another development, British imaging solutions company E2v was developing image sensors for Euclid, a dark-energy mapping mission currently under

[96] De Selding, Peter B. "Ka-Sat Enters Services as European Broadband Market Heats Up." 6 June 2011: 16.

[97] Ibid.

[98] Taverna, Michael A. "Remapping the Earth – New SSTL submetric smallsat concept could change economics for remote-sensing industry." Aviation Week & Space Technology 7/14 Mar. 2011: 58.

[99] Ibid.

development for ESA.[100] E2v was also awarded a contract with ESA to develop image sensors for Plato, a space telescope that searches for planets; this mission competed with two other space science missions, Euclid and Solar Orbiter, for two launch opportunities under ESA's CosmicVision program.[101] Euclid and Solar Orbiter were ultimately selected;[102] here, Euclid will be equipped with new front and back illuminated Charged Couple Device (CCD) image sensors. However, if Plato had been selected, it would have been equipped with 34 mini-telescopes, containing four back-illuminated charge-coupled devices on each of them. It would orbit the Sun 1.5 million kilometres beyond Earth for 6–8 years, and survey up to 40 % of the sky.[103]

Authorities at the Swedish Space Corp. report a near 100 % success rate for its Prisma formation flying mission.[104] Launched on a Russian-Ukrainian Dnepr rocket in June 2010, the two Prisma satellites conducted a year long mission to perform formation flying manoeuvres between the two spacecraft, while also testing several new technologies. One satellite operated in passive mode, while the other satellite operated around it autonomously, and made operational decisions on its own.[105] The Prisma mission paves the way for future formation-flying satellite missions, e.g. ESA's Proba-3. Analysis on Prisma also included tests on its High-Performance Green Propulsion (HPGP) system, developed by Ecological Advanced Propulsion Systems (Ecaps) of Sweden, which provided an alternative to conventional hydrazine fuel. At one third the cost of hydrazine, the HPGP system has a higher specific impulse, in addition to providing more thrust per fuel volume. It is also non-toxic, which allows fuelling to be conducted without the need for protective gear beyond standard clean room clothing.

The German DLR conducted its very first tandem mission, where ground controllers manoeuvred two radar satellites to within 350 m of each other in low Earth orbit.[106] Radar satellites TerraSAR-X, launched in 2007, and TanDEM-X, launched in 2010, originally had identical polar orbits 514 km in altitude, with a separation distance of 20 km. TanDEM-X's orbital period was reduced, letting it to move to a separation distance of 350 m from TerraSAR-X within 3 days. This separation enables the satellites to take simultaneous images of the same area, allowing for the production of three-dimensional digital evaluation models. This data provides a 12-m ground resolution, making objects this size or larger

[100] "E2v Tapped To Develop Image Sensors for Euclid." Space News 18 Oct. 2010: 8.

[101] "E2v Picked To Develop Image Sensors for Plato." Space News 25 Oct. 2010: 8.

[102] De Selding, Peter B. "Solar Orbiter, Euclid Chosen as ESA's Next Medium-class Missions." 5 Oct. 2011. Space News 6 Aug. 2012 http://www.spacenews.com/civil/111005-esa-selects-solar-orbiter-euclid.html

[103] Ibid.

[104] De Selding, Peter B. "Swedish Space Corp. Touts Success of Prisma Formation-flying Mission." Space News 6 Sept. 2010: 17.

[105] Ibid.

[106] "German Satellites Ready For Tandem Mission." Space News 18 Oct. 2010: 3.

distinguishable; the system also provides vertical accuracy of 2 m. The mission is managed by having one satellite transmit a radar signal to Earth, but both spacecraft would receive the reflected signal which is used to create the digital elevation models; this method also reduces power consumption and heat build-up, extending the transmission time for both satellites. While both satellites have a 5-year design life, the DLR is confident that TerraSAR-X will operate long enough to fulfil the tandem mission goal of 3 years of stereo collection.[107]

Across the Atlantic, Northrop Grumman Aerospace Systems has developed a pair of demonstration satellites that can track a ballistic missile launch through all phases of flight.[108] This unprecedented capability was achieved by the Space Tracking and Surveillance System (STSS) spacecraft, dubbed the "Holy Grail" for missile defence. Upon the launch of a missile, an STSS satellite will detect the heat signature of the launch with its acquisition sensor, and then use its gimballed tracking sensor to lock onto the boosting missile. The tracking data gets relayed to the other STSS satellite, which continues to observe the satellite as it re-enters the atmosphere toward its target. Future tests will involve attempts to cue the STSS system from the Defence Support Program missile warning satellites; determining if STSS satellites can produce missile tracking data good enough to cue the launch of ship-based interceptors; and, feeding data to the Aegis system while tracking a missile, to generate a "fire control solution" for an early interceptor launch.[109]

At the same time, a new squat, insect-like lander prototype underwent tests by engineers at NASA's Johnson Space Center.[110] They hope that the lander, named Morpheus, will 1 day enable manned missions to other planets or asteroids. The lander is equipped with an Automated Landing and Hazard Avoidance Technology (ALHAT) that uses lasers to spot dangerous craters or boulders that could make a landing spot unsafe. A safe landing spot can be determined quicker since the ALHAT lasers would image the surface of land area and identify hazards as it flies over. The lander is also being used to test a new lighter and safer mix of fuel; i.e. liquid oxygen and methane. While touted by NASA as a safer alternative to traditional propellants, it's also between 10 and 20 times less expensive, and potentially renewable on both the lunar surface and on Mars.[111] Current tests are being conducted using a crane and tether to maintain stability; despite some unexpected results, researchers are planning to test Morpheus in free flight up to an altitude of 30 m without a tether. If successful, Morpheus will have passed a significant milestone in its development.

[107] Ibid.

[108] Brinton, Turner. "STSS Satellites Demonstrate 'Holy Grail' of Missile Tracking." Space News 28 March 2011: 10.

[109] Ibid.

[110] Moskowitz, Carla. "Johnson Space Center Engineers Test New Lunar Lander Design." Space News 23 May 2011: 17.

[111] Ibid.

Intelsat's Galaxy 15 communications satellite, aka Zombie Sat, has been secured. The wayward satellite went out of control in April 2010, causing 6 months of alarm about interference to the operators of satellites within its path. These other satellites underwent complex unprecedented manoeuvres to limit serious interference and service interruption. By Dec. 17, Galaxy 15 had lost Earth lock, which caused it to lose enough power to shut down its C- and L-band payloads; and soon afterward, its battery drained completely, shutting off the radio beacons and removing the remaining interference risk.[112] Intelsat, with the contribution of SES, Telesat and SatMex expertise, developed a three-pronged strategy in dealing with Zombie Sat's interference threat. When approaching another satellite, Galaxy 15's transmission was minimized by using the narrow beam of a large antenna in Clarksburg, Maryland (in some cases, through the Beach Earth station facility in Hawaii). The energy available to Galaxy 15 was also minimized by changing the sensitivity settings of the satellite at risk of interference. Next, in the final stages of the fly-by, a "leap-frog" manoeuvre would be performed by the at-risk satellite, permitting the two spacecraft to stay in sync and maintain a minimum spacing of at least 0.2° until Zombie Sat began moving away again.[113] As a result, while Galaxy 15 approached a dozen satellites, these manoeuvres kept it from creating a serious impact.

Bigelow Aerospace conducted tests on the life support system of its inflatable habitats, destined to provide research facilities and hotel accommodations in space.[114] On March 31, the prototype environmental control and life support (ECLS) system underwent testing within a 180 cubic-meter test chamber, designed to replicate the interior volume and shape of the company's Sundancer inflatable module. The ECLS system is unlike other systems that are used for short duration trips between Earth and the International Space Station, because it is geared toward regeneration capability through recycling water from the environment and from any waste treatment process, and generating oxygen from water. The test involved locking three Bigelow engineers within that structure for 8 hours while performing tasks which demonstrated the system's "ability to control temperature, humidity, pressure, oxygen content and the removal of carbon dioxide and trace-gas contaminants from the environment." Bigelow will conduct a 30-h and week-long demonstration of the ECLS system in subsequent months. In parallel with the building of ECLS, Bigelow is finishing work on an analytical chemistry laboratory that will expand real-time monitoring and analysis of gases and liquids within the chamber to identify the components present.[115]

[112] Taverna, Michael A. "End of an Odyssey – Engineers review options, operators relieved as Intelsat regains control of Zombie Sat." Aviation Week & Space Technology 3 Jan. 2011: 26.
[113] Ibid.
[114] Svitak, Amy. "Bigelow Tests Life Support System." Space News 11 April 2011: 24.
[115] Ibid.

Boeing is developing a commercial crew capsule that is designed to ferry people to and from the International Space Station and future private space stations.[116] Meant to fill the void left by retirement of NASA's space shuttle, the Crew Space Transportation-100 (CST-100) spacecraft could carry up to seven people, and would be larger than the Apollo spacecraft, but smaller than the Orion capsule.[117] Bigelow Aerospace will assist Boeing with demonstrations and design work in areas where Bigelow has experience from the construction of its own orbital facilities and commercial space complex.

The U.S. Air Force launched two X-37B unmanned space planes into Earth orbit between 2010 and 2011. The first drone, Orbital Test Vehicle 1 (OTV-1), returned from a 7-months orbit on December 3, while the second, Orbital Test Vehicle 2 (OTV-2) was launched just over 3 months later; both missions were classified. OTV-1 spent more than 220 days in orbit, with some analysts speculating that the spacecraft served as an unmanned orbital spy platform. Expert Brian Weeden, former U.S. Air Force orbital analyst, suggests the X-37B's sensor payload can be reconfigured for each mission, i.e. to meet the emerging needs of military commanders or intelligence agencies.[118] The X-37B also carries enough propellant to change orbit in the middle of a mission, allowing it to change the coverage area on short notice. The X-37B's orbit duration can last up to 270 days, where the craft can rely on its solar array power system for much of its journey. The spacecraft has an automated re-entry and landing system, with flight controls that are driven entirely by electro-mechanical actuators instead of bulky hydraulic circuits. New thermal tiles were needed because the X-37B's steeper incline while in re-entry would create an increased amount of heat, while the spacecraft would have less surface area to bleed off that heat.[119] Weighing about 5,000 kg, the craft is about 9 m in length, with a 4 m wingspan, and 3 m in height. OTV-2 launched from Florida on March 5 on another classified mission.[120] While the secrecy surrounding these missions has raised concerns by Russia and China that the X-37B is a space weapon of some sort, the U.S. Air Force denies this charge, stating that it is merely testing sensor, guidance and navigation control hardware for future spacecraft.[121]

Earlier in the year, developers of the Mars Science Laboratory (MSL) rover, known also as "Curiosity," were forced to cancel plans to attach a 3-D zoom camera system, developed with the help of filmmaker James Cameron, due to a lack of time

[116] Chow, Denise. "Boeing Unveils Design for Commercial Crew Capsule." Space News 26 July 2010: 15.

[117] Ibid.

[118] Malik, Tariq. "X-37B Wraps Up 7-month Mission Shrouded in Secrecy." Space News 6 Dec. 2010: 17.

[119] Ibid.

[120] "USAF's X-37B Minishuttle Returns to Space on Atlas 5." Space News 14 Mar. 2011: 8.

[121] Ibid.

to get it ready for launch in November.[122] While in its final stages as of April 2011, there wasn't enough time to complete the units and integrate them with the rover; 'a number of technical problems' were cited as the cause for the delay. Fixed focal length cameras will replace the 3-D zoom camera system, and are expected to achieve all of the primary science objectives of the rover's Mastcam investigation. Had the zoom cameras been installed, they would have provided increased operational flexibility, improved stereo imaging, and a 3-D motion imaging capability that would enhance the public's visual exploration of Mars. The 3-D zoom developers, Malin Space Science Systems, will continue working on the cameras for potential application on future missions to Mars, the Moon, or on an asteroid.[123] By June 2011, Curiosity began undergoing tests at NASA's Jet Propulsion Laboratory before being shipped to Kennedy Space Center.[124] The rover is twice as long and five times heavier than its predecessor; and it carries ten science instruments for use on Mars' surface.[125] The rover was launched on 26 November 2011; and was expected to reach Mars by August 2012.

Space Exploration Technologies (SpaceX) has three Commercial Orbital Transportation Services (COTS) capsules, named Dragons, in varying stages of assembly. These Dragons, along with their Falcon 9 medium-lift launchers, will undergo separate missions to demonstrate their space-worthiness.[126] The first demonstration was conducted on Dec. 8 2010, where the Dragon capsule orbited Earth twice, and executed a pre-programmed series of manoeuvres and system checks needed for future rendezvous and dockings with the International Space Station.[127] Four of the capsule's 18 Draco thrusters were fired for 6 min to re-enter Earth's atmosphere, followed by the use of drogue chutes and three parachutes to slow the craft down to 8 m per second before landing in the Pacific Ocean. Future versions of the spacecraft will use a propulsive landing system, with parachutes as a backup.[128]

NASA's Orion space capsule, part of the Constellation programme that was terminated following the Obama administration's Fiscal 2011 NASA budget request, was recommitted as NASA's Multi-Purpose Crew Vehicle (MPCV).[129] This repurposed vehicle will be slimmed down to function as a crew lifeboat at the International Space Station. While NASA considered changing Orion, by including the substitution of a composite crew cabin for the capsule's aluminium structure,

[122] David, Leonard. "NASA Nixes 3-D Camera for Mars Science Laboratory." Space News 4 Apr. 2011: 15.

[123] Ibid.

[124] "NASA's Next Mars Rover Touches Down in Florida." Space News 27 June 2011: 8.

[125] Ibid.

[126] "SpaceX's Second Dragon Powered up for 1st Time." Space News 27 June 2011: 8.

[127] Klotz, Irene. "Shear Magic – SpaceX's Dragon soars through test flight into record books after last-minute fix." Aviation Week & Space Technology 13 Dec. 2010: 22.

[128] Ibid.

[129] Berger, Brian. "NASA Makes It Official: Orion To Be Multi-Purpose Crew Vehicle." Space News 30 May 2011: 5.

and the use of a different kind of launch abort system, it decided to stick with the existing Orion design since it already met the needed requirements. Orion could carry four astronauts for 21-day missions before landing in the Pacific Ocean; and the craft would have nearly 9 cubic meters of habitable space.[130] When rendezvousing with the International Space Station, Orion's Relative Navigation System will utilize its Sensor Test for Orion Rel-Nav Risk Mitigation (STORRM) during the docking stage.[131] STORRM consists of an eye-safe Light Detection and Ranging Vision Navigation Sensor, a high-definition camera, along with avionics and flight software.[132]

Russia has upgraded its Soyuz spacecraft with new digital enhancements. The Soyuz TMA-01M, launched for the first time on Oct. 7, features a variety of avionics and computer upgrades that are designed to be less operator intensive, and to make flying it easier.[133] The new avionics system weighs 70 kg less than its predecessor, allowing a small increase in cargo capacity. Also, when docked with the International Space Station, the spacecraft's computer systems allow it to interface with the onboard computers in the Russian On-Orbit Segment, allowing continuous monitoring of the Soyuz by Mission Control in Moscow.[134]

2.5.4 Suborbital Activities

Development of suborbital activity has steadily increased in the U.S. as well as parts of Europe and Asia. While perceived by many as a sector for tourism, it can also be used for experimental purposes.

Virgin Galactic conducted drop tests and landing tests of its SpaceShipTwo spacecraft between 2010 and 2011. SpaceShipTwo completed its first piloted glide test on 10 October, first released from WhiteKnightTwo at an altitude of 13,700 m, and gliding down to the Mojave Air and Space Port.[135] The flight tested for a clean release of the spaceship from its mothership and whether the spacecraft could free fly and glide back and land.[136] The flight also tested SpaceShipTwo's ventral, belly-mounted speed break, which was used to help slow down SpaceShipTwo while in

[130] Ibid.

[131] David, Leonard. "Orion Development Continues at New Lockheed Martin Center." Space News 28 May 2011: 14.

[132] Ibid.

[133] Harwood, William. "Realtime coverage of Soyuz TMA-01M launch." 7 Oct. 2010. CBS News – Space 29 Nov. 2011 http://www.cbsnews.com/network/news/space/home/spacenews/files/7f8aa558a8d86128b00b2b8b7c8c6c43-80.html

[134] Ibid.

[135] David, Leonard. "Virgin Galactic's Private Spaceship Makes First Solo Glide Flight." Space News 18 Oct. 2010:12.

[136] Ibid.

descent.[137] SpaceShipTwo's third drop test used water ballast to achieve an aft centre of gravity, and the craft was flown to a more aggressive stall indication to evaluate its stability and control.[138] After successfully increasing its top speed to 246 KEAS (knots equivalent airspeed) and loads to 3.5 g, the next test phase involved higher-speed subsonic flight with a short 15 s burst of power from its Sierra Nevada RM2 hybrid rocket.[139] By May 2011, five successful drop/glide tests had been conducted in addition to the sixth hot-fire test of a full-scale flight design rocket motor performed at ground test facilities.[140] The next milestones to come will involve short, medium, and long blasts from SpaceShipTwo's RM2 rocket motor.

In a parallel development in the US, XCOR Aerospace began finalising the structure of its Lynx suborbital spacecraft.[141] While the Lynx has undergone a number of design refinements since its first launch in April 2008, it was now reaching its last stage of structural refinement. After finishing its final round of wind tunnel tests, XCOR was expected to conclude the Lynx's design phase and begin assembly.

In Europe, the German aerospace centre (DLR) hopes to expand on the thermal protection technologies validated by the Mach 6 Shefex 1 in 2005 for the purpose of developing a small sub-orbital re-entry vehicle in 2020.[142] The Sharp Edged Flight Experiment (Shefex 2) demonstrator incorporates nine different thermal protection system tests, an actively cooled segment and a hybrid navigation system, while performing at a target top velocity of nearly Mach 11. It expands on the Shefex 1 by increasing its flight duration to 50 from 15 s, resulting in an increased re-entry flight distance of around 100 from 20 km. The Shefex 2 will be launched on a Brazilian VS-40 sounding rocket and will detach at an altitude of 200 km. Shefex 2 is expected to reach a temperature of 2,000 °C while descending between 100 and 20 km. It will then engage a parachute system near the end of its descent and land in shallow water near the Arctic Circle for recovery and data analysis. The Shefex 3 will operate at near orbital flight with a velocity of Mach 24 in 2016; and Shefex 4 is expected to be orbit-capable by 2020.

Also in Europe, Thales Alenia Space is preparing two atmospheric re-entry test vehicles for launches in 2011 and 2013 under ESA's Future Launcher Preparatory

[137] Norris, Guy. "Flying Enterprise – First glide test for SpaceShipTwo begins buildup to powered, suborbital evaluation." Aviation Week & Space Technology 18 Oct. 2010: 37.

[138] "SpaceShipTwo Conducts Third Gliding Flight Test." Space News 22 Nov. 2010: 3.

[139] Norris, Guy. "Flying Start – Low-speed SpaceShipTwo tests pave way for powered evaluation." Aviation Week & Space Technology 20/27 Dec. 2010: 37.

[140] David, Leonard. "Virgin Galactic's Tourist Spaceship Makes Longest Test Flight." Space News 2 May 2011: 14.

[141] Norris, Guy. "Stable Supersonic – Designers focus on suborbital Lynx airframe, after wind tunnel tests verify stability margin." Aviation Week & Space Technology 20/27 Sept. 2010: 36.

[142] Norris, Guy and Robert Wall. "Speed Trial – Experiments prepared for expanded hypersonic test from Norwegian site." Aviation Week & Space Technology 25 April/2 May 2011: 74.

Program.[143] These vehicles are the 450 kg Expert suborbital capsule and the 1,815 kg Intermediate Experimental Vehicle (IXV). Expert will be launched mid-summer, and its 17-min flight will follow a ballistic trajectory to reach an altitude of around 100 km, re-entering Earth's atmosphere at about 5 km per second.[144] The conical, blunt-nose spacecraft is manoeuvred through four flaps that are positioned at its base.[145] It will run ten experiments during its hypersonic descent through Earth's atmosphere. IXV will be launched in 2013 onboard Italy's Vega small satellite launcher; its 21-min flight will reach an altitude of 430 km, re-entering Earth's atmosphere at about 7.5 km per second. This larger demonstrator will perform in-flight verifications of structures, materials, guidance and control systems, and other critical re-entry technologies.[146] An inflatable balloon will emerge upon alighting in the water to ensure vehicle recovery. These tests are being conducted in an effort to further develop its understanding of re-entry issues for future international exploration programs.

EADS Astrium teamed up with Singapore-based companies in an effort to develop a fleet of spaceplanes at Singapore's Changi International Airport.[147] While the Singapore companies are currently designing and building an engineless small-scale demonstrator spaceplane to test aerodynamics and glide capability, the real spaceplane will be operated using both turbofan engines and a rocket power-plant. The turbofan engines will have about 10,000 lbs of thrust, whereas the spaceplane's rocket engine, a derivative of Astrium's Vulcain rocket engines, will have 40 t of thrust. EADS Astrium plans to apply for EASA certification as the use of conventional turbofan engines allows this spaceplane to takeoff and land from commercial airports. During takeoff, the spaceplane will use its turbofan engines to ascend to an altitude of 12 km, before its rocket engine takes over. After reaching a maximum apogee at an altitude of 100 km, it will glide for a period while in descent, before reengaging its turbofan engines to land. The four passenger spaceplane concept reached maturity since first being conceived in 2006, and is now seeking money for development.[148]

---

143 Taverna, Michael A. "From Dream To Reality – European hypersonic research shifts from drawing board to flight-preparation stage." Aviation Week & Space Technology 5 July 2010: 40.

144 De Selding, Peter B. "Europe Prepares Two Atmospheric Re-entry Test Vehicles." Space News 29 Nov. 2010: 7.

145 Taverna, Michael A. "From Dream To Reality...".

146 Ibid.

147 Francis, Leithen. "Launch Pad – Astrium hopes to base a fleet of spaceplanes at Singapore's Changi International Airport." Aviation Week & Space Technology 14 Feb. 2011: 52.

148 Morrison, Murdo. "PARIS: EADS Astrium looking for funds to launch Spaceplane." 22 June 2011. Flightglobal 30 Nov. 2011 http://www.flightglobal.com/news/articles/paris-eads-astrium-looking-for-funds-to-launch-spaceplane-358289/

2.5.5 Other Technologies

New developments in technology and science have occurred within the 2010–2011 period, which have implications both within and outside the space sector.

European automakers may save weight on vehicles by using spinoff fasteners developed for the International Space Station and NASA's Mars Pathfinder. MST Aerospace, the European Space Agency's technology broker, is pushing bolts containing an internal strain-measurement sensor into the auto industry for use in drive trains and axles.[149] A small ultrasound sensor imbedded in the bolt heads provides direct uniform readings instead of the scattered results obtained when using a coupling liquid and external meter. These new bolts have less than a 3 % margin of error, whereas current bolts may have as much as a 30 % error margin.

On another note, an atomic clock test program will be conducted onboard the International Space Station, using two new generation atomic clocks. EADS Astrium Space Transportation will oversee the development of the Atomic Clock Ensemble in Space (ACES) system which will involve two types of atomic clocks. The French Space Agency is developing the first atomic clock, Pharao; a laser cooled caesium clock designed for use in a microgravity environment.[150] The second clock is a hydrogen master clock, developed by the Observatory of Neuchatel, Switzerland; master clocks are already used by the European Space Agency on the four Galileo In-Orbit Validation positioning, navigation and timing satellites slated for launch in 2011.[151] Researchers will compare the performance of the ACES clocks with each other, as well as with ground-based atomic clocks.

Across the Atlantic, Canada's MacDonald, Dettwiler and Associates (MDA) is developing a servicing satellite that has the ability to grapple 75 % of the commercial communications satellites in geostationary orbit and either refuel them or make simple repairs with its robotic arm.[152] After using lidar ranging to approach a target satellite, the servicing satellite will then locate the target's apogee kick motor nozzle via video and target recognition software and draw itself up to its docking ring. Refuelling will be conducted in the same way satellites are fuelled on ground.[153] However, while Intelsat has already signed up for a refuelling service by MDA's Space Infrastructure Services (SIS) vehicle, agreeing to purchase about half of the 2,000 kg of fuel on the vehicle, other satellite manufacturers are sceptical and believe that some type of heavy government backing, or a public-private

[149] Morring, Frank Jr. "Space Spinoffs – ESA pushes high-tech bolts to auto industry." Aviation Week & Space Technology 20/27 Dec. 2010: 22.

[150] "Astrium To Manage Atomic Clock Demo on Station." Space News 26 July 2010: 9.

[151] Ibid.

[152] Morring, Frank Jr. "Changing the Games – Canadian robotics may stretch commercial satellite lifetimes." Aviation Week & Space Technology 21 March 2011: 23.

[153] Ibid.

partnership, is needed to make the servicing model viable for the entire duration of its 10 year operational lifespan.[154]

In another development, NASA's Gravity Probe B was able to demonstrate two key aspects of Albert Einstein's General Theory of Relativity, despite unexpected system noise that obscured some measurements.[155] While this noise obscured the extremely subtle space-time "frame-dragging" effect to within a 20 % margin of error, researchers achieved a 1 % or better margin of error with a second measurement known as the geodetic effect. General relativity predicts that massive rotating objects should drag space-time around themselves when rotating; i.e. as the Earth rotates, it pulls the space-time in its vicinity around itself, shifting the orbits of near-Earth satellites.[156] At its core, the satellite houses four nearly perfect spheres of fused quartz and silicon, cooled with liquid helium to 1.8°K and spun in helium gas to 5,000 rpm, creating super-conducting gyroscope rotors that generated a magnetic pointer along the axis of rotation allowing for precise measurements by digital magnetometers.[157] But for the effect of gravity on space and time, rather than constantly pointing in the same direction while the craft was in polar orbit around Earth, these gyroscopes experienced small but measurable changes in the direction of their spin while being pulled by Earth's gravity.[158] While this technology has very unique applications, some components were already used in NASA's Cosmic Background Explorer which confirmed the Big Bang Theory.[159]

The U.S. Defense Advanced Research Projects Agency (DARPA) has developed a telescope that will provide wide-field views of objects in geostationary orbit.[160] Currently, a network of space-based assets, optical telescopes and radar sites around the world are used to catalogue more than 20,000 space objects. Radar signals generally track satellites and debris in low Earth orbit, while optical systems, e.g. US Air Force's Ground-based Electro-Optical Space Surveillance (GEODSS) system, track more distant objects in the geo-stationary orbit. Most telescopes have spherical mirrors that have a single radius of curvature throughout, and require additional optics to project an image of a curved field of view onto a flat charged coupled device sensor. On the other hand, DARPA's optical design does not require

---

[154] Cf. De Selding, Peter B. "Intelsat Signs Up for MDA's Satellite Refueling Service." Space News 21 March 2011: 5, and De Selding, Peter B. "Satellite Builders Not Enthusiastic About In-orbit Servicing Project." Space News 21 March 2011: 5.

[155] Morring, Frank Jr. "Space – Gravity Probe B pushed spacecraft technology to the limit – and perhaps a little beyond it." Aviation Week & Space Technology 9 May 2011: 38.

[156] See Cain, Fraser. "Frame Dragging Confirmed." 22 Oct. 2004. Universe Today 23 Nov. 2011 http://www.universetoday.com/9984/frame-dragging-confirmed/

[157] Morring, Frank Jr. "Space – Gravity Probe B ...".

[158] "NASA's Gravity Probe B Confirms Einstein Theories." Space News 9 May 2011: 8.

[159] Morring, Frank Jr. "Space – Gravity Probe B ...".

[160] Brinton, Turner. "DARPA Space Telescope Will Track Objects in Geostationary Orbit." Space News 2 May 2011: 6.

the use of other refractive optics to create a flat field for projection onto a flat sensor.[161] With an aspheric mirror and curved charged coupled device sensor, DARPA's design is simpler and more compact, enabling it to conduct faster and more accurate and sensitive searches than the GEODSS system. While still in the alignment phase, once calibration is complete the system will move onto its technology demonstration phase to better track objects in geostationary orbit.

2.5.6 Innovation Policy

In the United States, as the Constellation programme was terminated following the Obama administration's Fiscal 2011 NASA budget request, Orion's long term future remains uncertain. Rather than returning to the Moon, the White House is more in favour of commercial orbital developments, and the human exploration of an asteroid in 2025.[162] While the Orion capsule has already been recommitted as NASA's Multi-Purpose Crew Vehicle (MPCV),[163] it had the potential for asteroid exploration flights as early as 2016 with an adequate build-up of test flights prior to the endeavour.[164]

Commercial suborbital spaceflight is getting a boost from the U.S. Congress. President Obama signed the 3-year NASA Authorization Act into law on 11 October 2010, allowing NASA to devote millions of dollars annually to commercial suborbital transportation projects.[165] In an effort to spur development of new technologies, as well as improve microgravity research and support the fledgling commercial reusable suborbital transportation industry, the Act authorizes NASA to invest $15 million annually on its Commercial Reusable Suborbital Research (CRuSR) program, allowing NASA to fly research payloads on commercial suborbital vehicles. "These flights are expected to reduce the risk for use of overall space operations by demonstrating application in a relevant environment." While initial test flights will carry NASA payloads designed to monitor the flight environment and the FAA's Automatic Dependent Surveillance-Broadcast payload, later flights will carry educational experiments, technology demonstrations, and research payloads. CRuSR's main goal "... is to provide researchers with the opportunities to validate components of developmental systems in a variety of areas, including propellant management, autonomous operation, communication and navigation,

---

[161] Ibid.

[162] Norris, Guy. "Orion Options – Planners craft fast-track launch, funding and exploration survival strategy for Orion." Aviation Week & Space Technology 23 Sept. 2010: 28.

[163] Berger, Brian. "NASA Makes It Official: Orion To Be Multi-Purpose Crew Vehicle." Space News 30 May 2011: 5.

[164] Norris, Guy. "Orion Options...".

[165] Werner, Debra. "Suborbital Spaceflight Gets a Boost from NASA, Congress." Space News 18 Oct. 2010: 12.

in-situ resource utilization, and space manufacturing." Masten Space Systems and Armadillo Aerospace were already awarded $475,000 for flight demonstrations.[166]

A new space race has begun within the U.S. in regard to the successor to the retired space shuttle. The second stage of the Commercial Crew Development (CCDev-2) initiative is underway, with federal funding going toward five separate contenders, i.e. Lockheed Martin, Blue Origin, Boeing, Sierra Nevada Corp., and Space Exploration Technologies (SpaceX).[167] The goal of the NASA's CCDev effort is to seed a commercial industry that can fly crews to the ISS within this decade; CCDev-2 is meant to mature designs that have a chance of growing into a full-scale system. Lockheed Martin's Orion crew exploration vehicle will be retooled as the new Multi-Purpose Crew Vehicle, and will be funded for 3 years under the NASA Authorization Act.[168] Blue Origin received just over $22 million in CCDev-2 funding to provide sub-orbital space flight to tourists and for science experiments in a three seat biconic capsule launched on an Atlas V rocket; it later aims to develop a reusable launch vehicle that will carry a seven-seat capsule to low Earth orbit. Boeing received the most funding of the recipients, i.e. $92.3 million for a pressure test article for its CST-100 capsule, and to evaluate a lighter-weight engine for its pusher-type launch abort system, to evaluate its parachute and airbag inflation systems for water landings, and to conduct full-scale tests of the capsule's pyrotechnics used for separating its service module before re-entry. Sierra Nevada Corp. has already put $20 million into its Dream Chaser spaceplane, and it received $80 million in CCDev-2 funding for preliminary design reviews, and drop tests. And SpaceX will use its $75 million CCDev-2 award to speed development of its side-mounted pusher-type launch abort system, in addition to preparing its initial design for crew accommodation evaluations.[169]

---

[166] Ibid.

[167] Morring, Frank Jr. "The New Space Race – With funding freed, NASA and its partners crank up work on possible shuttle successors." Aviation Week & Space Technology 25 Apr./2 May 2011: 24.

[168] Ibid.

[169] Ibid.

Part II
Views and Insights

Chapter 3
The Establishment of the UK Space Agency

Sa'id Mosteshar

3.1 Introduction

Unlike most countries active in space, the United Kingdom had no independent space agency until 2010. One of the government ministries, the Department for Business, Innovation and Skills (BIS), is primarily responsible for the development of UK space policy and the regulation of space activity.

3.2 History

Prior to the creation of the Agency, government oversight and regulation of space activity in the United Kingdom was entrusted to the British National Space Centre (BNSC), established in 1985. Before 1 April 2010, the United Kingdom discharged its responsibilities and regulatory oversight of space activity through the British National Space Centre. The BNSC was created to coordinate civil space activities across a number of government departments and research councils that have interests in space. It also acted as the UK point of contact with the European Space Agency (ESA), the European Commission, other space agencies and countries.

The main task of the BNSC was to coordinate the space activities of its partners. The BNSC was connected with ten government departments and non-departmental

S. Mosteshar (✉)
London Institute of Space Policy and Law and London University, London, United Kingdom
e-mail: Said.Mosteshar@Space-Institute.org

P. Hulsroj et al. (eds.), *Yearbook on Space Policy 2010/2011*,
Yearbook on Space Policy, DOI 10.1007/978-3-7091-1363-9_3,

public bodies (NDPBs),[1] with a remit to co-ordinate UK civil space activity, support academic research, nurture the UK space industry, and work to increase understanding of space science and its practical benefits.[2] The BNSC had no independent funding. The majority of its funding was provided by the Science and Technologies Facilities Council (STFC), the Natural Environment Research Council (NERC), the Technology Strategy Board (TSB), the Meteorological Office and the Ministry of Defence.[3]

In July 2009 the Government embarked on a review of the status and role of the BNSC, having invited and received comments from interested individuals and organisations on how UK civil space activities can best be funded and managed "to meet the challenges of the future for our civil space activities."[4] The main recommendation was the establishment of a space agency, but as matters moved so quickly following the consultation, a report was never published.[5] On 23 March 2010 the then Minister for Science and Innovation, Lord Paul Drayson, announced the establishment of the new UK Executive Space Agency, which would come into existence on 1 April 2010.

3.3 Rationale

In contrast to its previous structure and coordinating remit, the aim of the UK Space Agency is "to ensure the UK has a competence in space technology." An improvement was sought on the previous arrangement whereby BNSC worked within a structure based on a partnership across government. Consideration was given to the sort of structure that would ensure and encourage continued growth, and equally importantly, one that would ensure that the UK retains a core competence on space applications, technologies and systems in both industry and academia. The Agency is also to maximise UK benefits from space activities and to ensure that the UK remains at the forefront of global scientific excellence, as well as in the exploitation

[1] BNSC Partners were: Department for Business, Innovation and Skills, (BIS); Department for Children, Schools and Families, (DCSF); Department for Transport, (DfT); Ministry of Defence, (MoD); Foreign and Commonwealth Office, (FCO); Department for Environment, Food and Rural Affairs, (Defra); Natural Environment Research Council, (NERC); Science and Technology Facilities Council, (STFC); Met Office; and Technology Strategy Board, (TSB).

[2] UK Space Agency. 15 Nov. 2009. http://www.bnsc.gov.uk.

[3] See supra footnote 1.

[4] A Department for Business Innovation and Skills Consultation on the Funding and Management of UK Civil Space Activities, starting 22 July 2009, ending 14 October 2009. http://www.bis.gov.uk/assets/bispartners/ukspaceagency/docs/09-1170-consultation-funding-management-of-uk-civil-space

[5] The decision to form the Agency is also consistent with the Innovation and Growth Team report 2009–2010. The Team's recommendations are in the Innovation and Growth Strategy (IGS) Report, Feb. 2010. http://www.bis.gov.uk/assets/bispartners/ukspaceagency/docs/igs/space-igs-exec-summary-and-recomm.pdf

of space infrastructure, products and services. The Agency encourages interaction between industry and academia.[6]

The previous multi-partner approach meant that the BNSC could not take active responsibility for the overall delivery of the government's civil space strategy. The "bottom up" approach could have resulted in the UK unintentionally losing its strength and competence in key areas of space, if the decisions of individual partners did not take account of the wider strategic imperatives. Also, BNSC's structure had partner programme budgets operating on different timescales, militating against a unified long-term programmatic and financial outlook.

With a central agency, there can be an improved policy making process. For example, several government departments and agencies invest in Earth observation products. Under the previous structure it was not easy to get an overview of total government expenditure or to streamline products needed by the agencies. It is now possible to evaluate investment in a domestic Earth observation system that could serve various agencies and provide a reliable source of information. Such a system may not only save costs, but could be a source of income from international users. This could lead to innovative solutions being encouraged to meet identified needs with a potential anchor tenant. Industry will benefit from a uniform procurement process and from clearer direction, as will research institutions. These benefits are not confined to the particular example of Earth observation, but can flow from the new policy making environment.

The Agency also has greater amounts of information available to it about the space products and services needed nationally, which helps formulate a policy direction that can benefit the UK in all its space endeavours and investments. Much of the current budget is spent through ESA, where the Agency can better articulate and negotiate the UK position than was possible with the BNSC. The Agency has also the autonomy to speak for the UK on space matters and to initiate and conclude collaborative arrangements with other governments and agencies. An example of this is the Memorandum of Understanding that entered into force between the Indian Space Research Organisation (ISRO) and the Agency on 30 March 2011.[7]

[6] "How the UK Space Agency Makes a Difference." UK Space Agency 19 Nov. 2011. http://www.bis.gov.uk/ukspaceagency/who-we-are/how-we-make-a-difference.

The statement is made against the background of more than a decade in which the UK space industry has shown consistently high growth levels, that have been maintained even during the recession. Preliminary results from the ongoing biennial Study of the UK Space Industry indicate that growth over the last 3 years averaged about 11%. The UK's space sector contributes £7.5bn a year to the UK economy, directly employs 24,900 and supports a further 60,000 jobs across a variety of industries.

[7] "UK and India Agree Greater Collaboration in Space." 1 Apr. 2011. UK Space Agency 19 Nov. 2011. http://www.bis.gov.uk/ukspaceagency/news-and-events/2011/Apr/uk-and-india-agree-greater-collaboration-in-space.

A significant aspect of the new structure is regular interaction between the Agency and the wider space community in the UK. This takes place largely with industry, government, research and academic institution participation in the Space Leadership Council. Such interaction informs the Agency of the needs of the community, and the latter of Agency direction and priorities. The combined impact of focused government expenditure and strengthened international collaboration and negotiating position is intended to improve the competitive position of the UK within Europe and internationally.

3.4 Transition

The transition from the BNSC to the UK Space Agency that started on 1 April 2010 was a gradual process. This was completed by 1 April 2011, when the Agency became a fully functioning executive agency of BIS in keeping with its Structural Reform Plan.[8]

In the intervening period, from 1 April 2010 to 1 April 2011, the United Kingdom Space Agency operated as a shadow executive agency of BIS. To achieve the transition to a full executive agency, the Secretary of State for BIS issued directions[9] requiring the Natural Environment Research Council (NERC), the Science and Technologies Facilities Council (STFC) and the Technology Strategy Board (TSB) to transfer certain responsibilities for activities relating to scientific research in outer space to BIS, acting through the UK Space Agency. This Order makes provision for the transfer to the Secretary of State for Business, Innovation and Skills (in effect to the UK Space Agency) of all property, rights, liabilities and obligations held, acquired or incurred by:

- NERC, STFC and TSB in connection with certain agreements relating to activities of scientific research in outer space previously carried on by those research councils, and
- STFC in connection with grants awarded for developing scientific payloads and space mission operations.[10]

A full impact assessment was not produced for this instrument, as no impact on the private or voluntary sectors was foreseen.[11]

---

[8] Department for Business, Innovation and Skills. Structural Reform Plan: Monthly Implementation Update, published in July 2010. http://www.number10.gov.uk/wp-content/uploads/2010/07/BIS-Monthly-Update.pdf.

[9] United Kingdom Space Agency (Transfer of Property etc.) Order 2011, SI 2011/822 of 16 Mar. 2011. http://www.legislation.gov.uk/uksi/2011/822/made?view=plain.

[10] Ibid.

[11] Ibid.

3.5 Organisation

3.5.1 The Agency

The Agency's Chief Executive,[12] who is also the Accounting Officer for the Agency, is responsible for its financial affairs and reporting, and for supervising its day-to-day operation. A team assists the Chief Executive, with individual responsibilities and expertise in specific areas of space science and technology, as well as other aspects of the work of the Agency. The Chief Executive's responsibilities involve overseeing all UK civil space programmes, as well as acting as a liaison with government departments, industry, the European Space Agency (ESA), the European Union (EU), and with other space agencies. The Chief Executive also coordinates activities and consults with agencies responsible for national security, notably the Ministry of Defence (MoD). The Chief Executive reports to the Minister for Science and Universities,[13] within the Department for Business Innovation and Science.

3.5.2 Oversight and Advice

3.5.2.1 The Space Leadership Council

In March 2010 the Space Leadership Council[14] was formed at the invitation of the Minister for Science and Innovation to: provide advice to the UK Space Agency on its work plan and future opportunities; offer advice on the areas of space activity in which the UK should seek to develop and maintain global leadership; promote the UK's space industry and scientific excellence in space research, technology and applications; and oversee the implementation of agreed recommendations of the innovation and growth strategy (Fig. 3.1).

The Council addresses the full range of strategic issues and their impact on the domestic, European and global space sector. Its work includes identifying strengths and weaknesses to help target private and public investment. Establishing a unified UK position across government on space issues is also within the Council's remit. It takes a lead to ensure a productive dialogue between government industry and academia in order to identify and agree on shared priorities, in particular for R&D,

---

[12] Currently Dr. David J Williams.

[13] Currently Rt Hon David Willetts; also the Co-Chairman of the Space Leadership Council.

[14] The Council comprises 25 individuals who are senior industry, academic and government representatives. Other individuals with specific expertise in relevant fields, and government ministers or high officials also attend relevant meetings of the Council. *See generally* "Space Leadership Council." UK Space Agency 19 Nov. 2011. http://www.bis.gov.uk/ukspaceagency/who-we-are/how-we-work/space-leadership-council.

Fig. 3.1 The official announcement of the establishment of the UK Space Agency (*left* to *right*: Secretary of State for business, innovation and skills, Peter Mandelson, European Space Agency astronaut Tim Peake, science Minister Paul Drayson, and Logica chief executive Andy Green. Source: BIS)

skills, and training. The Council also oversees the Agency's implementation of agreed recommendations of the innovation and growth strategy report.[15]

Other functions of the Council include:

- Setting the timetable for the work undertaken in relation to the innovation and growth strategy.
- Providing advice to the UK Space Agency on its work plan and future opportunities.
- Offering advice on the areas of space activity that the UK should seek to develop and maintain global leadership in.
- Promoting the UK's Space industry and scientific excellence in space research, technology and applications.

The Council addresses the high-level objectives and policy direction of the Agency.

3.5.2.2 The Steering Board

The UK Space Agency has established a Steering Board,[16] comprising a Chairman and three members, to assist in providing strategic direction to the Agency. The Steering Board Chairman will work closely with the Chief Executive, who will

---

[15] The IGS Report contains the recommendations of the Innovation and Growth Team (IGT), established in 2009 to report on strategies to grow the space sector over a 20-year period.

[16] The appointment of the Chairman and three Members was announced on 26 August 2011. One Member will also be the Chairman of the Audit Committee.

represent the Board in public in relation to agreed policies and programmes. The Steering Board has collective responsibility to challenge and support the Executive Board of the UK Space Agency and exercise oversight on behalf of the sponsoring department, BIS.

The Steering Board's role is to:

- Advise on strategic direction and overall targets, within the framework of the UK Space Agency's objectives, wider government objectives and service to business.
- Bring an external perspective to provide strategic challenge on the business and corporate plans and endorse the annual report and accounts for submission by the Accounting Officer.
- Advise on and monitor performance, including setting of targets and performance against them, as well as provide an independent assessment of the targets and the underlying processes.
- Advise on major business developments that imply a significant change in the Agency's role or activities and on its exposure to and management of risk.[17]

The Steering Board will have at least an Audit Committee to assist in the oversight of the Agency.

3.5.2.3 Audit Committee

An Audit Committee is being established as a committee of the Steering Board. The purpose of the Audit Committee within this governance framework will be to help ensure a financially sound and efficient organisation, which makes full and effective use of resources in support of the UK Space Agency's objectives. The Audit Committee will support the Steering Board and the Chief Executive of the UK Space Agency as Accounting Officer, in their responsibilities for risk management, control and governance.[18]

3.5.3 Associated Entities

Not only because of its history, but also to enable the Agency to fully and effectively perform its functions, it collaborates and coordinates its activities with a number of other organisations and government entities. These include[19]:

[17] "UK Space Agency Opportunities." 22 Mar. 2011. UK Space Agency 19 Nov. 2011. http://www.bis.gov.uk/ukspaceagency/news-and-events/2011/Mar/uk-space-agency-opportunities.

[18] See also supra footnotes 16 and 17.

[19] "Who We Work with." UK Space Agency 19 Nov. 2011. http://www.bis.gov.uk/ukspaceagency/who-we-are/who-we-work-with.

- The Met Office,[20] which funds the UK's weather satellite programme.
- The Natural Environment Research Council (NERC),[21] which supports basic R&D for future Earth science missions and funds researchers to exploit the scientific data yielded by the Agency's Earth science missions.
- The Science and Technology Facilities Council (STFC),[22] which supports basic R&D for future space science and funds researchers to exploit the scientific data yielded by the Agency's space science missions.
- The Technology Strategy Board,[23] which funds a range of industrially-led developments across a large range of sectors, including future technology and applications of space systems and data.[24]
- UK Trade and Investment (UKTI),[25] with which the Agency works to support inward investment into the UK space sector and the export of UK space technology.
- The Science and Innovation Network, which represents UK science and research expertise in UK embassies across the world.

[20] A Trading Fund within the Ministry of Defence, the Met Office operates on a commercial basis under set targets. See Met Office 19 Nov. 2011. http://www.metoffice.gov.uk.

[21] NERC is the UK's main agency for funding and managing research, training and knowledge exchange in the environmental sciences. It is funded by the Treasury and income from external research, sale of data and consultancy. *See* Natural Environment Research Council 19 Nov. 2011. http://www.nerc.ac.uk

[22] The STFC is a non-departmental public body of BIS. It operates or hosts advanced experimental facilities, funded through mainly through BIS. The STFC tackles significant challenges such as meeting future energy needs, monitoring and understanding climate change, and global security. It has a broad science portfolio and works with the academic and industrial communities to share its expertise in materials science, space and ground-based astronomy technologies, laser science, microelectronics, wafer scale manufacturing, particle and nuclear physics, alternative energy production, radio communications and radar. See Science & Technology Facilities Council 19 Nov. 2011. http://www.stfc.ac.uk.

[23] The TSB is an executive non-departmental public body (NDPB), established in 2007 and sponsored by the Department for Business, Innovation and Skills (BIS). Its activities are jointly supported and funded by BIS and other government departments, the devolved administrations, regional development agencies and research councils. See Technology Strategy Board 19 Nov. 2011. http://www.innovateuk.org.

[24] A Service Level Agreement (SLA) defines how the Technology Strategy Board and the UK Space Agency work together. See "Technology Strategy Board and UK Space Agency Service Level Agreement." 11 Mar. 2011. UK Space Agency 19 Nov. 2011. http://www.bis.gov.uk/assets/bispartners/ukspaceagency/docs/service-level-agreement/technologystrategyboardsla2011march.pdf.

[25] UK Trade & Investment (UKTI) is a government agency, reporting jointly to BIS and the Foreign Office. It works with UK-based businesses to ensure their success in international markets, and encourage the best overseas companies to look to the UK as their global partner of choice. *See* UK Trade & Investment 19 Nov. 2011. http://www.ukti.gov.uk.

The Agency also works closely with many other government departments and agencies, as well as professional and industrial bodies. Some of these are reflected in the membership of the Space Leadership Council.[26] Others include the International Space Innovation Centre (ISIC), of which the Agency is a member.[27]

3.6 Functions

The UK Space Agency performs two related but distinct functions. First, it is the appointed regulatory authority for activities in outer space. Second, the Agency promotes the interests of the UK space sector and its contribution to the national economy. Each will be outlined in turn.

3.6.1 Regulatory Authority

The Outer Space Act of 1986 ("the Act") implements the international obligations of the UK and is the legal basis for the regulation of activities in outer space carried out by organisations or individuals established in the United Kingdom.[28] The Act confers licensing and other powers on the Secretary of State for Business, Innovation and Skills acting through the UK Space Agency. The Act ensures compliance with the UK's international obligations under treaties and principles covering the use of outer space, including liability for damage caused by space objects, the registration of objects launched into outer space and the principles for the remote sensing of the Earth.[29]

All space activities carried out by individuals or organisations established in the UK are required to be licensed under the Act. Licensees are obliged to:

[26] See further UK Space Agency 19 Nov. 2011. http://www.bis.gov.uk/ukspaceagency.

[27] ISIC announced in March 2010, was formally launched on 6 May 2011. It provides facilities for the public sector, private sector and academia to work together to develop new space applications. ISIC will support the £10 million National Space Technology Programme announced in the 2011 Budget. *See* Scientific & Technology Facilities Council – International Space Innovation Centre 19 Nov. 2011. http://www.stfc.ac.uk/ISIC/default.aspx

[28] The Act also applies to individuals and organisations established in United Kingdom Overseas Territories or Crown Dependencies.

[29] "The Outer Space Act 1986." UK Space Agency 19 Nov. 2011. http://www.bis.gov.uk/ukspaceagency/what-we-do/space-and-the-growth-agenda/uk-capabilities-for-overseas-markets/the-outer-space-act-1986.

Mosteshar, Sa'id. "Regulation of Space Activities in the United Kingdom." National Regulation of Space Activities. Ed. Ram S. Jakhu. Dordrecht Heidelberg London New York: Springer, 2010. 357–371.

- Permit reasonable access to documents, as well as the inspection and testing of their equipment and facilities by the UK Space Agency or its advisors as appropriate.
- Inform the UK Space Agency of any change in the licensed activity (e.g. change of orbit, change of owner) and seek approval prior to the change being made.
- Prevent the contamination of outer space and adverse changes in the environment of the Earth.
- Avoid interference in the space activities of others.
- Avoid any breach of the UK's international obligations.
- Preserve the national security of the UK.
- Insure themselves against third party liabilities, to a maximum of 60 million Euros, arising from the licensed activity. In this context, the UK Government should be named as an additional insured, and insurance should be for the launch and in-orbit phases of the mission.
- Dispose of the licensed space object appropriately at the end of the licensed activity and inform the UK Space Agency of the disposal and termination of the activity.

The reduction of the insurance obligation to 60 million Euros was introduced at the UK Space Conference in early July 2011.[30] Further changes to the Act, particularly regarding liability indemnity provisions,[31] are also likely to be made soon.

3.6.2 Promotion of the UK Space Sector

The current focus of the Agency is to promote and implement certain strategies recommended by the innovation and growth strategy report, published in February 2010.[32] The report advocates particular strategies to maintain and increase growth in the space sector, in order to capture 10% of the world market in the next 20 years.[33] In order to fulfil this and its wider remit to meet national needs, the

[30] See Outer Space Act 1986, Sec. 5; See also "Third-party Liability Reform for UK Space Projects." Rocketeers 9 July 2011. 19 Nov. 2011 http://www.rocketeers.co.uk/node/1618; and "The Outer Space Act 1986" UK Space Agency 19 Nov. 2011 http://www.bis.gov.uk/ukspaceagency/what-we-do/space-and-the-growth-agenda/uk-capabilities-for-overseas-markets/the-outer-space-act-1986.

[31] Outer Space Act 1986, Sec. 10. 22 Nov. 2011. http://www.bis.gov.uk/assets/bispartners/ukspaceagency/docs/osa/outer-space-act-1986.pdf.

[32] "UK Plan to Create £40bn Space Industry and 100,000 New Jobs." 8 Feb. 2010. UK Space Agency 19 Nov. 2011. http://www.bis.gov.uk/ukspaceagency/news-and-events/2010/Feb/uk-plan-to-create-40bn-space-industry.

[33] "The Space Innovation and Growth Strategy 2010 to 2030." UK Space Agency 23 Dec. 2011. http://www.bis.gov.uk/assets/bispartners/ukspaceagency/docs/igs/space-igs-main-report.pdf.

UK Space Agency is responsible for ensuring that the UK retains and grows a strategic capability in space-based systems, technologies, science and applications. The UK Space Agency therefore leads the UK's civil space programme, in order to win sustainable economic growth, secure new scientific knowledge and provide benefits to all citizens.

To support these goals, the Agency:

- Oversees the preparation of a corporate strategy and its effective delivery, monitoring and evaluation.
- Provides the central focus for civil space policy and programmes across government.
- Manages the space projects and programmes under the Agency's responsibility against schedule and budget.
- Works with departments, agencies and research councils, as well as with industry and academia, to ensure the effective and growing exploitation of space [across][by] government, business and civil society.
- Negotiates on behalf of the UK and manages the UK's relationship with other space agencies and trans-national organisations such as ESA, the European Commission, and international groups such as the Committee on Earth Observation Satellites (CEOS) and the International Space Exploration Coordination Group (ISECG).
- Manages the statutory duties of HM Government under the Outer Space Act and develops a space regulation policy that supports economic growth.
- Provides the face of the UK civil space programme to the UK general public through its communications activities and leads the delivery of the UK's space-related education, skills and outreach programme.
- Delivers efficient support services to support the central government's needs, including briefings and correspondence handling concerning the UK's civil space programme.
- Provides the financial management, procurement and operational tools needed to deliver the Agency's work.
- Meets the statutory financial and audit requirements of an executive agency of government.[34]

It will be noted that the promotion of the sector and industry are inherent in several of these Agency responsibilities and activities. Therefore, the Agency will be striking a delicate balance between its regulatory and supervisory role, and that of an advocate for and promoter of space sector participants.

[34] "What we do." UK Space Agency 19 Nov. 2011. http://www.bis.gov.uk/ukspaceagency/what-we-do.

3.6.3 Separation of Functions

The inherent friction between a regulator and promoter of activities is not unique to this field, or to the Agency. The Federal Communication Commission has often been criticised for its close relationship with those it regulates. In the frequency arena, Ofcom has the lead and can act as a check on the Agency's promotion of particular services or operators, should this become a concern. Nevertheless, in all areas of its regulatory activities the Agency will need to be perceived as impartial.

Within the Agency there are currently three directorates, all having a mix of responsibilities, including promotion of the sector. The regulatory function falls under the Director of Policy and Operations, who directs individuals not involved in the activities of the Agency other than in carrying out its regulatory function. As the structure of the Agency develops, more rigorous division of responsibilities may become necessary.

3.7 Policy Formulation

The UK Space Agency has recently[35] completed its policy consultation, seeking views on its draft strategies and their policy implications. These will be considered and reviewed by the Agency and its Policy Working Group to arrive at final recommendations for a coherent and long-term policy direction.

Security is an important aspect of space policy. This has two elements: security of the space environment, and terrestrial security (social, economic and environmental). It is therefore important to take account of all relevant issues in formulating a space policy. The Ministry of Defence is a critical participant and interested party in ensuring space security in both areas. The creation of the Agency has made it possible to more easily consider the civil and military security issues together, without exposing the needs of the MoD to a wide inter-departmental audience. A working group will address the needs of each and provide a channel for consideration of relevant factors in formulating a national policy.

More broadly, a centralised Agency will lead to a long-term and coherent policy that serves all constituencies. This was at best difficult under the BNSC regime, with each government department and agency focussing on its own needs and priorities.

---

[35] "UK Space Agency Strategy for 2011–2015 – Consultation Document." UK Space Agency 1 Apr. 2011. http://www.bis.gov.uk/assets/bispartners/ukspaceagency/docs/strategy/11-834-uk-space-agency-strategy-2011-2015-consultation.pdf.

3.8 The European Dimension

As the European Union develops its space policy,[36] the UK will need to respond with appropriate measures to fit within those policies. This will not entail policy harmonisation with those of other Member States,[37] but streamlining UK space policy with that of the EU, so as to maximise its effectiveness within Europe and internationally. In developing its space policy, the EU will also coordinate and cooperate with the European Space Agency.[38] To continue to secure a good return for the UK in its ESA participation, the Agency will also need to develop clear policies to strengthen and take a more pro-active role within ESA to ensure a positive impact on UK industry, technology and research.

3.9 Future Challenges

The UK Space Agency is still in its early stages of formation, having gained executive agency status in April 2011. Nevertheless, it has made a promising start with many of its intended structural and functional remits taking shape. Although it has commenced the formulation of a policy, it has not yet firmly settled on the direction that it will take.

A critical factor in the level of success the Agency will achieve in the future is funding. In the current economic environment this presents more of a challenge than it might otherwise. However, as one of the few strongly growing sectors in the economy, the challenge may be overcome with better and more focused outreach to the community at large. This is an area in which the Agency and industry have been active, but more needs to be done given the level of public awareness of the Agency and the role space plays in their everyday lives.

Associated with funding, the Agency will be challenged in identifying and supporting technology development. It will need to respond clearly to areas of particular expertise in the UK and resist the natural pressures from lobbyists and specific firms. This in turn necessitates recruiting the appropriate skills and expertise to serve the needs of the Agency.

[36] Article 189 of the Treaty of Lisbon on the Functioning of the European Union (TFEU). OJ 2010/C 83/01 of 30 March 2010. European space policy is not a new departure, but dates back several years, supporting GMES (Global Monitoring for the Environment and Security) and EGNOS (European Geostationary Navigational Overlay Service).

[37] Article 189(2) TFEU.

[38] Article 189(3) TFEU.

Chapter 4
The New German Space Strategy: How Strategic Is It?

Cornelius Vogt

4.1 Introduction

After almost 10 years, the German government finally released a new principal space policy document – "Making Germany's space sector fit for the future. The space strategy of the German Federal Government."[1] The spirit of progress and forward orientation has been encapsulated in its title. But does the document actually live up to its entire name? Evoking the term "strategy" entitles readers to have high hopes and expectations of the new space policy, especially given the fact that it came out only after a decade. Thus, it is an opportune time to examine the constituents of the new space strategy document, particularly by comparing its main features with those of its predecessor. Increased political efforts have been made towards space investments in recent years – Germany's public spending for space has been growing, a "High-Tech Strategy for Germany" was created in 2006, and the Office of the Federal Government Coordinator of German Aerospace Policy was established in 2007. These are key developments that we could expect to have contributed to the creation of the new space policy document that refers to itself also as a "space strategy." Perhaps, this is the latest milestone of the German government's plans to strengthen the country's position as a leader in technology development, innovation, and business. Therefore, assuming that the German government aims to increasingly exploit space for purposes in the realms of business, politics, science, and defence, the term "strategy" will be critically examined in order to discuss how "strategic" this new space policy document actually [is.][purports to be.]

The views expressed in this essay are those of the author and do not necessarily represent views held by the Future Forum Berlin of the Research Institute of the German Council on Foreign Relations.

[1] Federal Ministry of Economics and Technology, ed. Making Germany's Space Sector Fit for the Future. The Space Strategy of the German Federal Government. Berlin: BMWi, 2010.

C. Vogt (✉)
Research Institute of the German Council on Foreign Relations, Berlin, Germany
e-mail: sinjen@dgap.org

P. Hulsroj et al. (eds.), *Yearbook on Space Policy 2010/2011*,
Yearbook on Space Policy, DOI 10.1007/978-3-7091-1363-9_4,

4.2 Background

For a comprehensive understanding of the new German space strategy, this analysis will begin by looking at the process of its creation. The new German space strategy was publicly presented on 1 December 2010 in Berlin. Two things were particularly interesting about this presentation: first, the apparently little dissent on this policy document among the most relevant German actors in the space domain, i.e. politicians, ministerial staff, research institutions and the space industry; and secondly, as a result, the very limited media coverage that accompanied its announcement. One might ask why there was so little dissent. A possible explanation is that this had to do with the confidential and inclusive character of the process of the creation of the document. Consequently, criticism on the hard issues of the future of German space policy may have been expressed behind closed doors, prior to its publication.

Analysing a document on which everyone ostensibly agrees without running the risk of repetition is not the easiest task. The aim of this essay is to develop a useful set of criteria for the evaluation of the new German space strategy. Therefore, the guiding question will be how "strategic" the new strategy actually is. For the first time, the German government has published a space strategy. This sounds like there has been progress, but does this "strategy" live up to its title? The term "strategy" will be closely examined in this essay, in order to put the new German space strategy into a political science perspective. For this purpose, this chapter will begin with a few general comments on the concept of "strategy" that includes policy objectives, interests, means, uncertainties, the opponent's will, the reduction of complexity, and long planning cycles. Following this analysis, it will be possible to assess whether the basic steps and tools of strategy-making were sufficiently applied in the German space strategy document. In doing so, the shortcomings of the document will also be brought to the reader's attention.

4.3 Context and Evolution of the German Space Strategy

In recent years, the German government has placed more emphasis on space policy as evidenced in "The High-Tech Strategy for Germany"[2] of August 2006. Space technologies have played a significant role. In fact, public spending related to space technologies amounted to 3.65 billion euros.[3] Thus, space technologies constituted the highest public expenditure out of 17 fields in the technology sector, such as energy, information and communications, and nanotechnologies. The new German space strategy refers directly to the 2006 high-tech strategy[4] and sees itself in line with it. As a follow-up to the 2006 high-tech strategy, the German Federal Government

[2] Federal Ministry of Education and Research, ed. The High-Tech Strategy for Germany. Bonn, Berlin: BMBF, 2006.

[3] Ibid.: 104. See table: Funding for the High-Tech Strategy 2006–2009.

[4] Federal Ministry of Economics and Technology, ed. Making Germany's Space Sector Fit for the Future. The Space Strategy of the German Federal Government. Berlin: BMWi, 2010. 3, 5.

adopted the "High-Tech Strategy 2020 for Germany."[5] Space technologies are again prominently mentioned as one of the key investments for the future. The usefulness of space technologies is specifically outlined for areas such as Earth observation for climate and energy research, security, and communications.[6]

Another indication of the intensified political effort in space affairs was the establishment of the Office of the Federal Government Coordinator of German Aerospace Policy in 2007. Prior to that, no such office existed. Aerospace policy issues were subsumed as one of many other policy areas of the Federal Ministry of Economics and Technology. The creation of the office highlights the greater role the government wants to attach to aerospace policy. Parliamentary State Secretary Peter Hintze, who is one of the three Parliamentary State Secretaries in the Federal Ministry of Economics and Technology, has held this position since its establishment. In August 2009, he delivered his first report.[7] This document laid the groundwork for the space strategy that was subsequently published.

Since October 2009, Germany has been governed by a centre-right coalition under the leadership of Chancellor Angela Merkel. The coalition of the three governing parties agreed to develop a national space strategy within a year.[8] Throughout 2009, the German Aerospace Center had convened a group of experts from the German science community, as well as the German space industry, to draft a concept for the German space strategy.[9] This document was first published in November 2009.

In recent years, the German space budget has grown by about 10% annually. Space spending by the Federal Ministry of Economics and Technology alone rose from 764 million euros in 2004 to around 940 million euros in 2009.[10] Overall German public spending on space stood at approximately 1.2 billion euros in 2010. The central ministries in German space activities in terms of funding are the Federal Ministry of Economics and Technology, the Federal Ministry of Transport, Building and Urban Affairs, and the Federal Ministry of Defence.

There are two main drivers for this growing political attention to innovation, technology development, and particularly space utilisation: on the one hand, this is the result of the increasing international competition (and cooperation) in space; on the

[5] Federal Ministry of Economics and Technology, ed. Ideas. Innovation. Prosperity. High-Tech Strategy 2020 for Germany. Bonn, Berlin: BMBF, 2010.

[6] Ibid: 13, 17, 19.

[7] Bundesministerium für Wirtschaft und Technologie, ed. Bericht des Koordinators für die Deutsche Luft- und Raumfahrt. Berlin: BMWi, 2009.

[8] Growth. Education. Unity. The Coalition Agreement between CDU, CSU and FDP for the 17th Legislative Period. 26 Oct. 2009: 44 http://www.cdu.de/en/doc/091215-koalitionsvertrag-2009-2013-englisch.pdf.

[9] Deutsches Zentrum für Luft- und Raumfahrt, ed. Konzept für eine deutsche Raumfahrtstrategie. Köln. 2009.

[10] Bundesministerium für Wirtschaft und Technologie, ed. Bericht des Koordinators für die Deutsche Luft- und Raumfahrtpolitik. Berlin: BMWi, 2009: 38.

other hand, the German federal government has growing political ambitions in space and its related policy areas. These areas include activities that depend heavily on space assets, such as military operations in distant locations, with a view to achieving foreign policy objectives.

In conclusion, the new German space strategy is the result of an ongoing process that aspires to formulate a coherent and comprehensive German space policy within the larger framework of a national innovation and technology policy. The creation of the German space strategy involved contributions from (and consultations with) a large number of German space-related actors, including several federal ministries, the German Aerospace Centre, research institutes and the space industry. In short, the overall growing policy ambitions of the German government have led to an equally growing demand for the utilisation of space.

4.4 Strategy and Space Strategy: About Political Purpose, National Interest and Instruments

4.4.1 Steps Toward a Strategy

In a very general sense, all strategies attempt to balance political purposes with available means, while at the same time taking into account the strategic environment in which they have to operate. Strategy is terminologically and conceptually not limited to the domains of warfare or grand strategy.[11] It may also be applied to very specific political arenas, like space policy. The identification of one's political purpose is followed by the articulation of specific interests. The real difficulty is in finding a coherent fit. Without a thorough assessment of the strategic environment, this matching process is doomed to failure. Adding to the challenge of striking a good balance between political purpose and interests, one is obliged to match them by using specific instruments that are either readily available or under development.[12] This matching process, in turn, has to consider the instruments and interests of other (competing) players. In fact, the existence of at least one opponent is the element that conceptually permits the creation of a strategy.[13] It follows that the aim of a strategist is not to

[11] "Strategy comes into play in any contest of opposing wills even if not a shot is fired in anger. This is the true meaning of what the British call >grand strategy<, the French >total strategy< or >la stratégie de paix/guerre<, and the Americans >national strategy<." Buchan, Alastair. War in Modern Society: An Introduction. London: C.A. Watts. 1966. 82.

[12] Hart, B.H. Lidell. Strategie. Wiesbaden: Rheinische Verlags-Anstalt, 1960. 397.

[13] According to French general and strategist André Beaufre strategy is defined as "l'art de la dialectique des volontés employant la force pour résoudre leur conflit." Beaufre, André. Introduction à la stratégie. Paris: Librairie Armand Colin, 1963. 16.

overstretch, misguide, misuse, or underuse available instruments. Indeed, one might think that everything and anything seems to be interconnected, perhaps even chaotic. The question then is, why would anyone embark on a strategy in the first place if this endeavour appears so [straining][energy intensive], highly contingent, and fault-prone.

A potential answer could be that strategy is a useful tool to efficiently and effectively utilise limited resources by clarifying and prioritising among political objectives and the means required to accomplish them. Thus, a strategy helps to identify capability gaps and design a plan to fill such gaps in the future. Clearly, capabilities are not an end in themselves. They gain legitimacy from the political ends they are meant to meet. Prioritisation among different instruments means explaining why the one is chosen over the other. This implies making realistic assumptions on the relative costs and benefits of the different instruments at hand. The need for a strategy may also arise from the necessity to make decisions under complex and uncertain circumstances. This is often the case when long planning horizons are involved. In most cases, policy making occurs in an environment of uncertainty. A strategic approach may be helpful to manage uncertainty, particularly when multiple and diverse actors are involved, and when dynamic relationships among these actors are expected.

In short, a strategy needs to address the environment it is designed for, a country's goals and interests, and the desired and prioritised instruments in order to achieve political objectives. In addition, it needs to include awareness of its main competitors in the field. A strategy should also have an idea of how to deal with presumed uncertainties, while leaving room for flexibility in order to anticipate changes in its policy environment. The following remarks on the new German space strategy will focus on aspects that are specifically relevant to answering the question of how "strategic" it really is, as well as on policy issues of general importance to contemporary space affairs.

4.4.2 The German Space Strategy

The German space strategy begins with a few introductory remarks. It delivers a status update on German space activities, informs the reader about the country's space budget, its contribution to the European Space Agency, important preceding government documents that ought to be considered, basic facts about the German space industry, and finally, some general remarks on developments regarding the international environment of space affairs.

More specifically, the document states that space has become "an essential instrument for the achievement of economic, scientific, political and social goals."[14] Furthermore, it sees German space activities confronted with four major challenges:

(a) Growing international competition. Thus, the new space strategy can also be seen as a reaction to the ever changing international environment in space that is

[14] Federal Ministry of Economics and Technology, ed. Making Germany's Space Sector Fit for the Future. The Space Strategy of the German Federal Government. Berlin: BMWi, 2010. 3.

briefly described by the term "coopetition", meaning a situation where cooperation and competition are occurring simultaneously.
(b) Fundamental changes in European space policy, namely due to the provisions of the Lisbon Treaty that have assigned explicit competencies to the European Union with regard to space policy. In this respect, determining the future roles of the EU and ESA remains an important space policy issue. The strategy goes as far as calling these political changes within Europe "dramatic."[15]
(c) The emergence of new markets and commercial providers for space services, i.e. for private launch services.
(d) The growing dependence on space assets and services, coupled with their increasing vulnerability as space is becoming more crowded.

According to the document, its aim is to openly present Germany's current plans for space activities to a wider audience and to lay the groundwork for future German activities in space.[16] The latter is considered to be a facilitator of further coordination across different governmental bodies involved in German space activities. Nevertheless, there remains the unanswered question of why the document bears the title of a "strategy", when – according to its own objectives – it is rather a tool of external (information to the public) and internal (administrational management) communication (Fig. 4.1).

Utilising space technologies in order to face global challenges is defined as the overarching goal of the German space strategy.[17] With this formulation, it aligns itself with recent European efforts. More precisely, similar formulations have been employed in previous documents and meetings on the European level, such as the ESA Ministerial Council meeting in The Hague 2008 with its motto: "Space: A key asset for Europe to face global challenges."

The German space strategy then continues with a presentation of the most fundamental national goals of space policy for the coming years. It makes clear that "space is a means to an end [and that] all space projects must therefore be measured according to the contribution they make to achieving those [political] objectives. The Federal Government's strategy aims to use space to address key global challenges and, in so doing, exploit its economic potential to the utmost."[18] Four global challenges are identified as the areas of application of space-related instruments:

(a) Space and globalisation: globalisation and particularly flows of data, transport, and goods.
(b) Space and the knowledge-based society: the requirements and advancement of the modern knowledge-based society.

[15] Ibid: 4.
[16] Ibid: 5.
[17] Ibid: 7.
[18] Ibid.

Fig. 4.1 Federal Minister of Economics and Technology Rainer Brüderle with Peter Hintze, Parliamentary State Secretary and German Government Coordinator of Aerospace Policy (Source: BMWi)

(c) Space and the environment: climate change and the preservation of natural resources, where space technologies allow policy makers to base their decisions on concrete world scale evidence.
(d) Space and security: space assets can contribute to confronting contemporary security challenges that are often not limited to confined or proximate locations. The strategy specifically mentions that modern military operations rely heavily on space technologies. "Space systems in the domains of communication, navigation and Earth observation make a decisive contribution to Germany's ability to conduct an effective foreign and security policy and to achieve whole-of-government security preparedness."[19]

Following this section, the document presents what it calls the "Guidelines of the Federal Government's space policy."[20] Three basic principles are mentioned: (a) orientation toward benefits and needs, (b) orientation toward the principle of sustainability, and (c) intensifying international cooperation.

After these three relatively short and rather general chapters, the strategy goes further into detail on what it calls the "fields of action",[21] a section that spans through eight subchapters. The headlines of the subchapters are straightforward and give a fair outline of Germany's future focus in space policy. The fields of action include: the expansion of strategic space expertise; the establishment of a unified legal framework in space; the sustainable reinforcement of Germany's position in space research; the development of new commercial markets in space; the exploitation of space for civil and military security purposes; the distribution of roles of the main institutional actors in the European space sector; the definition of the roles of Germany and Europe in space exploration; and the safeguarding of technological

[19] Ibid: 8.
[20] Ibid: 9.
[21] Ibid: 11.

independence and independent access to space. Every subchapter follows a three-step approach: (a) explanation of the policy goal that is mentioned in its heading, (b) description of the international environment that this policy goal is surrounded by, and (c) specification of concise objectives in note form, i.e. bullet points. German interests, however, are not named. Finally, the German space strategy ends with a very condensed and straightforward summary of the previous chapters.

4.4.3 What Is So New About the New German Space Strategy?

The new German space strategy provides a mixture of important elements of continuity and much smaller traces of change or renewal in German space policy. There are some slight policy variations distinguishing the new space strategy from the 2001 German space programme. The biggest differences, however, are: (a) the new chapter about the exploitation of space for civil and military security, (b) the notion of sustainability in space usage that pervades almost the entire new document and makes a strong point in favour of national and international efforts for establishing a more comprehensive regulatory framework in space, and (c) the aim of technological independence.

4.4.3.1 Continuities: What Is Not So New?

The basic principle of German space policy continues to be the usefulness of space activities for the achievement of societal, political, scientific and economic goals. The 2001 space programme noted that "...actual demand and utility will be the governing criteria rather than a project's technical attractions."[22] Does this mean that only those projects that promise to be profitable on the commercial market will be considered for public funding? The unequivocal answer is no. According to the new space strategy, benefits of space projects with large public funding may result from their contribution to global environmental management, their provision of efficient support for government duties, their contribution to excellent research and advancements in science, and their added economic value.[23]

Germany's space activities will continue to focus on fields in which Germany has already attained a leading position, or where it has a real chance of doing so.[24] This is seen as an economic approach aimed at maximising the benefits derived from limited resources, avoiding resource overstretch, and achieving "...a substantial increase in

[22] Federal Ministry of Education and Research, ed. German Space Programme. Bonn, Berlin: BMBF, 2001. 13.

[23] Federal Ministry of Economics and Technology, ed. Making Germany's Space Sector Fit for the Future. The Space Strategy of the German Federal Government. Berlin: BMWi, 2010. 9.

[24] Ibid: 4. See also Federal Ministry of Education and Research, ed. German Space Programme. Bonn. 2001: 13f, 16.

German industry's share of global sales revenue in the commercial space sector."[25] Furthermore, it remains a general objective of Germany's space policy to capitalise on international cooperation in space as an additional means to further European integration and to uphold international partnerships.[26] In conclusion, several attributes of the new German space strategy were already mentioned similarly or identically in the 2001 space programme. Hence, numerous enduring objectives of the German space policy are being reaffirmed in the new space strategy.

4.4.3.2 Novelties: What Is New?

First of all, the new German space strategy points to the link between space and security. Space is seen as a means to improve the so-called whole-of-government security preparedness and therefore it fits into Germany's overall security policy approach. In fact, the document mentions that "military operations, in particular, are now inconceivable without the support of space-based systems."[27] As a result of the dependence on satellite imagery from foreign partners during the Kosovo conflict in 1999, the 2001 German space programme already called for a national core capability in satellite reconnaissance.[28] Thus, the reference to the link between space and security is not entirely new to German space policy. With the acquisition of Satcom BW and SAR Lupe, Germany now operates two military satellite systems. This wasn't the case when the 2001 German space programme was penned.

The new German space strategy not only points to the significance of space-based technologies for security purposes, but it also underscores the vulnerabilities that follow from a heavy reliance on space technologies. Moreover, the federal government advocates the "greater use of the potential for synergy between civil and military space research in the development of security-related technologies in the Earth observation and telecommunications sector, for example."[29] With regard to international cooperation in space activities for defence projects, the strategy clearly states that Germany favours bilateral or multilateral European arrangements over cooperation within ESA or the EU.[30] No reference, however, is made to the European Defence Agency.

Another feature of the German space strategy is the promotion of the principle of sustainability. In the eyes of the German government, the principle of sustainability converges with the need for comprehensive national and international legal

[25] Ibid: 18.

[26] Ibid: 3, 10. See also Federal Ministry of Education and Research, ed. German Space Programme. Bonn: BMBF, 2001. 11.

[27] Ibid: 8.

[28] Federal Ministry of Education and Research, ed. German Space Programme. Bonn: BMBF, 2001. 10.

[29] Federal Ministry of Economics and Technology, ed. Making Germany's Space Sector Fit for the Future. The Space Strategy of the German Federal Government. Berlin: BMWi, 2010. 19.

[30] Ibid: 21.

regulations regarding the utilisation of space. It is argued that a regulatory framework is required for future generations to use space to the fullest extent possible. The document specifically mentions the dangers to space-based infrastructure that stem from space debris or intentional interference.[31] In this context, the German government sees "a growing need for a national and international legal framework in the space sector."[32] Furthermore, the strategy specifically points to an urgent need for "regulations on preventing and disposing of space debris, liability in the event of collisions in space and other accidents and the protection of space-based infrastructures from attack."[33] The German government seems to consider it difficult to establish internationally binding rules and instruments that may help to avoid an arms race in outer space in the near future. Therefore, the strategy encourages confidence-building measures, such as the EU Code of Conduct for Outer Space Activities, to not give up on arms control in space.[34]

The newly obtained responsibilities of the EU in the realm of space policy, as they were formulated in the Lisbon Treaty, are addressed as well (for an analysis on recent developments from a European perspective, see the relevant chapter in this part). The 2001 space programme stated that "a longer-term aim is to bring ESA closer to the EU in organisational terms. This will provide scope for ESA to act under delegation from the EU."[35] Furthermore, the 2001 space programme proposed that ESA's geo-return mechanism should gradually align with the EU's industrial and support policy.[36] Current policies indicate that Germany has taken a step back from this position or that at least it wishes to delay the process of bringing ESA closer to the EU. In fact, it is Germany's policy to strengthen ESA as an "independent intergovernmental organisation."[37] Only in the long run, may the EU be in the position to assume new responsibilities and tasks in space.

According to the new strategy, Germany also aims to secure technological independence and guaranteed access to space. However, this independence has to come with a low price tag. The authors of the space strategy are aware of the fact that complete technological autonomy would be unaffordable, even on a European scale. It nevertheless advocates that "...it is essential for critical components and hardware to be readily available in Europe."[38] Reliance on single source technology providers may be obviated. The international traffic in arms regulations of the United States government are of particular concern, since certain hardware components can only be

[31] Ibid: 10.

[32] Ibid: 14.

[33] Ibid.

[34] Ibid: 14f.

[35] Federal Ministry of Education and Research, ed. German Space Programme. Bonn: BMBF, 2001. 15.

[36] Ibid.

[37] Federal Ministry of Economics and Technology, ed. Making Germany's Space Sector Fit for the Future. The Space Strategy of the German Federal Government. Berlin: BMWi, 2010. 21.

[38] Ibid: 24.

procured from the U.S. In cooperative international projects, like ESA's Ariane 5, Germany will continue to focus on bottle neck technologies, in order to stay involved in such projects.[39] For space transportation and the launcher sector, the German government emphasises that Ariane needs to be successful in the commercial market in order to be a cost-effective launcher. Technological independence and ensured access to space shall be achieved on a European and not on a national level.

To sum up, the novelties of the new German space strategy appear first and foremost as reactions or adaptations to recent international and European space activities and policy developments. They do not seem to express a genuinely new direction in German space policy that could be conceived as a challenging factor or an element of surprise to other space actors.

4.5 Conclusion: How "Strategic" Is Germany's Space Strategy? What Is Missing?

At a panel discussion on space applications and whole-of-government security at the Federal College for Security Studies in Berlin in March 2010, Johann-Dietrich Wörner, Chairman of the Executive Board of the German Aerospace Center, pointed out that in recent years German space policy has moved from a bottom-up to a top-down approach. Wörner said that the latter requires a thorough strategy.[40]

In light of this change, the question of whether the new German space strategy may be rightfully classified as a strategy arises. Indeed, several of the above-mentioned attributes of an ideal type strategy appear in the new document – even though sometimes in fragmented form. The mere fact that a new central policy document on German space policy has been published is something to be welcomed. Basic principles of German space policy are laid out in a single document, which can help with creating more cohesion among the different governmental bodies that are involved in German space activities. However, a number of (strategic) flaws are particularly striking. For example, no actual or assumed relative costs are mentioned. Therefore, the document appears as a general summary of political goals rather than a prioritisation of concrete political purposes, interests, and instruments. A rough estimate of relative costs and benefits of different purposes and instruments is required in order to make efficient use of a country's limited resources. In addition to this, some parts of the strategy are ambiguous due the document's simultaneous focus on 'benefits and needs' on the one hand, and visionary goals on the other. This blurs the

---

[39] Ibid: 23.

[40] Federal College for Security Studies. Panel Discussion. Raumfahrtanwendungen und gesamtstaatliche Sicherheit. 31 Mar. 2010, Berlin, Germany. Recording: 01:45-03:06. 11 Aug. 2011 http://www.baks.bund.de/DE/Veranstaltungen/Rueckblick/Rueckblick2010/Raumfahrt_ 2010/ Raumfahrt.html.

government's priorities.[41] Therefore, it appears that everything and anything seems to be in line with this strategy. On the other hand, this strategic vagueness leaves the political leadership enough room for manoeuvre.

The Federal Government does not see its space strategy as a tool to enhance solely German interests. It rather aims at fostering the utilisation of space in the name of humankind, i.e. the improvement of human living conditions. The German government intends to improve human living conditions on Earth using space applications: "space for the benefit of the Earth"[42] is the strategy's overriding motto. This approach could have served as a reference point for the entire strategy, but it does not.

Readers of the new German space strategy should be able to draw something like a 'strategic pyramid' with different levels of goals, means and costs, as well as an image of the strategic environment that surrounds German space policy. Unfortunately, some pieces of the puzzle are missing. For example, as in the case of German space related interests, strategic gaps are hardly addressed throughout the entire document. This omission adds to the impression that the document for the most part is a response to external developments. Indeed it avoids exerting pressure on competitors in the space field through planning unexpected actions or policy shifts.

Ultimately, what will matter most to the success or failure of the new German space strategy will be its implementation. In other words, it needs to be translated into actual programmes and projects. The German Aerospace Centre is authorised to draft a German space programme that follows the strategic guidelines. With regard to the above-mentioned legal regulations on space utilisation, the German government will come forward with a national German space Act. In conjunction with the satellite data security Act that came into effect in December 2007, a comprehensive national legal framework will thus be established.

In recent years, the German government has made considerable efforts to reinforce and improve the country's position in space. It remains to be seen whether German space policy is really altering to a strategic top-down approach, as the head of the German Aerospace Centre observed. By and large, the new German space strategy constitutes a continuation of German space policy, which demonstrates a strong (but not exclusive) focus on the economic, industrial, and commercial potential of developments in space.

---

[41] "The Federal Government focuses its space policy strictly on benefits and needs while, at the same time, targeting visionary goals." Federal Ministry of Economics and Technology, ed. Making Germany's Space Sector Fit for the Future. The Space Strategy of the German Federal Government. Berlin: BMWi, 2010. 9.

[42] Ibid.

Chapter 5
"Towards a Space Strategy for the EU that Benefits Its Citizens": The EU's Declaration of Intents for Space

Matxalen Sánchez Aranzamendi

5.1 Introduction

Commission Communication COM(2011) 152[1] (the Communication) had been long awaited since Commissioner Tajani announced his intention to elaborate a new space policy, shortly after taking office as new Commissioner for Enterprise and Industry in February 2010. This unilateral initiative came as a surprise to the European space community. Fourteen months and a Space Council later, the same European space community witnessed the release of a European Commission Communication entitled "Towards a Space Strategy for the EU that Benefits its Citizens" at a press conference[2] held at the ESA site in Frascati (Italy). The several months of separation between the initial declaration by Commissioner Tajani, a year of space active EU Presidencies and a Space Council in November 2010, among others, blurred the connection between Commissioner Tajani's declaration and the Communication that eventually ensued. Still the Communication was received nervously and raised some questions as to the meaning of the word "strategy", the role of the Communication in the context of the European space policy, and its conception as an apparently unilateral by the European Commission.

The following chapter aims at providing an analytical insight into the Communication by presenting the role of the Communication in the European space policy, its potential to set a space strategy and its nature as an EU initiative in the context of the

[1] European Commission. Communication to the Council, the European Parliament, the European Economic and Social Committee and the Committee of the Regions Towards a Space Strategy for the European Union that benefits its Citizens. COM (2011) 152 final of 4. April 2011. Brussels: European Union.

[2] "A New Space Policy for Europe: Independence, Competitiveness and Citizen's Quality of Life". Europa Press Release (IP/11/398) 4 Apr. 2011.

M.S. Aranzamendi (✉)
ESPI, Brussels, Belgium
e-mail: matxalen.sanchez@espi.or.at

P. Hulsroj et al. (eds.), *Yearbook on Space Policy 2010/2011*,
Yearbook on Space Policy, DOI 10.1007/978-3-7091-1363-9_5,

EU policy arena. With this purpose, this chapter will first attempt to determine the meaning and role of a space strategy in Europe by briefly examining the development of the European space policy and some of the most recent examples at national level. It will then briefly look at the contents of the Communication, and it will assess the nature and role of the Communication. Finally, based on this analysis the chapter will draw conclusions on the nature, role and potential of the Communication.

5.2 Definition of a Space "Strategy" in the European Context

As indicated by its title, the Communication seeks the eventual formulation of a space strategy, but it does not lay down such a strategy in itself. In fact, the European space policy making process has been evolving around the formulation of a space strategy as the pivotal element of the European space effort since its beginnings in 2000,[3] endowing it with a particular European meaning. During the same period, the space scenario has evolved, posing new challenges and bringing new perspectives. These changes have triggered the adaptation of national space strategies and policies, in order to face the new challenges and the call by the Space Council[4] to draw a joint European strategy to address these issues. Given the reference to a strategy and the unavoidable role of the EU in European space policy making, the Communication must be analyzed in the light of its relevance to the European space policy making process, as well as of its capacity to serve as a space strategy *stricto sensu*.

5.3 The European Quest for a Space Strategy

The word "strategy" has been present in the making of the European space policy since its inception in 2000.[5] On 27 September 2000 the European Commission published a Communication entitled "Europe and Space Turning a New Chapter,"[6] which responded to the calls of the ESA Ministerial Council and of the EU Council

---

[3] Council of the European Union. Resolution on the Reinforcement of the Synergy Between the European Space Agency and the European Community. OJ EC (98/C 224/01) of 22 June 1998. The resolution was adopted in parallel by both the ESA Council and the EU Council.

[4] Council of the European Union. 7th Space Council Resolution "Global Challenges: Taking Full Benefit of European Space Systems." 25 Nov. 2010. Inv. 1. Brussels: European Union.

[5] Council of the European Union. Resolution on Developing a Coherent European Space Strategy. OJ (1999/C 375/01) of 2 Dec 1999. Brussels: European Union Note: It mentions for the first time the objective of formulating a strategy.

[6] Commission of the European Communities. Communication to the Council and the European Parliament on Europe and Space: Turning a New Chapter. COM (2000) 597 final of 27 Sep. 2000. Brussels: European Union.

to "prepare, together with the European Space Agency (ESA), a communication on a European space strategy."[7] The 2000 Communication reflected the joint position of the EU and ESA and set the basis for creating a task force that would elaborate that strategy:

> ...the strategy as proposed in this communication is a further milestone in establishing a coherent approach for space in Europe. It is not intended as a detailed road map for programming space activities in the coming years.

The groundwork of analysing the European space panorama, and setting the basis for action areas and designing a *modus operandi* was carried out by a task force[8] that issued a set of recommendations "that should lead to" the definition of "a genuine European Space Policy":

> The document covers in particular the three components of the European strategy for space: (i) strengthening the foundations of space activities; (ii) enhancing scientific knowledge; (iii) reaping the benefits for markets and society; as well as horizontal issues mainly addressing industrial policy, international aspects and the future relation between the EU and ESA.
>
> It should lead to the definition and establishment of a genuine European Space Policy, which would combine eventually the strategy per se, a European Space Programme (an essential element due in particular to its structuring effect on the space constituency), as well as a set of implementing rules.

While beginning to define the substance of a future European Space Policy, the task force identified three fundamental pillars for it: a space strategy that would capture the substance of European action in the space domain, a space programme that would provide for a functional structure, and implementing rules. At that stage, a strategy was one of the three elements of the Space Policy, albeit a crucial one, as it would set the objectives for its realisation.

The task force set the basis for a Green Paper and consultation that led to a White Paper in 2003.[9] The White Paper crystallised the result of 3 years' work on the elaboration of a final European Space Policy. The publication of the White Paper coincides roughly with the signature of the EU-ESA Framework Agreement[10] and the creation of the Space Council, which took over the task of guiding the space

---

[7] Ibid.

[8] Commission of the European Communities. Communication from the Commission to the Council and the European Parliament "Towards a European Space Policy." European Commission and the European Space Agency Joint Task Force Report. COM (2001) 718 final of 7 Dec. 2001. Brussels: European Union.

[9] Commission of the European Communities. Green Paper. European Space Policy. COM (2003) 17 final 21 Jan. 2003. Brussels: European Union; European Commission. White Paper. Space: A New European Frontier for an Expanding Union. An Action Plan for Implementing the European Space Policy. COM (2003) 673 final 11 Nov. 2003. Brussels: European Union.

[10] Council of the European Union. Framework Agreement between the European Community and the European Space Agency. RECH 152 (OC589) final of 7 Oct. 2003, entered into force in May 2004. 13 Dec. 2011. http://ec.europa.eu/enterprise/policies/space/files/policy/european_communityandeuropeanagency frameworkagreement_en.pdf.

policy making process. The latter culminated in 2007 with the joint formulation of the European Space Policy (ESP) by ESA and the EU. The ESP was embodied in a Commission Communication[11] and submitted to the Space Council.

The ESP Communication captures the status of European space activity to that date and builds on the work carried out by the task force and the first three Space Councils to identify areas of action and potential objectives.[12] The structure of the ESP Communication does not differ from previous proposals for a strategy, but it does not comprise the three pillar structure initially envisaged by the task force. In essence, the ESP solely captures the strategic part, while it is accompanied by a document setting the preliminary elements for a European Space Programme.[13] The latter was only given the status of a working document and it was not considered by the fourth Space Council.

In short, the ESP failed to set the ground for the adoption of the European Space Programme and for creating the appropriate governance schemes. This fundamental weakness of the Communication has remained in European space policy making during recent years, while new challenges have emerged in terms of increasing global competition in space activities and industry as well as in the area of space security. The April 2011 Communication addresses both the gaps left open by the ESP and the emerging challenges by following a style similar to that of the ESP. The fact that it covers previously uncovered questions, that it follows the orientations of the latest Space Councils and that it is formally captured in a Commission Communication, triggers the question of the role of the Communication beyond the EU scope enshrined in its title.

5.3.1 Recent Developments in National Space Policies and Strategies

Efforts to formulate a space policy or a space strategy are not an exclusive domain of the EU. Recent years have seen intense activity in adopting national space policies and strategies that may shed some light on what a European space strategy could aim at. The most recent among these strategies is the German space strategy, which is contemporary with the recent establishment of a British Space Agency and

[11] Commission of the European Communities. Communication from the Commission to the Council and the European Parliament "European Space Policy." COM (2007) 212 final 26 Apr. 2007. Brussels: European Union.

[12] Marboe, Irmgard. "National Space Legislation: The European Perspective." Nationales Weltraumrecht/National Space Law. Development in Europe. Challenges for Small Countries. Eds. Christian Brünner and Edith Walter Böhlau. Wien, Köln, Weimar, 2008: 31.

[13] Commission of the European Communities. Commission Staff Working Document. European Space Programme-Preliminary Elements. SEC (2007) 504 final 26 Apr. 2007. Brussels: European Union.

its ongoing work on the formulation of a Space strategy for the coming 2 years.[14] In addition, the international scene has also recently witnessed initiatives of this kind. Most notably, the Obama Administration conducted a reorientation of the U.S. Space Policy in 2010, followed by the adoption of a new National Security Space Strategy in 2011.

On 28 June 2010 the U.S. announced its new Space Policy[15] which adds to the number of policy changes brought in by the Obama Administration. The new U.S. Space Policy was followed in January 2011 by the U.S. National Security Space Strategy (USNSSS),[16] which reflects the change in the approach[17] to global challenges adopted by the U.S. Space Policy.[18] Together, these documents mark the goals of U.S. space activity and set guidelines for their achievement. The U.S. Space Policy is the vehicle to convey principles inspiring the U.S. activity in the space field, to enumerate the goals to be pursued by U.S. space programmes and to set the guidelines for the achievement of these goals. While the principles contained in the policy set the political goals to inspire all space activity, inter-sector guidelines act as a code of conduct to guide decisions in the achievement of sector objectives laid out by sector guidelines. On its side, the USNSSS "builds on the strategic approach laid out in the National Security Strategy"[19] and focuses on the security aspects of space. It marks strategic objectives and sets a roadmap to achieve them.

While the U.S. National Space Policy sets the overall objectives and inspiring principles for the U.S. Space Sector, the U.S. strategic objective for space focuses on the use of space for its security objectives and on ensuring the protection of U.S. strategic space assets. This approach contrasts with the latest reviews of national

[14] See relevant chapters in this part.

[15] "National Space Policy of the United States of America." 28 June 2010. The White House, Washington, D.C. 13 Dec. 2011 http://www.whitehouse.gov/sites/default/files/national_space_policy_6-28-10.pdf.

[16] "National Security Space Strategy." Jan. 2011 U.S. Department of Defense and Office of the Director of National Intelligence 13. Dec 2011. http://www.defense.gov/home/features/2011/0111_nsss/docs/NationalSecuritySpaceStrategyUnclassifiedSummary_Jan2011.pdf.

[17] Pagkratis, Spyros. "International Cooperation in the New U.S. Space Policy: Opportunities for Europe." ESPI Perspectives 42. 22 Dec. 2010. The author argues that the change in the approach is characterised by a softer language and the change towards an attitude based on cooperation. According to the author this is a natural change to be attributed to the change of administration. However, he argues that the Policy remains the same in content.

[18] The new U.S. space policy was motivated by the changes in the global space scenario and reacts to three main challenges: the increasing use of space to provide services for citizens, increasing global competition caused by the growth in the number of space faring nations and changes in the concept of space security.

[19] The strategy points out at a number of challenges such as the increasing international competition, the opening up of new markets together with the growing importance of private enterprise business models and the greater dependence on space both for the delivery of public and consumer services as well as international cooperation and cooperation in the EU context.

space strategies in Europe that focus on the strategic value of space technologies as the central element in achieving the key goals of their national policies.[20]

"The Space Strategy of the German Federal Government"[21] is the most recent European space strategy document. In similar fashion to the U.S. Space Policy, the German strategy seeks to adapt to the changes in the space scene. The German strategy refers to the paradigm shift in space as no longer seeing space as a goal in itself, but as the means to provide for services to citizens. Furthermore, it focuses on the strategic role of space to preserve a leading role in key policy areas of the German economy.

The core of this strategic direction is comprised by a set of policy orientations and a catalogue of action items. While policy orientations provide for general objectives to inspire the entire space activity,[22] the German strategy lists a number of objectives grouped in major interest areas of the German space sector that include: furthering regulatory and legislative action, ensuring and progressing the highly specialised technology base (e.g. in system integration and remote sensing), creating appropriate synergies between civilian and military technologies and anchoring Germany's presence in basic space science and exploration. Overall, the German strategy recapitulates emerging space policy challenges, Germany's wider economic and policy interests in space, and the emerging need of its space sector to set the objectives of its future programmes and the guidelines to be respected during their realisation.[23]

The UK has taken a similar approach in determining the strategic value of space. In April 2011 the British government launched a consultation[24] on a new national space strategy for the period 2011–2015, which is to succeed the current one for

[20] Other countries like Japan have also turned to a more security based conception of the word strategy. In this sense, Japan reformed its basic Space Law to include security aspects as it was perceived that it could not meet its international commitments without incorporating the security dimension. Aoki, Setsuko. "Challenges for Japan's Space Strategy." AJISS-Commentary No. 34. 26 June 2008; Suzuki, Kazuto. "A Brand New Space Policy or Just Papering over a Political Glitch? Japan's New Space Law in the Making." Space Policy 24 (2008): 171.

[21] "Making Germany's Space Sector Fit for the Future. The Space Strategy of the German Federal Government." Nov. 2010. Federal Ministry of Economics and Technology 13 Dec. 2011. http://www.bmwi.de/English/Redaktion/Pdf/space-strategy,property=pdf,bereich=bmwi,sprache=en,rwb=true.pdf.

[22] According to the Strategy, space projects will be judged and funded according to their efficiency in addressing societal issues in competition with terrestrial technologies.

[23] Remuß, Nina-Louisa. "Space as a Key Driver for Innovation Improving People's Living Conditions on Earth." German Journal of Air and Space Law 60.2 (2011): 231.

[24] "UK Space Agency Strategy 2011–2015: Report on public Consultation." 1 Nov. 2011. UK Space Agency 14 Dec 2011 http://www.bis.gov.uk/assets/bispartners/ukspaceagency/docs/strategy/11-834-b-uk-space-agency-response-to-the-agency-consultation-2011-2015.pdf.

"Strategy to Grow UK Space Sector." 10 Feb. 2010. BBC News 14 Dec. 2011 http://news.bbc.co.uk/2/hi/science/nature/8506141.stm.

2008–2012.[25] This proposal coincides with the entry into operation of the recently set up UK Space Agency.[26] The proposed space strategy has a twofold objective: on the one hand to provide guidance for the UK Space Agency that will guide all civil space activities; on the other hand, to create a link between space and society by setting up a strategy that fundamentally supports the overall UK national growth strategy. In fact, the proposal refers explicitly to the UK growth strategy. The contribution of space to each of these growth connected aspects is explained per heading and accompanied by a list of objectives. The proposed UK strategy ends with a heading on the UK Space Agency, its functions and the cooperation it seeks to carry out with the country's space industry, ESA and the EU.

In conclusion, there is no single concept of the strategic nature of space. Each country defines the meaning of the word strategy according to its general policy interests. However, the trends in space strategies during recent years show certain common characteristics regarding the form and purposes of a space strategy. In all of the abovementioned examples, the review of national space strategies was motivated by two main factors: the emergence of changes in the global scene and the value of space in sustaining national strategic objectives. Consequently, space strategies can be defined as a catalogue of broad orientations and more concrete objectives aspiring to face these challenges and meet national interests.

5.4 The EC Communication of April 2011

5.4.1 The Communication in a Snapshot

The Communication presented by the European Commission on 4 April 2011 is a 12-page document that captures the essence of the EU's space portfolio under Art. 189 TFEU. In its seven sections, the Communication elaborates on the political background and legal basis that support the EU's action in space, lists its programmatic priorities for space, and describes the needs for further action on the functional side (industrial policy and governance).

The contents of the Communication build on the major EU political events of recent years and are inspired by the orientations of the Space Councils that have taken place since the adoption of the ESP Communication. The style is markedly policy oriented as both priorities and functional elements are mainly presented in

[25] "UK Civil Space Strategy. 2008–2012 and beyond." 14 Feb. 2008. British National Space Centre 14 Dec. 2011. http://www.bis.gov.uk/assets/bispartners/ukspaceagency/docs/ukcss2008-2010.pdf.

[26] De Selding, Peter B. "New Government Space Agency To Centralize British Space Efforts." Space News 23 Mar. 2010.

Amos, Jonathan "'Muscular' UK Space Agency launched." 23 Mar. 2010. BBC News 14 Dec. 2011 http://news.bbc.co.uk/2/hi/8579270.stm.

Fig. 5.1 Antonio Tajani, vice-president of the European Commission (Source: ESA)

the framework of broader EU policies, such as climate change, security and competitiveness. In this sense, the Communication does not address challenges to the space sector, nor does it create a roadmap to address them.

5.4.1.1 Politico-legal Basis and Policy Background

The Communication is an effort to embed space in the general EU policy objectives, while providing a specific policy rationale for the EU's space priorities. In this sense, the Communication starts with an introduction that highlights the social, economic and strategic value of space, and refers to the contribution of space to the major policy strategies of recent years where space is explicitly mentioned, such as the Europe 2020 Strategy and the EU Industrial Policy. The Communication underpins the support of all relevant institutions (European Parliament, Council and the Space Council) for the EU taking action in the space field. In addition, it underlines Art. 189 TFEU and the EU-ESA Framework Agreement as the legal basis of such actions (Fig. 5.1).

5.4.1.2 Priorities

The substance of the EU's action in space according to the Communication is captured in a set of priorities that comprise the following:

- Satellite Navigation: the Galileo and EGNOS Programmes.
- GMES and climate change.
- Security: SSA and space for security.
- Exploration.

Satellite Navigation

Contrary to the classic European tendency of focusing on the completion of Galileo, the Communication brings EGNOS to the forefront as the currently existing European Satellite Navigation system. This priority area describes the EU's approach to SatNav. Although the completion and immediate operationalisation of Galileo continues to be the most immediate priority, action in this area is to develop along the lines of downstream market development and the creation of business opportunities.

GMES and Climate Change

GMES is embedded in the wider Climate Change objective. While the full completion and operationalisation of GMES is the top and most immediate priority in this area, the core of the priority area should be the development of GMES into more than a research activity. In this sense, EU efforts in this area would focus on the development of downstream services and the creation of business opportunities. The other side of the EU's efforts for GMES focus on the development of its climate change nature and its potential to respond to security needs, including the exploitation of the dual use opportunities offered by GMES.

Security: SSA and Space for Security

Space and security is the third priority area. This priority area reflects the relevance that security aspects have adopted in EU affairs. It is also the reflection of the bigger openness to incorporate security aspects in the space discussion in Europe. Action under this priority area should develop on three levels; the release of the full potential of dual use possibilities of current civilian assets such as GMES or commercial non-military satellites, coordination of existing national assets or development of European assets to fulfil the requirements of the European external action, and development of the European Space Situational Awareness (SSA) capacity.

Space Exploration

Space exploration is the fourth priority. The Communication does not present clear-cut actions in this area but describes the role of the EU in exploration as an enabler of the link between exploration and challenges in the social, political and economic areas. According to the Communication the EU should pursue actions on cross-fertilisation in essential technologies, participation in ISS cooperation, fostering independent access to space and the set-up of a high-level international platform.

The Communication recognises there is a political dimension to space exploration that goes beyond solely research and development issues.

5.4.1.3 Functional Elements

After listing the priorities that should constitute the core of the EU's action in the space field, the Communication deals with other aspects of a more functional or strategic character. These aspects are:

- Competitiveness (industrial policy, research and satellite communications for fostering innovation).
- The international dimension of space policy.
- Governance.
- Basis for a European Space Programme.

Competitiveness (Industrial Policy, Research and Satellite Communications for Fostering Innovation)

This section contains some of the most awaited and meaningful messages of the Communication as it deals with industrial policy. The Communication briefly describes the strong points and main characteristics of the sector. Space generates innovation and requires a highly qualified work force while it strongly depends on public procurement and has to contend with increasing competition. The European space sector also includes the high participation of SMES. The Communication describes the main objectives of the space industrial policy as the steady and balanced development of the industrial base, greater competitiveness on the world stage, non-dependence for strategic sub-sectors and the development of the market for space products and services.

According to the Communication, these objectives should inspire an industrial space policy that should be elaborated in close cooperation with ESA and the Member States ("MS") and should reflect the needs of each sub-sector. The Communication also mentions tools for the industrial policy such as an appropriate regulatory framework, adequate financial tools for research and innovation and the most appropriate procurement procedures. The Communication also mentions the "possibility to adopt specific provisions under particular legislative acts." As the latter shows, the language of the Communication is vague and anything could be read into that wording. In general it could be interpreted as the recognition of the special nature of space and a will to adopt specific measures that differ from ordinary mechanisms. This is particularly relevant in the area of procurement as it could mean greater flexibility in EU-ESA relations.

The reference to research and development and [posterior][subsequent] attention to the "Common Strategic Framework for Research and Innovation" (now

"Horizon 2020" after November 2011)[27] seems to have predicted intentions not to fund space under a specific heading but again under the Research Framework Programme.

International Dimension

The definition of objectives in the international arena is a highly strategic task. Although international relations should not be an end in itself, the definition of objectives in this area is crucial to defining the EU's position as a global actor. However, the Communication is rather weak in this task. No objectives are listed in terms of cooperation with other space faring countries. The Communication stays on a political level by reiterating the intention to broaden Space Dialogues to other countries and the need to better integrate space related matters into the EU's external policy, while mentioning Africa specifically. Supporting the development of the Code of Conduct for Outer Space activities and Europe's involvement in GEOSS are the two multilateral initiatives mentioned by the Communication.

Governance

The strongest language of the Communication is found in the section dealing with governance. The section on governance defines the role of the EU in relation to the MS and ESA. The Communication highlights its role as coordinator under Art. 189 TFEU and aims at stronger cooperation with MS in order to ensure better synergies between space and other policy areas. While the language relating to MS is in cooperative terms, the tone regarding EU-ESA relations is more demanding. The Communication uses a straight-forward approach to addressing ESA, when it calls for:

- The creation of an EU-ESA industrial policy.
- The reassessment of the ESA-EU framework agreement.
- The reconsideration of ESA's procurement practices regarding EU programmes.
- The increased involvement of ESA in space security issues.

Basis for a European Space Programme

The last functional element deals with the creation of a European Space Programme. The Communication refers to its competence under Art. 189 TFEU

[27] European Commission. Communication to the Parliament, the Council, the European Economic and Social Committee and the Committee of the Regions. Horizon 2020 – The Framework Programme for Research and Innovation. COM (2011) final 30 Nov. 2011. Brussels: European Union.

to adopt a European Space Programme. The Communication seems to suggest that, despite its name, this European Space programme adopted by the EU would not be "the" European space programme affecting all European space actors but an EU complementary specific programme (an EU European Space Programme) based on multi-sector actions. What the European Commission might have meant with such complementarity remains unclear.

5.4.2 Deciphering the Communication

The contents of the Communication reveal that its primary purpose is to map out the material scope of the EU space competence as much as it is to define the role and competence of the EU in European space policy making. As mapped out by the Communication, the EU's approach to space is marked by three major characteristics:

- Alignment with the paradigm shift in space policy.
- Recognition of the space and security dimension.
- Tackling of governance and industrial policy issues.

Space policy has experienced a "paradigm shift" whereby space activities by themselves are no longer a sufficient justification for governmental support in the space field.[28] For the last two decades, national space policies have been shifting towards an approach that is based on the development of applications and provision of space-based services. As seen above, this has been reflected in the most recent space policy reforms. In this sense, the 2010 U.S. Space Policy explicitly mentions that "Space systems allow people and governments around the world to see with clarity, communicate with certainty, navigate with accuracy, and operate with assurance."[29] Along this line, the German Space Strategy is crafted to meet wider policy demands. The Communication is not extraneous to the "paradigm shift"; moreover, it fully embraces an approach based on applications and space services to serve other policy objectives. In fact, the Communication covers all the usual elements comprising the European space policy and reflected in all the ESP Communications and the Space Councils (Galileo, GMES, launchers, space foundations, SSA, space security and space exploration). However, the Communication adopts a different systematic approach, policy actions and objectives are grouped along three main policy goals that reflect the three main policy trends of

[28] Change in the conception of space activities has long been identified. Prof. von der Dunk coined the term "paradigm shift" and elaborated its significance in: von der Dunk, Frans. "As Space Law Comes to Nebraska, Space Comes Down to Earth," Nebraska Law Review 87 (2008): 498.

[29] Supra 15 p. 1.

the EU's policy of the last decade: climate change,[30] security integration[31] and competitiveness.[32] In addition to the policy-based approach, the Communication strongly emphasises the development of downstream services and business opportunities, which are major objectives in the fields of SatNav and GMES.

This cross-policy and market based approach is not only the consequence of the "paradigm shift"; it also comes as a natural development in the EU's integration mechanisms. On the one hand, market integration has been the major driver of the EU's evolution since its inception. On the other hand, the EU has enjoyed implicit competences to develop and take action as long as this has been necessary for the completion of the internal market and the achievement of its constitutional goals.[33] A look into the development of the EU's space competence reveals its strong focus on the development of applications and market opportunities rather than on the exploration and fostering of space science and technology.

It must be noted that despite the prevailing policy oriented approach, the Communication gives an important role to the historical motive for space action. In spite of the marginal capacity of the EU in these areas, the Communication elaborates on space exploration where development of critical technologies and maintenance of independent access to space are listed among the prime objectives. The vague objectives in this area and recognised shortcomings of the EU[34] structures for these two areas seem to suggest that the inclusion of exploration may be a placeholder for future unforeseen developments. Most importantly, the inclusion of exploration plays a psychological role in the European citizen's mind,

[30] The EU has taken a strong commitment towards climate change in the recent years, with a policy package on reducing carbon emissions by 20 % by 2020, a Directorate General dedicated to Climate Action and a roadmap for 2050, the EU aims at playing a leading role in the global arena. Connie Hedegaard, Member of the European Commission 14 Dec. 2011 http://ec.europa.eu/commission_2010-2014/hedegaard/index_en.htm.

[31] One of the major changes in European integration brought about by the Lisbon Treaty pertains to External Action and the Security and Defence areas that are going through a communitarisation process and [suppose][constitute][imply] a major milestone in European integration. An analysis of these changes is provided in: European Union External Action 14 Dec. 2011 http://www.consilium.europa.eu/eeas/security-defence.

[32] The search for competitiveness has been at the forefront of the EU political strategy for the past decade. Labelled as the Lisbon strategy initially and Strategy EU 2020, competitiveness has been the benchmark for European Policies for the past 10 years. Europe 2020 Strategy 14 Dec. 2011 http://ec.europa.eu/europe2020/index_en.htm.

[33] Case C-22/70 *Commission of the European Communities v Council of the European Communities* (AETR Case) European Agreement on Road Transport. AETR case/implicit competences and Lisbon TEU.

[34] Fourth Conference on the European Union Space Policy. "A Space Policy for Society and Citizens." 8–9 Nov. 2011, Brussels, Belgium. 14 Dec. 2011 http://www.spaceconference.eu/2011/welcome_en.shtml.

as it is a political symbol.[35] In this context, the Communication is full of references to independent access to space. Independent access to space is the motto for each of the applications covered in the Communication as well as for the space industrial policy and international cooperation.

The security related strategic value of space technologies is self-evident, particularly if we look back to the origins of humanity's quest to space and the development of other national policies. Europe's efforts in space policy can be regarded as somewhat anomalous in this respect as security aspects have been deliberately excluded from European joint space efforts (a good example is the interpretation of the ESA Convention clause on peaceful purposes as excluding any security related activity). The extensive attention dedicated by the Communication to security represents a remarkable step in European space policy making while it breaks with taboos such as the full development of the dual use nature of civilian assets like GMES.

Security occupies a central role in both U.S. and German space strategies and even Japan has revisited its Basic Space Law to allow security purposes in order to be able to meet its international commitments.[36] The Communication resembles to some extent the Japanese case. The tackling of security aspects reflects the debate and the calls of the Space Councils for extended action in space security and goes hand in hand with the "communitarisation" of external relations and security decisions since the Lisbon Treaty. The changes brought by the Lisbon Treaty to the EU architecture call for a more visible participation of the EU in international cooperation and crisis management. In this vein, strong European participation in crisis management, humanitarian action and emergency relief are heavily dependent on reliable space infrastructures. Action proposed by the Communication on this level is focused on the development of common space capabilities to meet those goals.

The development of such capabilities not only responds to realising common policy goals, it also aims at providing infrastructure that MS could not develop on their own for their own security purposes that fall outside the EU's remit. By exploiting the dual use capacities of European systems, coordinating national assets or by developing new assets, the Communication depicts an EU with the role and duty of achieving goals that MS are not equipped to achieve, either by themselves, or in the context of ESA.

Lastly, the Communication covers industrial policy and governance issues, which have been subject to intense discussion in the European policy making arena. The Communication sends a number of messages in both areas. The European Commission has often been criticised for lack of appropriate industrial policy

[35] European Space Policy Institute Conference, ed. European Identity through Space, 13 Nov. 2009. 14 Dec. 2011 http://www.espi.or.at/index.php?option=com_content&view=article&id=424:13-November-2009-espi-holds-pioneering-conference-on-european-identity-through-space&catid=39:news-archive&Itemid=37.

[36] Supra 19.

to foster space competitiveness and for a procurement policy that fails to adapt to the specificities of the space sector. The Communication is disappointing regarding the former as it only dares to announce the elaboration of an industrial policy in consultation with the MS and ESA while referring to the European Industrial Policy of 2010 that includes space as a specific area but provides little more than a description of the European space sector.[37] Instruments such as regulations are mentioned but no objective or action is mapped out and financial resources are once again referred to the Research and Innovation framework that has been deemed inadequate by industry. Procurement rules and an appropriate regulatory framework are among the most relevant tools for crafting a space industrial policy. The Communication's language referring to "a better use of the European regulatory framework, regarding trade in particular" and to the definition of "the most appropriate type of procurement procedures" may sound promising but throws little light on the direction the EU should take.[38]

The most daring messages are contained in the section related to governance. This section contains the key to interpreting the meaning of the parallel competence of Art 189 TFEU as it defines the role of the EU in relation to its MS. The document depicts the EU's role as a coordinator of the Member States' space policies. While this seems to confer a rather weak role to the EU, it nevertheless respects MS competence to manage national space policies and programmes. Yet, it must be noted that the EU has competences in other policy areas that may affect space, such as environment. Decisions taken by the EU under those competences may lead to further coordination of national space policies. Finally, the EU's upcoming role in European space policy making is reaffirmed by the position the Communication takes with respect to EU-ESA relations. Using carefully balanced language, the document invites ESA to create a procurement scheme exclusively suitable for EU programmes, and to adapt its internal rules to accommodate space security oriented activities. In spite of the apparently harsh tone of the Communication (as we are not used to hearing demands from the Commission to ESA), such invitations have already been voiced in the context of the Space Council. The Communication remains on role sharing by endowing the EU with the political lead and coordination tasks while it reserves technology development tasks for ESA. However, the Communication remains silent on more crucial governance issues such as programme governance, where ownership and responsibility need to be allocated.

[37] Communication from the European Commission to the European Parliament, the Council, the European Economic and Social Committee and the Committee of the Regions "An Integrated Industrial Policy for the Globalisation Era Putting Competitiveness and Sustainability at Centre Stage", COM (2010) 614, Brussels.

[38] Hobe, Stephan, Hofmannová, Mahulena and Jan Wouters, eds., A Coherent European Procurement Law and Policy for the Space Sector: Towards a Third Way. Cologne Studies on International and European Law. Vol. 22. Berlin: LIT Verlag, 2011.

5.5 Conclusions

The Communication COM (152) “Towards a Space Strategy for the EU that Benefits its Citizens” is the first Commission Communication on Space Policy since the adoption of Art. 189 TFEU and the first of similar nature to the ESP Communication since the latter’s adoption. The Communication’s title refers to the adoption of a strategy and comes at a time when national space policies and strategies are under review, while a new approach is also required at European level. The role of the Communication must be assessed according to three criteria; its nature in the context of Art. 189 TFEU, its place in the European space policy-making process, and its value as a strategy for space.

The first criterion relates to the interpretation of the EU space competence. The recognition of the EU space competence as an explicit shared competence by the Lisbon Treaty triggered questions on the nature and the practical application of a peculiar competence that neither abides by the pre-emption principle nor allows harmonisation but yet endows the EU with wide powers to adopt a European Space Policy and a European Space Programme. In this context, the Communication acts primarily as the instrument to map the material scope of the competence and define the powers of the EU under this competence. It must be noted that the Communication refers constantly to the EU instead of using the wider term “Europe.” It is clear from this wording that the Communication does not intend to deal with joint European space policy making but with the remit of the EU.

In this sense, the Communication describes the political grounds that legitimate the EU’s action in the space field including the adoption of a space policy and a space programme. The description of priorities defines the material scope of the competence while the considerations on industrial policy and financial means describe the EU’s tools to achieve those policy objectives. The Communication is of great value in understanding the “parallel competence” nature of the space competence as it describes the role of the EU in its relation to the MS. In short, the Communication acts as the first step to an EU Space Policy.

It provides the political grounds that legitimate the EU’s action in the space field, it determines the reach and boundaries of the EU space competence, it describes the potential actions and intentions of the EU within those boundaries, and opens the door for concrete action in the form of an EU European Space Programme. In addition, the Communication is of great value as it defines the position of the EU in the European space policy-making process.

The second question relates to the strategic capacity of the Communication. The strategic nature of the national space strategies and policies presented in this chapter lies in their capacity to ensure the continuity and success of space technologies and assets to support major national policy objectives and strategic goals. The latter often refers to defence purposes and national positioning towards international counterparts. All abovementioned examples set objectives and general principles to guide future space programmes towards the achievement of national strategic goals.

The title of the Communication suggests a first step towards the establishment of a space strategy but how suited is it to lay the grounds of a strategy? The main feature of the Communication is the strong recognition of space as a strategic element to achieve Europe's policy goals and the consequent prioritisation of independence in access to space and availability of space based services. In this respect, the powers of the EU in the space field are determined in the context of pursuing other major policy objectives (i.e. climate change, competitiveness and security). However, other than describing broad action areas such as SatNav, GMES or exploration, the Communication fails to identify concrete objectives or to define the direction of future EU space action, and to act as an inspiring element for the future EU space programme. The Communication is also vague in the definition of the contribution of space to European security and external action objectives. While vague language regarding defence objectives is explained by the fact that the EU does not have an integrated defence policy, the Communication could have elaborated more concrete objectives for cooperation with other international space partners. In conclusion, the Communication lacks the determination of a space strategy to mould the EU's position as a global actor or to map out the EU's strengths and weaknesses and fix a direction accordingly.

The last criterion to assess the role of the Communication refers to the place of the latter in the context of the European space policy-making process. The first section of this chapter has dealt with the evolution of the European Space Policy and its search for a strategy. As mentioned, the current ESP Communication was aimed at constituting the strategic pillar of the European joint endeavour in space. Despite its remarkable work in mapping out and describing joint programmes and action areas, it remains on a rather political level. This generalist character and the emergence of new policy and governance needs have triggered calls for the adoption of a new space strategy. Is the Communication such strategy? Can it substitute for the ESP Communication? As mentioned before, the Communication is primarily a solely EU document aimed at setting the first step in defining the EU's space competence. Although it follows the recommendations of the Space Council, the Communication has neither been prepared jointly by ESA and the Commission nor endorsed by the eighth Space Council. In addition, its value as a strategy is limited and it is doubtful that it aimed at providing the strategic basis of the future European Space Programme. Therefore, the Communication cannot substitute for the ESP Communication. However, it does incorporate elements that complement the ESP Communication and may serve as input to the joint European Space Policy discussions.

Chapter 6
Space-Based Capabilities for Internal Security Operations: A Critical Assessment of the Case of Land Border Surveillance

Alexander Kolovos

6.1 Introduction

On 28 January 2011, at the invitation of the European Council, the European Commission published a staff working paper entitled "Determining the technical and operational framework of the European Border Surveillance System (EUROSUR) and the actions to be taken for its establishment."[1]

EUROSUR is a system that should be operational on a pilot basis as of 2011. It is expected to reinforce the control of the external border of the Schengen area, especially its southern maritime and eastern land borders, and step up the fight against irregular migration and cross border crime. Using modern surveillance technologies (satellites included), EUROSUR should support the European Member States in reaching full situational awareness at their external borders and increase the reaction capability of their law enforcement authorities.

The creation of EUROSUR is as a response to one of the emerging challenges the EU has to face: the need to protect its citizens at home. Threats such as this call for European responses that can combine civil and military means at national as well as European level. Border control and surveillance contributes to the management of migratory flows at the external borders of the EU. According to the European Commission, the purpose of border surveillance is to prevent unauthorised border crossings, to counter cross-border criminality and to apprehend or take other measures against persons who have crossed the border illegally.

---

[1] European Commission. Commission Staff Working Paper. Determining the Technical and Operational Framework of the European Border Surveillance System (EUROSUR) and the Actions to be Taken for its Establishment. SEC(2011) 145 final 28 Jan. 2011. Brussels: European Union.

A. Kolovos (✉)
Hellenic National Centre for Space Applications (HNCSA, Ministry of Defence), Athens, Greece
e-mail: alexanderkolovos@yahoo.com

P. Hulsroj et al. (eds.), *Yearbook on Space Policy 2010/2011*,
Yearbook on Space Policy, DOI 10.1007/978-3-7091-1363-9_6,

The 2011 Working paper is the last of a series of relevant documents that were initiated in 2008, when the European Commission presented the Communication on examining the creation of EUROSUR, a decentralised system of (existing) systems, which should be established gradually in three phases, consisting in a total of eight interlinked, but separate steps.[2] These steps were not identified as concrete actions, but rather as a roadmap providing the main parameters for the development of EUROSUR, and they were adopted by the European Council in June 2008.

Following this decision, in February 2010 the European Council called on the Member States to implement the steps set out in the 2008 roadmap, in order to reinforce interagency cooperation and enhance the Member States' border surveillance capabilities as soon as possible. Phases one and two should cover the maritime and land external borders, while Phase three should focus on the maritime domain. The first two phases contain a series of measures that should be developed and implemented at EU level:

- Phase one deals with upgrading and extending national border surveillance systems, and interlinking national infrastructures in a communication network.
- Phase two targets the necessary research and development to improve the performance of surveillance tools and sensors (e.g. satellites, Unmanned Aerial Systems etc.), and develop the common application of surveillance tools.

6.2 The Migration Trend as a Threat to EU Internal Security

The last decade found thousands of desperate refugees fleeing poverty, drought, oppressive regimes, or civil war in North Africa and the Middle East, and seeking refuge to EU territory. Consequently, European external borders are facing considerable pressure from irregular migration, which is expected to continue in the future. This trend started in the early 1990s and has increased since 2000.[3] The European Council recognised that the booming population growth on its borders posed problems for Europe, first with the adoption of the EU Security Strategy[4] and later, in November 2010, with its Internal Security Strategy. The latter identifies the

[2] Commission of the European Communities. Communication from the Commission to the European Parliament, the Council, the European Economic and Social Committee and the Committee of the Regions. Examining the Creation of a European Border Surveillance System (EUROSUR), COM(2008) 68 final 13 February 2008. Brussels: European Union.

[3] This pressure in general is fuelled by the world's population that is increasing by 80-million persons each year. World population had increased fourfold in a 100 years, growing from 1.75 to 6.9 billion people. To put this into perspective, the world is adding a quarter million more persons each day, after deaths are taken into account. The world's production of food is increasing only +1.5 % per year, while total world population is currently increasing at approximately 4% each year.

[4] Solana, Javier. "A Secure Europe in a Better World. European Security Strategy." 12 Dec. 2003. European Union Institute for Security Studies 15 Nov. 2011 http://www.iss.europa.eu/uploads/media/solanae.pdf.

main crime-related risks and threats that Europe faces today, including activities relevant to border control such as organised crime, drug trafficking, human trafficking, trafficking in arms and cross-border crime. The development of the EUROSUR system is recognised as one of the key initiatives under the European Commissions' Communication on Internal Security Strategy[5] (namely in its fourth objective: "strengthen security through border management", and especially in Action One: "exploit the full potential of EUROSUR").

6.2.1 Quantitative and Qualitative Nature of the Threat

In line with the definitions given in the Schengen borders code, the control of the external borders of the EU consists of border checks at border crossing points and border surveillance. Border surveillance should be effective in preventing and discouraging persons from circumventing the checks at border crossing points, and in detecting unauthorised crossing of the external borders. The land borders are divided into regulated areas and unregulated areas. The regulated areas are located along paved roads, railways, etc. and are easy to monitor since they are normal border control checkpoints. Migrants often travel in acute conditions facing extreme hardship, with the consequence of risking the loss of their own lives. They take different routes and use various methods to reach European borders, such as small boats reaching coasts in a clandestine manner, moving on foot or on vehicles (cars, trucks, motorcycles, or bicycles on mountain roads and off road forest paths), or even riding horses, always trying to take advantage of the morphology of the terrain.

Furthermore, a significant problem that needs to be tackled more effectively in the course of border management is cross-border crime, such as terrorism, trafficking of persons, drug smuggling, smuggling of weapons, etc.[6]

Consequently, it can be said that the illegal immigration problem has two dimensions: one is of a quantitative nature while the other is qualitative and related to crime. Obviously, the case of cross-border crime, particularly organised crime, includes higher risks, both to the people involved in border control and to the internal security of the host country, but also on an EU level in general.

---

[5] European Commission. Communication from the Commission to the European Parliament and the Council. The EU Internal Security Strategy in Action: Five Steps Towards a More Secure Europe. COM (2010) 673 final 22 Nov. 2010. Brussels: European Union.

[6] European Commission. "Legislative Proposal on the Establishment of a European Border Surveillance System (EUROSUR)." Oct. 2010. Roadmaps 2011 – Impact Assessment. 15 Nov. 2011 http://ec.europa.eu/governance/impact/planned_ia/docs/2011_home_008_eurosur_en.pdf.

6.2.2 *Border Surveillance Pillars: Situational Awareness and the Management of the Reaction Capability*

Border surveillance is carried out by border guards in stationary or mobile units, who perform their duties by patrolling or stationing themselves at places known or perceived to be sensitive. The main purpose of border surveillance through patrolling measures is to prevent and discourage unauthorised border crossings, to counter cross-border criminality, and to take measures against persons who have crossed the border illegally. As such, the concept of border surveillance activities is focused on two main and interrelated areas[7]: the provision of situational awareness, and the management of the reaction capability.

It is in this framework that the European Council conclusions of February 2010[8] invited the European Commission to report before the end of 2010 on how surveillance tools, such as satellites, could be involved in monitoring EU land and sea borders.[9] From the following analysis, it seems that utilising space systems to monitor terrestrial borders has so far gained more attention than surveillance of maritime borders.

6.3 Space System Needs for EU Internal Security Missions

According to the Schengen Catalogue, the management of each Member State's external borders regarding border control resides with the civilian public authorities that are the police forces. So border control constitutes a form of civilian operations, which institutionally falls under the jurisdiction of EU's third Pillar (cooperation in Justice and Home Affairs).[10] The use of space assets for such civilian crisis

[7] "European External Border Surveillance System (EUROSUR)." 14 Mar. 2008. Europa – Summaries of EU Legislation. 15 Nov. 2011 http://europa.eu/legislation_summaries/justice_freedom_security/free_movement_of_persons_asylum_immigration/l14579_en.htm.

[8] Council of the European Union. Council Conclusions on 29 Measures for Reinforcing the Protection of the External Borders and Combating Illegal Immigration. 2998th Justice and Home Affairs Council meeting, 25–26 Feb. 2010. Brussels: European Union.

[9] Under conclusion 14, the Council invited the Commission to report before the end of 2010 on how the conclusions of the Global Monitoring for Environment and Security (GMES) border surveillance group on common application of surveillance tools, such as satellites, could be implemented in the EU land and sea borders.

[10] Under the Maastricht Treaty (1991), the European Union consists of a three pillar organization that includes the old EC (the economic pillar), a pillar for a new Common Foreign and Security Policy (the second pillar), and a third pillar on Justice and Home Affairs, which covers issues of internal security like immigration and criminal activity. Each pillar would have its own rules, with the latter two more inter-governmental in design. The Treaty of Lisbon, which came into force on December 1, 2009 amends existing EU treaties, formally abolishing the old pillar structure, with the explicit recognition that the Common Foreign and Security Policy will remain subject to its own rules. Currently the actual policies are still pursued by the old three pillars competencies.

management operations has already been acknowledged and a framework has been formulated. A list of generic space needs requirements has been already defined by the EU Civilian Committee (CivCom), in the context of the second pillar (European Security and Defence Policy, ESDP).[11]

It should be noted that the overarching European Space Policy (ESP, of 2007) formally establishes a link between space activities and the European Security and Defence Policy (ESDP).[12] ESP also recognizes as a priority the need to improve coordination and synergies between defence and civilian space programmes and technologies, in a user-driven approach. CivCom has underlined that many of the user needs and requirements for space systems in the planning and conduct of military crisis management operations are equally applicable to civilian crisis management operations.[13]

As such, the relevance of space policy to the ESDP area and the generic needs and requirements are outlined in the following European Council documents, which constitute the existing institutional framework in this area[14]:

- European Space Policy: ESDP and Space (no. 11616/3/04 of 16 Nov. 2004).[15] This document provides the identified and agreed upon ESDP requirements to be reflected in the general EU Space Policy and its corresponding European Space Programme, and is recognised as a cornerstone regarding space in the ESDP domain.
- Generic Space Systems Needs for Military Operations (no. 6920/06 of 7 Feb. 2004).
- Space Systems Requirements as per RC 05.[16]

---

[11] In accordance with the Lisbon Treaty, the former European Security and Defence Policy (ESDP) was renamed as the Common Security and Defence Policy (CSDP).

[12] The EU is developing a common security and defence policy, covering all questions relating to its security, including the progressive framing of a common defence policy. The European security and defence policy allows the EU to develop civilian and military capacities for international crisis management. The ESDP is part of the common foreign and security policy (CFSP).

[13] Space as a contributor to Internal Security was analysed in a special ESPI workshop in 2009. Molard, Bernard. "Space and Internal Security in Europe – The Setting." Conference on Space and Internal Security. European Space Policy Institute, Vienna, Austria. 28 May 2009 http://espi.or.at/images/stories/dokumente/Conference2009/molard.pdf.

[14] For the development of this institutional framework see a detailed analysis by Alexandros Kolovos. See Kolovos, Alexandros. "The European Space Policy – Its Impact and Challenges for the European Security and Defence Policy." ESPI Perspectives 27. 2 Sept. 2009 http://www.espi.or.at/index.php?option=com_content&task=view&id=399&Itemid=1.

[15] Council of the European Union. Note from General Secretariat to COREPER/Council. European Space Policy. "ESDP and Space." 11616/3/04 REV 3 of 16 Nov. 2004. Brussels: European Union.

[16] "Generic Space Systems Needs for Military Operations" (6091/06) and "Space Systems Requirements as per Requirements Catalogue 2005" (RC 05) approved by the GAERC (doc. n. 13732/05 dated 7 November 2005).

http://register.consilium.europa.eu/pdf/en/05/st13/st13732-ex01.en05.pdf.

- The Report of the Panel of Experts on Space and Security (SPASEC Report) of March 2005 also provides useful insights in that respect.[17]
- Outline of Generic Space Systems Needs for Civilian Crisis Management Operations (2006).

Finally, according to the latter EU Council decision,[18] "relevant space system services for civilian crisis management operations include in particular Earth observation, communications (CIS) and Positioning-Navigation-Timing (PNT). Civilian crisis management operations may also require intelligence information. One source of this is signals intelligence."

6.3.1 Generic Space System Needs Adopted by the European Council

Looking at the survey of the entire range of space-related technologies needed for military and civilian ESDP missions that have been addressed in the above EU documents, one can summarise the following capabilities:

- **Earth observation**. Earth observation is a key area in all kinds of operations, in order to collect and process data into information and make it available for its timely exploitation by a wide range of users. Intelligence gathering, surveillance and reconnaissance are the pre-requisites of early warning, advance strategic and operational planning, conducting operations in the area of responsibility including the preparation and employment of friendly forces.
- **Communications**. Secure and reliable communications are essential for exercising command and control of any operation. They are also vital on an operational and tactical level. These capabilities may often depend upon space-based assets, especially where local infrastructure is insufficient.
- **Positioning, navigation, and timing**. The availability of real-time accurate position determination, precise navigation and timing data from satellite systems has become increasingly vital for the effective conduct of any type of operation required in the framework of an EU crisis response capability.
- **Signal intelligence**. Signal intelligence satellites are designed to detect transmissions from broadcast communications systems such as radios and

[17] "Report of the Panel of Experts on Space and Security." 1 Mar. 2005. European Commission – Enterprise and Industry 15 Nov. 2011
http://ec.europa.eu/enterprise/policies/space/files/article_2262.pdf.

[18] Council of the European Union. Note from the Committee for Civilian Crisis Management to the Political and Security Committee. Outline of Generic Space Systems Needs for Civilian Crisis Management Operations. 10970/06 of 27 June 2006. Brussels: European Union.

other electronic systems, from which intelligence can be gathered on the perpetrators' location and intentions.
- **Weather, oceanography and mapping**. The analysis of weather, terrain and other environmental factors is a critical step in the preparation and conduct of operations. Space assets have been extensively used for mapping and producing 3D terrain models suitable for training, mission simulation and planning.

6.3.2 Earth Observation Satellites Chosen in the EUROSUR Framework

Following a technical study carried out in 2009,[19] the European Commission has so far given priority to Phases one and two of EUROSUR, in order to meet the pressing operational needs facing the EU, and in particular at its southern and eastern borders.[20] The European Council[21] has agreed to call on the Member States to implement the phases and steps laid down for the development of EUROSUR as soon as possible, in order to reinforce cooperation and their respective border surveillance capabilities. Space assets appear at the following steps of EUROSUR Phase two, under the title "development and implementation of common tools and applications for border surveillance at EU level":

Step 4: research and development to improve the performance of surveillance tools. Earth observation (EO) satellites appear as one of the two emerging tools of interest for border surveillance purposes (along with UASs).

Step 5: common application of surveillance tools. The application of new surveillance tools (such as EO satellites) could provide Member States' authorities with surveillance information on their external borders and on the pre-frontier area. Consideration should be given to how the EU can support Member States by setting up mechanisms enabling the shared use of investment intensive tools,

[19] See *Cf.* point 23 of Council of the European Union. Council conclusions on Public Deliberations. 2873rd JUSTICE and HOME AFFAIRS Council meeting of 5 and 6 June 2008. Brussels: European Union. The study was implemented between January 2009 and January 2010.

[20] Commission of the European Communities. Commission Staff Working Paper. Report on Progress Made in Developing the European Border Surveillance System (EUROSUR). SEC (2009) 1265 final 24 Sept. 2009. Brussels: European Union.

[21] Council of the European Union. Council Conclusions on 29 Measures for Reinforcing the Protection of the External Borders and Combating Illegal Immigration. 2998th Justice and Home Affairs Council meeting of 25 and 26 Feb. 2010. Brussels: European Union.

such as satellites.[22] FRONTEX could act as a facilitator, e.g. liaising with service providers in order to procure satellite imagery on behalf of several Member States.[23]

Step 6: common pre-frontier intelligence picture (CPIP). CPIP facilitates targeted intelligence reaction: on the basis of intelligence received from third countries' authorities, a vehicle used for a criminal activity is identified abroad and tracked by using satellites until its interception in EU territory.

In the subsequent two reports of 2009 and 2011, the European Commission presented the progress made under the different steps in terms of the goal of having EUROSUR operational by 2013. In these two documents, space assets were assigned the task of improving the technical ability to identify and track objects of interest, such as:

- Step 4, aimed at improving the performance of surveillance tools, was fulfilled through the implementation of the sixth and seventh Framework Programmes for research and development (aimed at improving the technical ability to detect, identify and track small boats used for illegal migration and related cross border crime).
- Regarding Step 5, the GMES border surveillance group prepared a concept study for the common application of tools for border surveillance, which included a detailed technical overview of surveillance sensors, capabilities and platforms, such as satellites. A number of projects was to be covered under the FP7 2011 work programme (space theme).
- Finally, the objective of the CPIP (Step 6) was identified as supporting the Member States to improve their situational awareness of the pre-frontier area and providing them with additional time for reacting adequately. One of the CPIP components should include basic geodata, such as topographic and thematic maps and nautical charts that should be managed by FRONTEX on behalf of the Member States.

---

[22] Global Monitoring for Environment and Security (GMES) observation capacities can contribute to improve the mapping and monitoring of border areas. GMES aims to make earth observation a tool for governance, it is concerned with developing applications for satellite observation, but also takes into account collateral airborne and ground-based observations. Information is expected to be collected from the meetings of the GMES expert group on EUROSUR, which has elaborated a GMES strategy to support EUROSUR and which has prepared a concept of operations on the use of GMES for border surveillance with exploratory supporting actions in 2012. "Application of Surveillance Tools to Border Surveillance 'Concept of Operations'." 7 July 2011. European Commission – Enterprise and Industry 8 Jan. 2012.
http://ec.europa.eu/enterprise/policies/security/files/doc/conops_gmes_en.pdf. This document does not implicitly or explicitly represent either the views of the European Commission or those of the Agencies mentioned in it. The purpose is to serve as orientation material to guide proposals for the FP calls 2012 with respect to activities related to Border Surveillance, taking into account the role of relevant stakeholders, lessons learned from R&D projects and other activities in the area.

[23] FRONTEX, which was established in 2004 as a Community Agency under the first pillar of the European Union, is responsible for the coordination of operational cooperation between Member States to facilitate the application of Union law, including with regard to border surveillance.

6.4 Bring Space Back to Earth: Bridging the Gap Between Research and the Operational Shortfalls

6.4.1 The Institutional Deficit

6.4.1.1 The Missing ESDP Framework

Cooperative approaches are the main tools selected for building a European Space Architecture. On 25 June 2003, the second Pillar's Policy and Security Committee[24] contributed a position paper (SN 2211/1/03) to the European Commission's consultation process on the Green Paper on European Space Policy. An interesting extract of the paper was the following: "in the light of the on-going development of ESDP, further and regular trans-Pillar reflection is needed to ensure that the security and defence aspects of CFSP and ESDP are taken into account during further deliberations on an EU Space Policy and possible programmes."

This well-balanced position included all the main principles that were subsequently incorporated into the European Space Policy of 2007, issued by the European Commission and ESA, such as bi-pillar cooperation, synergies, and dual use of systems, which are also keywords of the ESP. In order for these to be achieved, the European Council has instituted a "permanent inter-pillar dialogue" between the European Commission and the second and third pillars, dealing with the development of a European space programme.

However, these provisions were not taken into account in the space dimension of EUROSUR. In all of the European Commission's EUROSUR communications, there are no references to the already existing institutional framework that was built under the second pillar and approved by the European Council. In particular, there is no reference to the ESDP or its Space Policy, nor to any generic space system requirements for civilian operations that have been already adopted by the Council (Fig. 6.1).

It appears that the EUROSUR framework failed to integrate the whole range of identified users' needs. The lack of a focal point within the second pillar on the one hand, and the transfer of the space responsibilities decided in 2005 within the European Commission (from DG Research to DG Enterprise and Industry) on the other, may have played a role in this discontinuity. It should also be noted that the new European Commission relevant authority (Space R&D unit [H3]), deals mainly with the GMES services, based on user views and requirements federated by the relevant GMES bureau. Nevertheless, we have seen that GMES is only a part of the developing EU space programme.

[24] PSC: Policy and Security Committee at the Ambassadorial level. Established also in 2001, it deals with all aspects of CFSP including ESDP.

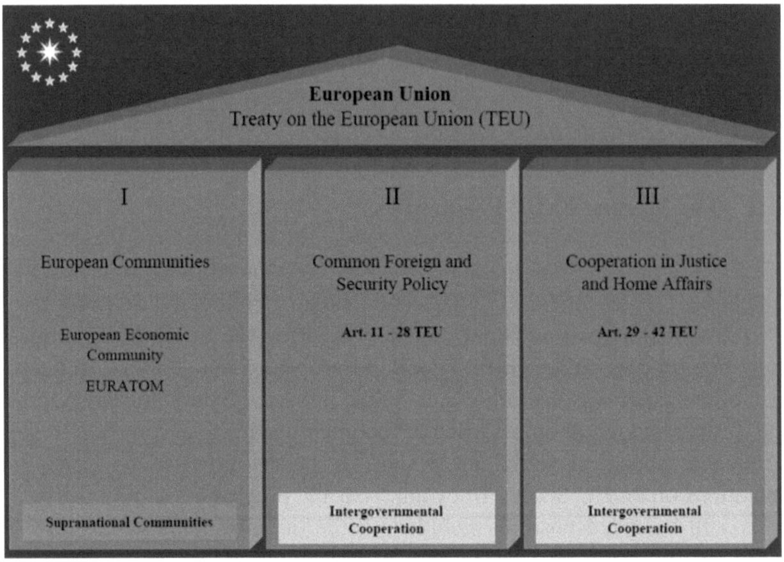

Fig. 6.1 The "permanent inter-pillar dialogue" between the European Commission and the second and third pillars failed in the EUROSUR case to take into account the existing institutional framework, or to integrate the whole range of identified users' needs and requirements, although the Commission had been invited to do so (Picture taken from page 10 of the presentation: Space and Internal Security in Europe – The Setting, by Bernard Molard, ASD EUROSPACE Security & Defence WG Chairman, Vienna – 28 May 2009, available at: http://www.espi.or.at/images/stories/dokumente/Conference2009/molard.pdf)

6.4.1.2 FRONTEX Expansion and EUSC Competencies

Another issue that arises from an institutional point of view is the European Space Policy recommendation regarding the use of existing government structures. It is expected that by 2013, FRONTEX will provide selected services, such as the common application of surveillance tools, including satellite imagery, and information and intelligence on the pre-frontier area, to the National Coordination Centres (NCCs).

This raises the question of whether FRONTEX or the NCCs will have the necessary capabilities, knowledge and expertise to do so, as well as the role and involvement of the present structure – namely the European Union Satellite Centre (EUSC) that currently collects satellite imagery and turns it into geospatial information products. EUSC is the only relevant CFSP/ESDP operational tool that supports the EU and Member States' Earth observation needs related to crisis management operations.

EUSC is already involved in providing GMES services and it has demonstrated its capability to meet internal security applications' demand.[25] The importance of continuing to strengthen the ability of FRONTEX to contribute to the management of the external borders of European Member States is well understood, but this must be done with respect to the competences of the institutions and bodies like the EUSC, which has over 20 years of experience in this domain.

GMES Concept of Operations document (see footnote 22) identifies EUSC along with the European Maritime Safety Agency (EMSA) as actors in the EUROSUR system, without specifically addressing their roles and responsibilities. But, as is clearly stated, this document does not implicitly or explicitly represent the views of either the European Commission or those of the Agencies mentioned in it.

6.4.2 EUROSUR Focuses on R&D in EO Systems

It is a fact that Earth observation is an essential element of any security-related application, and that a significant number of related dual use systems currently exists, or is under development in Europe, on a national and a multi-national level. The European Commission's arguments link EO satellites mainly to the surveillance of Europe's external borders, as well as the adjacent pre-frontier zone (i.e. for monitoring ports used as points of departure for illegal migration).

In order to satisfy this "surveillance term", an all-weather operational capability able to function day and night is required. EO space-based systems equipped with electro-optical, infrared and RADAR sensors can generally satisfy such requirements, with varying degrees of accuracy.

According to the EUROSUR concept "a key operational objective should be to [...] pay[ing] attention to geographical circumstances and differences between types of borders, in particular between land and maritime borders", which may have different needs.

6.4.2.1 Land and Maritime Borders Have Different EO Needs

Arguments like the one iterated in Step 6,[26] that"a vehicle or vessel used for a criminal activity is identified abroad and is being tracked by using satellites",

---

[25] Ussorio, Alessandra and Juan Luis Valero. "Earth Observation Applications and Critical Infrastructures in the Field of CFSP/ESDP – The Experience of European Union Satellite Centre." Presentation. European Space Policy Institute Workshop on Space and International – Developing a Concept for the Use of Space Assets to Assure a Secure Europe, 28 May 2009, Vienna, Austria http://www.espi.or.at/images/stories/dokumente/Conference2009/ussorio.pdf.

[26] See 5 fn.10. of Commission of the European Communities. Communication from the Commission to the European Parliament, the Council, the European Economic and Social Committee and the Committee of the Regions. Examining the Creation of a European Border Surveillance System (EUROSUR), COM(2008) 68 final 13 Feb. 2008. Brussels: European Union.

may give the impression that EO satellites can be used in all time-critical situations (even to monitor actual land border crossings).

In the sea domain, the need for better performance of search and rescue mechanisms thereby saving more lives has been clearly expressed. As such the capacity to detect small boats in the open sea must be enhanced, and difficult mission requirements, such as tracking a vehicle by satellite in real time, could be regarded as important enough to mobilise R&D on improving satellite capabilities in the maritime domain.

However, this might not be the case for the land border monitoring mission area. It is true that EO satellites are not generally used as a tactical means in land border surveillance due to their limited revisit time compared to the moving speed of immigrants using tracks. In these situations, aerial systems work best, despite the geopolitical constraints that their use might face. Time consuming parameters, like the entire operational chain from mission tasking to information delivery to the hands of the end users-especially in the theatre of operations, including the processing and fusion of information, complicate the situation. This is an area on which R&D for land border monitoring missions could focus best. This is verified both by the 2008 European Commission Report which recognises that EO satellites' use for tracking is limited and by a GMES Report, according to which a "continuous efforts, in particular for the tasking of satellites and the retrieval of archive images, should be made to improve the timeliness as some products are still generated too late to be of use in an operational phase."[27]

6.4.2.2 Focus on the Southern and Eastern External Borders Reveals Other EO User Needs

Since 2008, the EUROSUR network has the mandate of involving as many Member States of the southern and eastern external borders as possible. The organisational-technological level of sophistication among these Member States varies considerably.

Border guards in some of these countries may have more simple needs, such as the availability of printed maps,[28] the use of GIS systems, or portable PCs. Such simple needs like 2D or 3D border mapping that represent the low-time "strategic" applications of EO satellites were absent from the original [arguments][justification] of the three-step approach adopted by the European Commission. It was only in 2011 that the issue arose when, for the first time, the latest European Commission staff working paper made a reference to low-time EO needs, such as basic geodata information.

---

[27] European Commission. Space, Security and GMES. Enterprise and Industry Directorate General. GMES Bureau. GSB-GMES-2011/05 of 30 June 2011. Brussels: European Union.

[28] An ESA study, describing the Lessons Learned from various EU operations, concluded that Maps/imagery in most cases, were insufficient or non-existent. See "European Space and Human Security Working Group Report." Oct. 2006. European Space Agency, 21 Nov. 2011 http://www.agoria.be/upload/belgospace/ESHS_Report_Complete_%20Final%20091006.pdf.

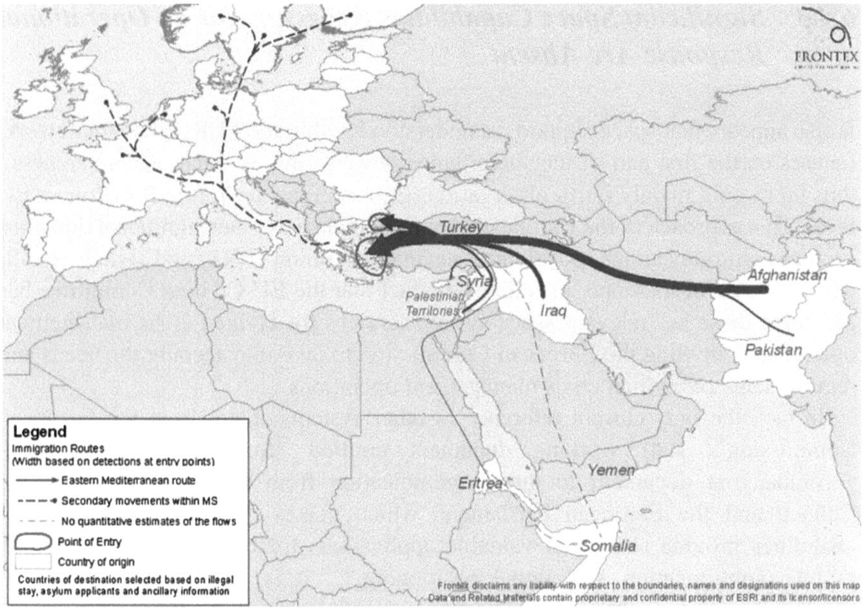

Fig. 6.2 The EUROSUR network is mandated to involve as many Member States of the southern and eastern external borders as possible, especially in regions where the EU faces the greatest immigration flows (Picture taken from page 7 of the presentation: Situation Regarding Illegal Migration at the EU External Borders, by Mari Kalliala, Head of Analysis and Planning, risk Analysis Unit, FRONTEX, Vienna – 28 May 2009, available from: http://espi.or.at/images/stories/dokumente/Conference2009/kalliala.pdf)

This fact creates the impression that it is only now that the 'bring space down to Earth'[29] motto is being implemented. This assumption is further confirmed by a November 2010 Joint Task Force Report,[30] according to which the revisit and delivery times that are currently feasible through the GMES Space Component (GSC) are not expected to fully satisfy tactical information requirements, despite the fact that GSC is in a position to provide up-to-date topographic and cartographic information in support of long term intelligence requirements and operations planning (Fig. 6.2).

[29] European Parliament. European Parliament Resolution of 20 Nov. 2008 on the European Space Policy: How to Bring Space Down to Earth. P6_TA(2008)0,564 of 20 Nov. 2008. Strasbourg: European Union.

[30] "Civil-Military Synergies in the Field of Earth Observation – Final Report." 26 Nov. 2010. European Union, European Defence Agency, Secretary General of the Council, European Space Agency Joint Task Force 21 Nov. 2011.

http://ec.europa.eu/enterprise/policies/space/files/research/civil_military_synergies_2010_11_24_en.pdf.

6.4.3 Significant Space Capabilities Necessary for an Operational Response Are Absent

It also appears that space support, as understood within the EUROSUR framework, focuses on the first part of the surveillance process (that is situational awareness), thus promoting mainly Earth observation (EO) satellites for future R&D on an EU level. This approach of the European Commission leaves other significant domains, such as communications, positioning-navigation-timing (PNT) and signals intelligence outside of its scope, in spite of the fact that the EU Civilian Committee has identified these as "relevant space system services for civilian crisis management operations", inviting the European Commission to take into account the needs and requirements of civilian crisis management operations.

In fact, the only current reference to other systems comes from the European Commission's staff working document entitled "Impact Assessment", an accompanying document to the Communication from the Commission to the Council and the European Parliament, which makes the following reference: "Satellites provide also other valuable applications for border surveillance such as telecommunication and global positioning."[31] It is interesting to note that according to the mandate of the European Commission space R&D unit (H3),[32] GMES services also include the integration of satellite communication and navigation. Nevertheless, no such development has taken place so far for the EUROSUR system.

One would assume that these two space domains would be also addressed and explored further. However, satellite capabilities needed for the management of reaction capability, such as satellite communications or PNT, are not included in the measures adopted so far in relation to the establishment of EUROSUR. Nevertheless, both functions are needed in order to implement the Integrated Border Management (IBM) concept at a national, regional and local level, which the EU Schengen Catalogue has recommended.[33] These capabilities are vital because border guards face communications shortfalls in mountainous areas, not permitting real time data flow to all border control command levels, and creating a major problem for patrol management and the conduct of operations. Border guards

---

[31] The 2008 Communication, and the accompanying Impact Assessment (IA), (SEC (2008) 151) were based on an analysis of the work done for setting up the European Patrols Network (EPN), and on the results of two studies (MEDSEA and BORTEC). See: Commission of the European Communities. Commission Staff Working Document. Examining the creation of a European Border Surveillance System (EUROSUR). SEC (2008) 151 of 13 Feb. 2008. Brussels: European Union http://www.statewatch.org/news/2008/feb/eu-com-68-08-eurosur-ia.pdf.

[32] "Space – What we do." 9 Dec. 2010. European Commission – Enterprise and Industry 21 Nov. 2011 http://ec.europa.eu/enterprise/policies/space/research/what-we-do/index_en.htm.

[33] "EU Schengen Catalogue. External Borders Control. Return and Readmission. Recommendations and Best Practices." 7864/09 LB/mdc of 2009, available at:

http://www.schengen.mira.gov.ro/English/Documente/utile/catutil/Updated%20EU%20Schengen%20Catalogue.pdf.

usually use GSM phones or even VHF transmissions (which sometimes can be intercepted by unauthorised third parties), which have operational limitations areas according to ground morphology.

Interestingly, the need for satellite communications has also been identified by the European Commission. The SPASEC Panel (2005) confirmed the relevance of using data relay satellites in support of EU border surveillance operations along with Earth observation systems, in order to fight against illegal activities. Furthermore, other systems like ESA's TDRSS, GEO (Inmarsat) and LEO (Iridium) SatCom constellations, together with other national systems can be used for such purposes. Having in mind the criminal activity problems aforementioned, a critical step in any border control scenario is the time when a border guard takes an unidentified person's biometric data (e.g. fingerprints). If he doesn't find that person's name in the relevant databases, he will simply let him pass the checkpoint. With the use of traditional communication means (GSM phones, internet if the available capacities allow it), transmitting and receiving biometric data could take several hours at least.

Another important need is the capability of border guards to know and make known their position, along with the real time location of other patrols in the area. The location information that satellite navigation systems like GPS/EGNOS and GALILEO provide is necessary to keep senior commanders permanently informed of the positions of the various patrol teams. Furthermore, these signals need to be superimposed upon digital maps. In particular, the ability to provide users with knowledge of the physical environment by utilising Geographical Information Systems (GIS) with embedded informed cartography and/or space based imagery, in conjunction with localisation and navigation capabilities supported by communicating portable devices, has become a key plank in consolidating the actual role of space systems on the ground.[34]

Finally, another area of interest is the field of signals intelligence (interception and jamming). Few countries currently have the means to intercept and listen to GSM communications that would permit the detection and localisation of all kinds of electromagnetic activity attributed to illegal purposes. This mission area would also merit a future R&D effort.

6.5 Conclusion

2010–2011 was a significant period for the EUROSUR System. First, in November 2010 the European Council identified border management as one of the key initiatives within the EU Internal Security Strategy. Second, in January 2011 the Commission published the framework and the actions to be taken in order to establish EUROSUR. Third, EUROSUR is scheduled to achieve initial operational capabilities on a pilot basis by the end of 2011.

[34] Pasco, Xavier. "Space Capabilities for Crisis Management: Reducing Gaps, Improving Action." Fondation pour la Recherche Stratégique 21 Nov. 2011. http://www.frstrategie.org/barreCompetences/espace/doc/SCCMsummary.pdf.

Space is an important "force multiplier" for operational and tactical missions, such as land border control. It is therefore a domain that the EU must exploit further, and this is the objective of the EUROSUR steps, through instigating research and development in the space observation domain. The analysis above found that there are some institutional issues that deserve consideration, especially since the Treaty of Lisbon foresees space as a shared competency, covering the needs of all end users, while coordinating and integrating different domains and policies. On the other hand, current experience has demonstrated that operational actors are faced with various deficits, not only in plain EO products, but in other areas vital to border control as well. They consequently deserve better support, especially in terms of the related R&D effort. The current challenge is to bridge the gap between R&D and the actual operational end-user shortfalls. Widening space applications to include other mission areas, while deepening R&D in these new domains and integrating the entire range of space assets, would constitute a balanced, as well as indispensable future step. On a political level, Europe answered the dilemma of deepening vs. widening operational capabilities a long time ago, by simultaneously striving for both. It seems that the role space assets could play in the EUROSUR operational structure must follow the same path.

Chapter 7
European Access to Space: Challenges and Perspectives

Anna Veclani, Nicolò Sartori, and Rosa Rosanelli

7.1 Introduction

Today, the EU finds itself confronted by two imperatives regarding its space policy: first, the completion of Galileo and GMES; and second, the maintenance of independent, reliable and cost-effective access to space capability. The latter is the condition for any other European space activity. Indeed, all the major space powers dispose of an autonomous launch system. The lack of such an asset in Europe [could have][has] serious consequences in both political and economic terms. On the one hand, all the institutional activities that characterise the European space policy (like Galileo and GMES), as well as national space programmes [are] [will be] entrusted to third countries, creating dependence on them, including for missions of a strategic nature. On the other hand, the European aerospace industry and its niches of excellence [could][will] be weakened in favour of the growth and development of extra-European industries.

For 30 years, the European launcher Ariane has allowed both civilian and military European institutional satellites to be placed in orbit, and thanks to the launch company Arianespace it has been the global point of reference for commercial missions. Furthermore, over the past decade ESA has begun a process of diversification with the adaptation of the Russian Soyuz rocket and the development of Vega, the inaugural launches of which took place at the end of 2011[1] and the beginning of 2012 respectively.

[1] Soyuz launcher's inaugural flight took place on October 21, 2011, carrying two Galileo satellites.

A. Veclani (✉) • R. Rosanelli
Istituto Affari Internazionali (International Affairs Institute, IAI), Rome, Italy
e-mail: a.veclani@iai.it

N. Sartori
Paris-Sod University, ■, ■

P. Hulsroj et al. (eds.), *Yearbook on Space Policy 2010/2011*,
Yearbook on Space Policy, DOI 10.1007/978-3-7091-1363-9_7,

These successes are not enough to guarantee independent European access to space for the future. ESA, the EU and their respective Member States must meet two challenges to guarantee the sustainability and effectiveness of European policy on space access. The first concerns the modernisation of the available family of rockets, as the current generation of launchers will require updating to remain competitive on the global level and respond to the most complex European institutional missions. The second challenge derives from the growing international competition in the sector, which involves both public and private actors, and which sees the traditional space powers and the emerging new actors trying to capture large shares of the global market.

To ensure the sustainability of R&D programmes and launcher production, it will be necessary to find a solution to the problems of finance and governance affecting the launch company Arianespace. Despite its commercial successes, high losses have beset the European colossus, forcing shareholders and governments to adopt exceptional and costly measures. For example, in the face of losses of 135 million euros over the years 2009 and 2010, the company was re-capitalised to the amount of 127 million euros, of which 80 million was in cash. Furthermore, at the proposal of the French government, the March 2011 ESA Council meeting approved the allocation of 212 million euros over a period of 2 years for the exploitation of Ariane 5.[2]

The aim of this chapter is to evaluate the different policy options for the maintenance of a European launch system that is independent and competitive on a global level. To this end, the first part of the chapter will analyse the two challenges referred to above, while the second will address the issue of governance and funding of Arianespace, which is one of the most important issues related to a credible and effective answer to these challenges.

7.2 The Challenge of Renovation

Today, the European family of launchers includes the heavy-lift Ariane 5, the medium-lift Soyuz and the lightweight Vega. The synergy between these different categories should ensure performance and flexibility, allowing for the entire range of launch requirements to be covered. That notwithstanding, the evolution of satellite technologies and characteristics, as well as the global demand for launches, requires an appropriate policy of diversification and modernisation of the fleet of European launchers.

[2] "ESA's Council Decisions." 22 Mar. 2011. ESA 20 Nov. 2011 http://www.esa.int/esaCP/SEMGN71U5LG_index_0.html.

Ariane 5 carries out its missions in two configurations. The first, with a launch capacity of 10 tonnes to geostationary transfer orbit (GTO), is used for the launch of heavy telecommunications satellites. The second, with a capacity of 20 tonnes to low-Earth orbit (LEO), is used for the launch of the Automated Transfer Vehicle (ATV) that supplies the International Space Station.[3] Ariane 5 can perform both double and single payload launches. Double payload launches enable lower prices to clients, but they also require greater efforts in terms of the pairing of satellites and the scheduling of launches, two problems currently affecting the competitiveness of Ariane 5 on the international market.[4]

The successor to Ariane 5, the Next-Generation Launcher (NGL), is the subject of debate and will not enter service before 2025. The main issue to be resolved concerns the strategic choice between the development of another heavy-lift rocket for double payload launches, or of a medium-lift rocket for single payload launches. The second option, which could reduce costs by adapting itself to medium-weight satellites, appears to be the favourite and, in the long term, could be a replacement for the European version of the Russian rocket Soyuz.[5] The 2012 ESA Ministerial Council will not only have to decide on this, but also on whether to launch and finance a related preparatory programme in the short term.

Meanwhile, work has begun on the development of a medium-term solution, namely the Ariane 5 Midlife Evolution (ME). The new motor VINCI will increase the Ariane 5 launch capacity by about 20% (or two more tonnes). Furthermore, this re-ignitable engine will permit double launches over a wider range of orbits, once a satellite is in position.[6] Ariane 5 ME would gather in a single launcher the [functionalities][capabilities] of the two current configurations of Ariane 5, thus reducing costs. The inaugural flight is planned for 2017,[7] provided that the ESA Member States currently participating in the preparatory programme approve its financing during the upcoming ESA Ministerial Council. Otherwise, Ariane 5 will run into serious difficulties vis-à-vis international competition.

[3] "Launcher Family." Arianespace Service & Solutions 20 Nov. 2011 http://www.arianespace.com/launch-services/launch-services-overview.asp.

[4] "2010 Annual Report." Arianespace Service & Solutions 20 Nov. 2011 http://www.arianespace.com/about-us-corporate-information/Annual-Report-2010.pdf.

[5] Bigot, Bernard, Yannick D'Escatha, and Laurent Collet-Billon. "L'enjeu D'une Politique Européenne de Lanceurs: Assurer Durablement à l'Europe un Accès Autonome à l'Espace." 18 May 2009. La Documentation Français 20 Nov. 2011 http://www.ladocumentationfrancaise.fr/rapports-publics/094000223/index.shtml.

De Selding, Peter. "Astrium To Lead Studies of Successor to Ariane 5." 7 July 2010. Space News 20 Nov. 2011 http://www.spacenews.com/contracts/100707-astrium-studies-ariane-successor.html.

[6] "Vinci Rocket Engine – Thrust Chamber – Cryogenic Propellant Rocket Engine for the ESC-B Upper Stage of Ariane 5." Astrium – An EADS Company 20 Nov. 2011 http://cs.astrium.eads.net/sp/launcher-propulsion/rocket-engines/vinci-rocket-engine.html.

[7] "Ariane 5 Development." 6 Jan. 2010. Astrium – An EADS Company 20 Nov. 2011 http://www.astrium.eads.net/en/news2/ariane-5-development.html.

The Russian-manufactured rocket Soyuz joined the family of European launchers following a 2005 ESA-Russia agreement.[8] The European version was adapted to launches of satellites of up to 3.2 t from equatorial latitude into medium (MEO) and low (LEO) Earth orbits, including the launch of constellations. It is used for orbiting medium-weight telecommunications, navigation, Earth observation and scientific research satellites.[9] Soyuz will probably not be able to replace Ariane for missions having a higher political and strategic value.[10] For example, since Soyuz is not a fully European launcher, it would be preferable to avoid transfers of information and technology related to the launch of military payloads. Furthermore, the fact that the contract with Russia to provide the launcher is limited in time facilitates its replacement by NGL in the long term.[11]

The small launcher Vega, the result of an Italian initiative, became an optional ESA programme in 1998. It is intended above all to serve the needs of the institutional small satellite market and to partially respond to the growing interest in launching microsatellites,[12] given the fact that it can place in LEO several light weight payloads simultaneously.[13] Vega would constitute the ideal solution for launching Earth monitoring and scientific satellites up to 1.5 t in weight, or for replacing satellites within constellations, avoiding the delays associated with double launches.[14]

The three European launchers are developed, produced and managed according to an ESA system of tenders that foresees a prime contractor responsible for the development and production of each launcher, which it then hands over to Arianespace. The latter is responsible both for commercialising launchers and conducting operations. The prime contractor industries take part together with Arianespace's technical staff in the integration of components at the European space centre of Kourou, in French Guiana. The prime contractor for Ariane 5 is EADS-Astrium, for Soyuz the Russian space agency Roscosmos, and for Vega the European Launch Vehicle (ELV), an Italian company owned 30% by the Italian

[8] "International Cooperation." ESA Launchers 20 Nov. 2011 http://www.esa.int/esaMI/Launchers_Home/SEMCDI1PGQD_0.html.

[9] "Soyuz Overview." Arianespace Service and Solutions 20 Nov. 2011 http://www.arianespace.com/launch-services-soyuz/soyuz-introduction.asp.

[10] Lasagni, Pier Giuliano. "Politica Industriale e Ricerca Italiana per lo Spazio nel Quadro della Politica Spaziale Europea." 10 June 2005. Attività del VAST 20 Nov. 2011 http://vastxiv.camera.it/attivita/31/192/193/194/203/programma.asp.

[11] Bigot, Bernard, Yannick d'Escatha, and Laurent Collet-Billon. "L'enjeu d'une Politique Européenne de Lanceurs: Assurer Durablement à l'Europe un Accès Autonome à l'Espace." 18 May 2009 http://www.ladocumentationfrancaise.fr/var/storage/rapports-publics//094000223/0000.pdf.

[12] The category includes satellites ranging from 10 to 500 kg.

[13] Vega will be able to carry out single launches of the heaviest satellites in the category (400–500 kg), and multiple launches of the lightest satellites.

[14] "Da Kourou a Vega: Lavori in Corso." 21 Oct. 2004. ESA Italia 20 Nov. 2011 http://www.esa.int/esaCP/SEMGD21A90E_Italy_0.html.

space agency ASI and 70% by Avio. Both EADS-Astrium and ELV are at the same time shareholders in, and suppliers of, Arianespace, a problem which will be examined in greater depth below.

Even when Vega and the European version of Soyuz become operational, it will be still necessary to establish a precise strategy for modernising the European launcher family. The speedy evolution of launch and satellite technologies will turn Ariane 5 into an inefficient and expensive launcher. Global launch demand will follow trends in space services that are mainly obtained through large scale satellites placed in GTO for telecommunications, and medium or small scale satellites in LEO for Earth observation and scientific missions.[15] Furthermore, the growing demand for space services does not coincide with the development of a greater number of satellites or with an increase in launches, but rather with the development of heavier, more durable and technologically advanced satellites, organised in small constellations. For these missions' purposes, the performance of the single launch configuration of Ariane 5 is poor, while the double-launch is inefficient. Only the decision to produce Ariane 5 ME and start a preparatory programme for NGL will allow Europe to meet launch requirements and market demand for the next 15 years. As for Vega, Italy has already set in motion a national programme (known as Lyra) aimed at increasing the performance of the launcher, including for military missions.[16]

7.3 The Challenge of International Competition

National governments play a central role in maintaining independent access to space through investment, regulations limiting foreign competition, and orders placed with national space industries.[17] In 2009 and 2010, institutional demand accounted for around 70% of the global market. Most orders came from the US, Russian and Chinese governments. Although institutional European launches decreased, Europe maintained a significant international influence in terms of commercial launches (see Fig. 7.1).[18] In addition to this, while during the Cold War the launcher sector was dominated by the US and the USSR, in recent years the market has opened up to

---

[15] Berteau, David J., Gregory Kiley, Guy Ben-Ari, Joshua T. Hartman, Gary Powell, Stephanie Sanok, and Brian Green. "National Security and the Commercial Space Sector – An Analysis and Evaluation of Options for Improving Commercial Access to Space." 26 July 2010. CSIS – Centre for Strategic & International Studies 20 Nov. 2011 http://csis.org/publication/national-security-and-commercial-space-sector.

[16] "Piano Triennale Delle Attività 2010–2012." Agenzia Spaziale Italiana 20 Nov. 2011 http://www.asi.it/files/ASI_PTA_2010-2012_3.pdf.

[17] Ibid.

[18] "Commercial Space Transportation: 2010 Year In Review." Jan. 2011. Federal Aviation Administration http://www.faa.gov/about/office_org/headquarters_offices/ast/media/2010%20Year%20in%20Review.pdf.

Fig. 7.1 Vega on the launch pad (Credits: ESA – S. Corvaja 2011)

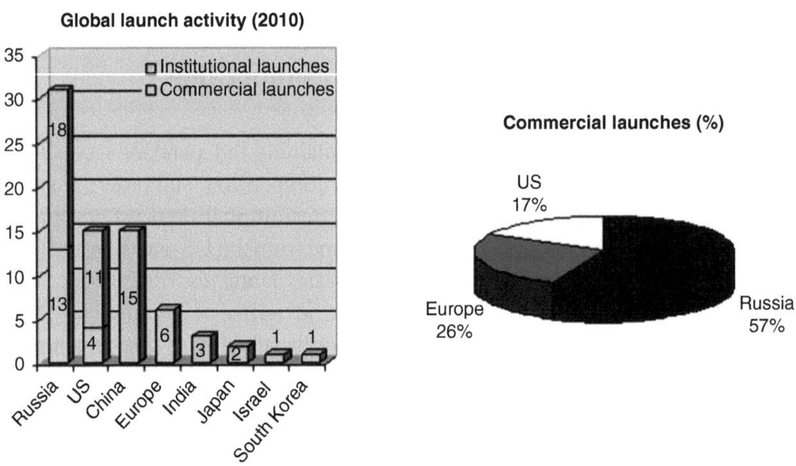

Fig. 7.2 Global launch activities. Focus on commercial launches (2010) (Commercial Space Transportation: 2010 Year In Review." Jan. 2011. Federal Aviation Administration http://www.faa.gov/about/office_org/headquarters_offices/ast/media/2010%20Year%20in%20Review.pdf)

competitors such as China, India, Brazil, Japan, Israel and South Korea. Global competition depends on a certain number of factors, such as price, reliability, availability and security, as well as on the nature (public or private) of the client (Fig. 7.2).

The **price** of the launch, which is often reserved and negotiated in the purchase phase, is not in and of itself a determining factor, especially if the purchaser is a government whose main objective is the success of the mission. Other factors are equally, and sometimes more, important.

A launcher's **reliability** is assured by a required minimum of five to six launches per year, allowing for the funding of research and development activities, fuelling efficient industrial cycles and guaranteeing the security and continuity of operations.[19] All of these aspects contribute to the probability of success of the launch. Reliability is a fundamental element: the owner, either public or private, of a satellite whose launch fails will incur very high additional costs in terms of manufacturing a replacement and/or suffering an interruption of services. The latter complication is particularly serious in the case of military satellites, bearing in mind the possible consequences for national security. Although Ariane 5 can count on 46 consecutive successful launches and is considered to be one of the most reliable launchers in the world, it must be underlined that while in past years it maintained an average of six flights per year, it dropped to five in 2011.[20]

Availability is connected with the scheduling of launches on useful and predetermined dates and without delays. For governmental clients who have to launch military satellites, availability at short notice is a priority. This happens sometimes to the detriment of commercial customers, who could be forced to wait for a long time or to postpone the launch of their own satellites. Availability is becoming an issue for Ariane 5, as pairing two satellites is proving more and more difficult in terms of keeping the launch schedule and of abiding by strenuous technical requirements, considering the increasing weight of telecommunication payloads that overstretch the launcher's limits.

Finally, **security** is related to technology transfer, considering that the operator of the launch has to gain access to a certain amount of information regarding the payload, in order to integrate it with the launcher. A government that does not dispose of the capability to launch military satellites must necessarily turn to a country that it considers reliable. For commercial satellite operators, the possibility of access to foreign launch services will also depend upon national legislation.[21]

The principal commercial launch operator in the US market is the United Launch Alliance (ULA), a joint venture by Lockheed Martin and Boeing that dominates

[19] Bigot, Bernard, Yannick d'Escatha, and Laurent Collet-Billon. "L'enjeu d'une Politique Européenne de Lanceurs: Assurer Durablement à l'Europe un Accès Autonome à l'Espace." 18 May 2009 http://www.ladocumentationfrancaise.fr/var/storage/rapports-publics//094000223/0000.pdf.

[20] The last Ariane 5 flight was performed on September 21, 2011; the next is planned for early 2012; See "Arabsat-5C and SES-2 are Orbited on Arianespace's Fifth Ariane 5 Flight of 2011." 21 Sept. 2011. Arianespace Service and Solutions 20 Nov. 2011 http://www.arianespace.com/news-mission-update/2011/846-va204-success.asp.

[21] Being characterised by a noteworthy technological proximity to intercontinental ballistic missiles for military use, launch products and technologies are particularly sensitive. Some countries have for this reason adopted regulations precluding their transfer, or subjecting it to particular authorisation procedures.

the US institutional launch market for heavy and medium weight launches, consequently distorting market competition rules in commercial launches. However, the emergence of Orbital, SpaceX, and the international consortium Sea Launch, could encourage greater competition in the US institutional market, as well as in the international commercial market in general.[22] The Russian Proton and Soyuz launchers provide extremely reliable services at a relatively low cost. The country's launcher family will soon be expanded with the new Angara rockets[23] and the lightweight Rockot, commissioned by ESA.[24] Ukrainian launchers are also active in the commercial launch market, with the rockets Zenit[25] and Dnepr-1.[26]

Since the Chinese government made independent space access a strategic objective, the country's launch capability has expanded considerably and is constantly gaining ground, thanks to its increased reliability[27] and competitive price. Today, an increasing number of international players, both institutional and commercial, are seeking launch services from the Chinese Great Wall Industry Corporation[28] (CGWIC), which is developing the heavy-weight launcher Long March 5, able to compete directly with Ariane 5 and the American rockets Atlas V and Delta IV.[29] India is also demonstrating growing entrepreneurship in the launch sector.[30] However, its ability to capture the commercial market is limited by the prominent role of the public sector on the supply side, and by the absence of a

[22] Berteau, David J., Gregory Kiley, Guy Ben-Ari, Joshua T. Hartman, Gary Powell, Stephanie Sanok, and Brian Green. "National Security and the Commercial Space Sector. Initial Analysis and Evaluation of Options for Improving Commercial Access to Space." 30 Apr. 2010. Center for Strategic and International Studies. 20 Nov. 2011 http://csis.org/files/publication/100430_berteau_commercial_space.pdf.

[23] Khrunichev State Research and Production Space Center 20 Nov. 2011 http://www.khrunichev.ru/main.php?id=44.

[24] "Business Profile." Eurockot Launch Services 20 Nov. 2011 http://www.eurockot.com/joomla/index.php?option=com_content&view=article&id=20040650&Itemid=58.

[25] Made up of the American company Boeing (40%), the Russian company RSC Energia (25%), the Norwegian company Aker Solutions (20%) and the Ukrainian company SDO Yuzhnoye/PO Yuzhmash (15%).

[26] International Space Company Kosmotras 20 Nov. 2011 http://www.kosmotras.ru/en/company/.

[27] In 2010, CGWIC carried out all its 15 planned institutional launches with success.

[28] Berteau, David J., Gregory Kiley, Guy Ben-Ari, Joshua T. Hartman, Gary Powell, Stephanie Sanok, and Brian Green. "National Security and the Commercial Space Sector. Initial Analysis and Evaluation of Options for Improving Commercial Access to Space." 30 Apr. 2010. Center for Strategic and International Studies. 20 Nov. 2011 http://csis.org/files/publication/100430_berteau_commercial_space.pdf.

[29] "Long March 5 Will Have World's Second Largest Carrying Capacity." 4 Mar. 2009. Space Daily 20 Nov. 2011 http://www.spacedaily.com/reports/Long_March_5_Will_Have_World_Second_Largest_Carrying_Capacity_999.html.

[30] "Rapport du Groupe Sectoriel sur le Transport Spatial." L'Académie Nationale de l'Air et de L'Espace 8. 20 Nov. 2011 http://www.academie-air-espace.com/upload/doc/docComm/Transport Spatial.pdf.

true space industrial production chain. By contrast, private players are fundamental in the Japanese launcher market, where the public sector is active in the development and use of the new launcher H2-B.[31]

Emerging space actors such as Brazil, Israel and South Korea are also developing their own capabilities. Brazil is involved in the development of the Veículo Lançador de Satélites (VLS).[32] Israel has an independent launch capability thanks to the lightweight Shavit rocket.[33] Finally, the efforts of **South Korea** have been so far frustrated by two failed launches of the lightweight launcher Naro-1.

The dynamism of the international launcher market underlines the need for the constant improvement of European launch capabilities. While a significant international market presence in the heavyweight launch sector for Ariane 5 can be foreseen, a certain degree of technological flexibility is also needed in order to satisfy diverse commercial demands. In the lightweight launch sector, Vega will win a significant share of the global commercial market only if it is able to compete in terms of reliability and price. In this respect, the number of launches, both institutional and commercial, carried out every year will be fundamental. Although institutional launches are experiencing a decline, the upcoming deployment of the GMES and Galileo constellations could offer the opportunity to use European launchers to place more than 30 satellites in orbit in the medium term. In order to remain competitive, Arianespace will have to continue to gain a significant share of the global commercial market. Between 2011 and 2017 Arianespace is expected to carry out the launch of only six institutional customers' satellites per year, out of an average of 27–28,[34] which represents only 20% of the forecast for global satellite launches in that period.[35]

7.4 The Governance and Financing Aspect

The most urgent policy priorities that the main European space players must address concern Arianespace's financing and governance. Indeed, only a rigorous and long-term investment strategy, as well as a clear division of tasks between the

[31] Japanese Aerospace Exploration Agency 20 Nov. 2011 http://www.jaxa.jp/projects/rockets/h2b/index_e.html.

[32] Agencia Espacial Brasileira 20 Nov. 2011 http://www.aeb.gov.br/indexx.php?secao=lancadores.

[33] Israel Aerospace Industries 20 Nov. 2011 http://www.iai.co.il/14471-15689-en/BusinessAreas_SpaceSystems_Launchers.aspx.

[34] European Space Agency Launcher Programme Board. Assessment of the Launch Service Market Demand (2010 Release). March 2010 Paris.

[35] Ibid.

Fig. 7.3 Ariane 5 flight VA202 (Credits: ESA/CNES/Arianespace/Photo Optique vidéo du CSG – S. Martin)

main actors, political and industrial, will be able to render the planning of programmes for launcher development and production clearer, while strengthening the presence of Arianespace in the international market.

As previously mentioned, both Arianespace's shareholders and the ESA Council, faced with the company's losses, were forced to intervene to recapitalise the undertaking and to support Ariane 5 prices respectively. Nevertheless, considering their consistent contributions in launcher programmes, many member states are reluctant to write off Arianespace's losses. For example, in 2010 the European Guaranteed Access to Space (EGAS) programme was concluded, having financed for 6 years part of the production costs of Ariane 5 in order to make Arianespace self-sufficient, but without success. For this reason, the 2012 ESA Ministerial Council will have to make decisions concerning long term financing mechanisms for Arianespace.[36]

Another cause of reluctance to support Arianespace was the lack of transparency in its financial management. ESA launched an external audit to shed light on Arianespace's management and on the reasons for the poor economic performance of the consortium. This led to the agreement that three ESA governments, Italy, Germany and Sweden, would be represented at Arianespace shareholder meetings to guarantee greater transparency (Fig. 7.3).

Arianespace's economic difficulties can be ascribed to different factors, such as:

- The inefficiency of the industrial cycle, due to ESA's principle of "geographic return." The latter allows companies in the different ESA countries to receive industrial contracts proportional to the financial contributions channelled by their respective governments to ESA's programmes. As a result, prime contractors and subcontractors are not necessarily selected on the basis of cost and efficiency.

[36] De Selding, Peter. "ESA Putting Arianespace Finances Under the Microscope." 14 Jan. 2011. Space News 20 Nov. 2011 http://www.spacenews.com/civil/110114-esa-arianespace-finances.html.

- The conflict of interest involving certain companies that are simultaneously shareholders in, and suppliers to, Arianespace. This is true for both prime contractors and subcontractors.
- The need to maintain competitive prices on the global market, even if they are too low to cover the production cost of the launcher.
- The presence of competitors in the international market that benefit from the financial support of their respective governments.
- The relatively elevated production costs in Europe and the additional expenses related to the integration of components overseas, in French Guiana.
- The unfavourable Euro/Dollar exchange rate, given that launch services are denominated in Dollars, whereas Arianespace's costs are borne in Euro.

The first four elements are closely connected to the governance and financing of Arianespace. It is, therefore, above all the public-private nature of the company's structure that is the object of serious reflection within ESA, from both a political and an industrial point of view. Arianespace is an international undertaking, the capital of which is owned by 21 shareholders from 10 different European countries.[37] The largest share is held by France that controls through its seven shareholders 64.1% of the total shares,[38] followed by Germany with 19.84%.[39] The eight other countries, among them Italy,[40] control the remaining 16.05%, with their respective share percentages not exceeding 4% per shareholder. From the point of view of the companies involved, the European aerospace giant EADS owns more than a third of the capital of Arianespace, through no less than six of its subsidiaries.[41] Next are the French company Safran (10.57%) and the German MT Aerospace AG (8.26%). Finally, the relative majority holding (of 34.68%) is owned by the French space agency, CNES.

The economic difficulties encountered could be overcome by opting for a different form of governance, ranging from a system of shareholding that would be entirely public and dominated by the presence of national space agencies, to one that would be entirely private. Both solutions, however, pose problems deriving from the nature of the space transport sector and from the limited European institutional market, which indeed make it extremely difficult to have break-even launch activities.

[37] The countries with an interest in Arianespace are, in order of the size of the holding, France, Germany, Italy, Belgium, Switzerland, Sweden, Spain, Holland, Denmark and Norway.

[38] CNES (34.68%), Astrium SAS (16.85%), Safran (10.57%), L'Air Liquide SA (1.89%), Clemessy SA (0.11%), EADS France SAS and Compagnie Deutsch SAS.

[39] Astrium GmbH (11.59%) and MT Aerospace AG (8.26%).

[40] The Italian share of 3.38% is controlled by Avio.

[41] The French firms Astrium SAS and EADS France SAS, the German company Astrium GmbH, the Spanish firms EADS Casa and Crisa and the Dutch company Dutch Space BV.

The first option would be to make Arianespace an entirely public entity. This would ensure better control over the company's accounts by the national space agencies. Greater transparency, however, would not necessarily imply a reduction of the company's costs, because of the "geographic return" principle. In such a case however, given the strategic value attributed to independent space access, the allocation of public money to cover the financial losses of Arianespace would be easier to justify. Nevertheless, it is not clear whether European governments would be willing to invest in the company, or if they even dispose of the necessary resources. A partial alternative could be to include the EU as a shareholder. This would emphasise the strategic importance of access to space for Europe, while ensuring a collective commitment on the part of all member states to support the progress made so far by the major European space faring countries, through their national space agencies.

The second option would be the privatisation of Arianespace, which could address the economic and commercial problems and lead to the reduction of the extraordinary costs induced by the "geographic return" principle. However, under the European single market law, a completely privately owned company would not be entitled to receive public financing to cover losses, and it would be difficult for those governments that might be willing to have recourse to the appropriate instruments to derogate from this regime.[42] Finally, governments might be reluctant to pay the high costs of launches managed by Arianespace only to maintain the productive capacity of the companies involved. On the contrary, they would have an incentive to turn to cheaper third country operators whenever possible. A possible privatisation could nevertheless be carried out in different ways. There are three alternatives, as follows:

1. Floating of the company on the market. In light of the consortium's known financial problems, the possibility of attracting private investors willing to acquire the 34.68% share currently held by the French space agency can only be considered remote.
2. Total acquisition of Arianespace by the largest European aerospace group, EADS. This would leave complete control of European launch capabilities in French hands, with obvious political and strategic implications. In addition, it would represent a step back with respect to European cooperation in space transportation inaugurated in 1973 on a French proposal to transform its "Lanceur à Trois Etages de Substitution" into an ESA programme, Ariane.

[42] Based on the European Commission's Interpretative Communication on the application of Art. 346 TFEU, member states can exceptionally derogate from internal market rules if it is proved that there are essential security interests at stakes.

3. The formation of a new company, able to bring together the core of current European industrial capabilities in the space access sector, i.e. these of EADS Astrium Space Transportation and Avio. This would ensure a rationalisation of the European technological and industrial base, consolidating in a single entity, apart from the production of Ariane and Vega, additional activities that today are broken up, such as the commercial activities of Eurokot and Starsem,[43] or the industrial activities of Europropulsion and Regulus.[44]

It should be noted that the third approach would allow future technological and industrial challenges to be addressed in a unified way, in particular those connected with the development of NGL and the evolution of Vega. The new company should be responsible both for all industrial activity (research, development, industrialisation and production) and for all commercial exploitation (institutional and commercial demands). Furthermore, it should be organised in national "subsidiaries" specialised in their current competences and subject to supervision and control by their respective governments, in order to ensure the protection of the different national interests.

The new company created in this way would take on the current role of Arianespace, simplifying and improving the efficiency of the European launch sector. Its shareholders would be EADS (in which the French and Spanish governments have interests; in addition, the German government could be obliged to buy Daimler's shares in EADS[45]) and Avio (in which Finmeccanica, itself 30% controlled by the Italian government, has an interest). In this way, although the new company would be a private one, the interests of the countries with a greater involvement in space access activities would be protected. The final decision on a new form of governance (regardless of whether it endorses one of the options outlined above or not) that was expected to be taken by the end of 2011, [might be][has been] postponed until 2012. A final decision might be reached at the occasion of the next ESA Ministerial Council.

[43] Eurockot Launch Services GmbH, a joint venture between EADS Astrium (51%) and Khrunichev space centre (49%), operates the launches of the Russian launcher Rockot. Starsem is a Russo-European consortium (35% EADS Astrium, 25% the Russian space agency, 25% Samara Space Centre, 15% Arianespace) which operates the launches of the Russian rocket Soyuz from the cosmodrome of Baikonur.

[44] Through the company Europropulsion (owned 50% by Avio and 50% by the French company Snecma) and the company Regulus (controlled 60% by Avio and 40% by the French company SME), Avio has designed and specified, and is currently producing, solid propellant motors for the launcher Ariane 5.

[45] "German Government and Daimler AG Plan Sale of 7.5% of EADS Shares by Daimler to KfW Bank Group." Daimler 10 Nov. 2011 http://www.daimler.com/dccom/0-5-7171-1-1439170-1-0-0-0-0-0-87-7164-0-0-0-0-0-0-0.html.

7.5 Conclusions

Maintaining an independent space access capability is one of the hinges of European space policy, without which EU and Member States' space programmes would have to depend on third countries for providing launch services. However, the rapid technological evolution in progress and the redefinition of the global politico-economic balances could undermine Europe's capacity to maintain independent, effective and sustainable access to space. In the wake of the successes achieved by ESA and Arianespace in recent decades, it is necessary to respond quickly to the two main challenges in the European launcher sector, in order to be able to continue to play a leading role in that sector. The first challenge is the need to modernise the family of European rockets currently available in order to anticipate the new commercial and technological demands of the satellite market and to continue to guarantee an autonomous launch capability for European institutional missions, civilian as well as military. The second challenge, closely connected to the former, is the ability to compete in the international launcher market together with established powers, such as the US and Russia, and new players, such as China, who are ready to invest in launch technologies in order to strengthen their own strategic independence and international prestige, as well as to capture greater shares of the global market.

The specificities of the European space transportation sector, such as the reduced number of institutional launches programmed for the coming years, render these challenges particularly difficult and require a renewed and shared approach to the European space access policy. In particular, it appears essential to improve the corporate and financial framework in which the European launcher industry operates. The lack of transparency in the management of Arianespace and its recurrent economic losses limit the capacity for industrial programming and technological development, leading to a potential reduction in competitiveness in the face of increasing international competition.

Only through a clear and definite governance reorganisation, coupled with a precise definition of the main political and industrial actors' roles, will it be possible to maintain a European launch capacity that is essential to increasing the EU's scientific and technological prestige, strengthening its independence and positioning it as a global space actor. In the absence of such progress, any solution envisaged to address the governance and financial problems of the space transportation sector will be influenced by the reorganisation of the relations between the EU, ESA and their respective Member States, within what is commonly referred to as the "institutional triangle." It is therefore a matter of redefining the whole European governance in the field of space, not only in the segment of launchers. This is an urgent necessity, all the more in light of the new role foreseen for the EU in space activities by the Lisbon Treaty (Arts. 4[3] and 189 of the Treaty on the Functioning of the European Union [TFEU]).[46]

---

[46] Art. 4(3) specifies furthermore that "... the exercise of that competence shall not result in Member States being prevented from exercising theirs."

Finally, the competitiveness and efficiency of the space transportation sector will be also influenced by the upcoming space industrial policy, to be proposed by the European Commission. The latter should recognise the strategic value of the aerospace domain and address its specificities, namely its high-technology nature, its limited competitiveness, the modest size of its market, the different procurement rules applied by single member states, ESA and the European Commission (geographic return vs. competition), and the role of public funding inside and outside Europe. Such a policy should be able to stimulate a certain degree of integration in the European space market, while strengthening its industrial and technological base.

Chapter 8
Cooperation in Space Between Europe and Israel in Light of the Recent ESA-ISA Agreement

Daniel Barok

8.1 Introduction

The beginning of 2011 will be remembered as a period of significant political relations' warming between the state of Israel and Europe. One of the symbols of this trend was the Cooperation Agreement that was signed between ESA and the Israeli Space Agency (ISA) on 30 January 2011.[1] This outcome was the fortunate culmination of earlier contacts and discussions started in 2007, which the Israeli space community hopes may eventually "open the door" to full ESA membership.[2] It is expected that cooperation between Europe and Israel on agreed missions could benefit both parties by complementing their capabilities, satisfying their needs, sharing their development costs, and mutually improving their technological know-how and industrial efficiency. Since the signing of the agreement, ESA experts and their Israeli counterparts have been jointly identifying areas of mutual interest, notably in the field of space science and exploration, followed by reciprocal introductory meetings and a comprehensive survey of the Israeli space industry and technology basis, currently conducted by ESA.

This article reflects the personal perspective of the author and does not represent any official policy of the State of Israel nor the Israeli Space Sector.

[1] "Israel signs Cooperation Agreement." 31 Jan. 2011. ESA 19 Nov. 2011 http://www.esa.int/esaCP/SEMKE3Y1LJG_index_0.html.

[2] In the 1960s Israel was invited to join ESRO and ELDO as a full member. Despite the interest and support of the Israeli scientific community, the Israeli government declined the offer.

See further Paikowsky, Deganit. "From the Shavit-2 to Ofeq-1- A History of the Israeli Space Effort." Quest (2011 forthcoming).

D. Barok (✉)
BINGO, Business Development and Consulting Services, Tel Aviv, Israel
e-mail: danielbarok@inter.net.il

P. Hulsroj et al. (eds.), *Yearbook on Space Policy 2010/2011*,
Yearbook on Space Policy, DOI 10.1007/978-3-7091-1363-9_8,

The aim of this paper is to scrutinise the motives underpinning ESA-ISA cooperation and to review both the positive and adverse factors influencing its development. This chapter is composed of four main parts. The first part outlines the history of Israeli space programmes, presenting their constraints, achievements and current activities. The second part reviews the Israeli and European motivation for collaborating in space. The third reviews current Israeli civil space collaborations. Finally, the fourth part evaluates the underlining factors that could potentially support or hinder the fulfilment of this agreement. As a last minute update before publication, comes the epilogue (written a few months after the original article). It describes the most recent encouraging and supporting events that improve the situation of the Israeli space sector and increase the prospects for fruitful collaboration.

8.2 Israel in Space: A 30-Year Legacy

"Israel is a small state with the security challenges of a major power", Rear Admiral Ophir Shoham[3] once said. Nevertheless, and perhaps as a result of the security challenges with which it is forced to contend, Israel is not generally perceived as a small state. An example of the kind of power which to some extent alters Israel's status is the fact that it is one of the few relatively small countries that builds and launches satellites into space.

Already in the early 1960s, not so long after the successful launch of Sputnik, Israel launched its first meteorological rocket.[4] The official Israeli space programme, however, was only established in the early 1980s. Its establishment was motivated mainly by defence and security needs, as well as by the desire for self-sufficiency in cutting-edge technologies. The Egypt-Israel Peace Treaty (1979) and the withdrawal from the Sinai Peninsula distanced the Israeli Defence Force (IDF) from Egyptian territory on the other side of the Suez Canal, and it consequently lost much of its intelligence-gathering capabilities, including the ability to carry out manned reconnaissance over-flights. Such flights were considered a violation of Egyptian sovereignty and were a very sensitive issue for the nascent friendly relations between the two countries.[5]

---

[3] Rear Admiral Ophir Shoham. Current head of the Defence Directorate for Research and Development (DDR&D) in the Israeli Ministry of Defence (IMOD), described the State of Israel at a conference held in March 2004 at Tel Aviv University. Shoham, O. "Economics and Defence." Tel Aviv.

[4] The Shavit-2 sounding rocket was launched on July 5, 1961. The launch took place as part of a "small-scale space race between Israel and Egypt and was used to show Israel's superiority over Egypt and to boost Israeli morale." Burleson, D. Space Programs Outside the United States. North Carolina and London: McFarland & Company Inc. 2005: 153.

[5] Paikowsky, Deganit. Clubs of Power: Why Nation-States Join the Space Club. Tel Aviv University, 2010.

Paikowsky, Deganit. From the Shavit-2 to Ofeq-1- A History of the Israeli Space Effort. Quest (2011 forthcoming).

In order to establish an Israeli space programme, a suitable infrastructure was developed under the leadership of the Israeli Ministry of Defence (IMOD). Feasibility studies were drawn up by the Israeli defence industries concerning the production of launch vehicles, satellites and payloads. The studies (completed by December 1980) came out positive and work began in Rafael, Israel Aerospace Industry (IAI) and ELOP. In January 1983, the Israeli government authorised the creation of the Israeli Space Agency (ISA) headed by the Minister of Science, Prof. Yuval Neeman, under the authority of the Ministry of Science.[6] As a government agency, ISA coordinates and supervises the civil national space programme. This includes national space policy, as well as external relations and collaboration with foreign countries and other space agencies. It also coordinates scientific research activities and space exploration initiatives that involve Israeli academic institutions and industries.

In contrast to other states, Israel specialises in developing lightweight satellites. This particularity stems, among other reasons, from its limited launching capabilities. It is also related to the country's overall strategic situation that does not allow launching satellites eastward (in the direction of the rotation of the Earth), because of the danger that parts might fall onto the territory of hostile states, as well as because of the strategic threat involved in firing a projectile in their direction. Thus, Israel can only launch satellites in a westward direction – opposite to the rotation of the Earth. This limitation has forced Israel to develop satellites which are as light as possible and which minimise the load on the launcher.[7]

Along with this "space odyssey", the initial Ofeq project suffered from a severe budget shortage. In a report published at that time by the State Comptroller, criticism was levelled at the decision-making process and the lack of willingness to grant the project an appropriate budget. The report determined that the benefit to the state from implementing such a project was great: "the project in question makes a contribution to the state as a whole in a variety of respects, including influencing employment levels in the economy, the technological advance of industry and the maintenance of critical information centres."[8] The improvements made possible by space activities not only concern developing technologies or manufacturing hardware. Above all, they represent a new dimension that has never been utilised before. Therefore, they should be viewed and understood as a change of perception and philosophy, which is concomitant to mankind's extension into space, and the use of its expanses for civilian and military purposes. One can ask: what can be gained from an independent space programme? The contributions which are part and parcel of such a programme can be divided into three categories: security, economic and social.

---

[6] Government Secretariat. Decision number 1255 (4/49) of the Ministerial Committee for Science dated January 23, 1983 was approved as a government decision, entitled Israeli Space Agency (ISA Archives).

[7] Report of the State Comptroller for 2001. Annual Report 52a. Project Ofeq: 252–3.

[8] Ibid.

While struggling with budget constraints, the Israeli space industry couldn't rely on the domestic market in order to maintain it's know how, and therefore it had to seek new markets. Marketing attempts required overcoming a series of obstacles, starting with the Israeli Export Control Regime and followed by several market penetration difficulties abroad that required massive governmental support. In spite of the difficulties, Israel has gradually expanded its national space expertise by developing commercial communication satellites (the Amos series, operated by the Spacecom company) and commercial Earth observation satellites (the EROS satellites, owned by ImageSat International N.V), as well as an experimental satellite built by university students.[9]

Furthermore, in 2008 Israel launched TECSAR, a sophisticated Synthetic Aperture Radar (SAR) satellite. Up to now, 13 satellites have been successfully launched, and at the time this chapter was completed (September 2011) 10 of these satellites were operational, performing in full compliance with their respective design goals. Several more satellites are currently being developed and expected to be launched in the coming years, including a nano-satellite for technology demonstration purposes. Israel's first astronaut Ilan Ramon, who attracted much attention in Israel, tragically perished onboard the Space Shuttle Columbia STS-107 mission, in February 2003. The Astronaut mission was an exception to the overall pragmatic security and economic benefits-oriented approach of Israel to space activities. On top of these projects, ISA is involved in a variety of space exploration projects in which it cooperates with other space agencies. Finally, Israeli space industries compete in the global space market by promoting their satellites and other sub-system products (Fig. 8.1).

As a result of the conclusions of a special committee nominated by the President of Israel, Mr. Shimon Peres, and with the involvement and approval of Prime Minister Netanyahu, a significant focus on space activities was created within government policies. Israel has crafted a new national space plan, which was not yet publicly released, formally approved, or officially funded by the time this chapter was completed.[10] If fully implemented, Israel's new national space plan could partially address the issues of limited domestic demand and small manufacturing scale. In addition, a greater budget allocation for civilian space could increase the country's involvement in international cooperation in space activities and boost Israeli exports to the international space market.[11]

While finalising this chapter, an exclusive interview was published by the Jerusalem Post,[12] marking the retirement of Brig.-Gen (res.) Haim Eshed, who

[9] TECHSAT, which was launched on a Russian launcher in 1998.

[10] "Israel's Space Capability and Competitiveness." 2011 Space Competitiveness Index. Futron. 2011: 51.

[11] Paikowsky, Deganit, and Isaac Ben-Israel. "Trends in Shaping Space Policies Around the World." Yearbook on Space Policy 2009/2010: Space for Society. Eds. Kai-Uwe Schrogl, Spyros Pagkratis and Blandina Baranes. Vienna: SpringerWienNewYork, 2011. 260.

[12] Katz, Yaakov. "Former Military Space Head: Israel is Losing its Edge." 2 Sept. 2011. The Jerusalem Post 19 Nov. 2011 http://www.jpost.com/Defense/Article.aspx?id=236395.

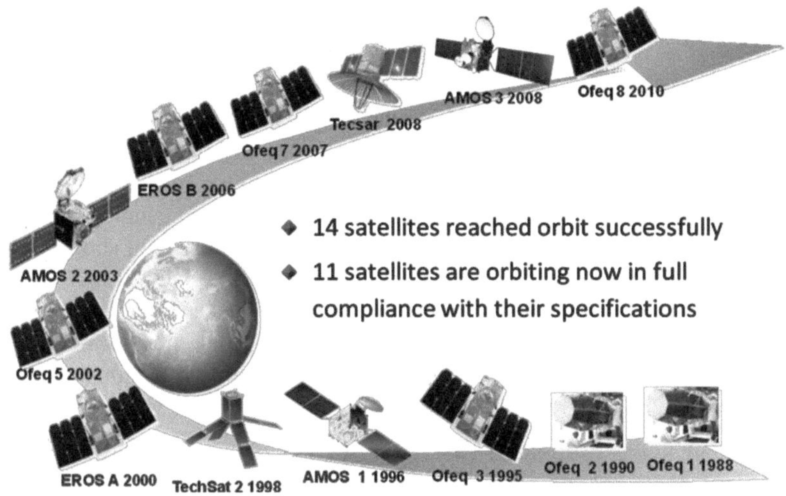

Fig. 8.1 Israel's space heritage

founded and led the Defence Ministry's Space Division for more than 30 years. Some quotes from the interview:

> Israel stands to lose its qualitative military edge in space as the government refuses to increase financial support for research and development.

This warning came in response to the government's refusal to release funds that it had pledged last year as part of the implementation of the recommendations of the above-mentioned committee, including the allocation of 1.5 billion New Israeli Shekels[13] over 5 years. Eshed stated that over the past 20 years the Israeli Space Industry had achieved total sales of less than 2.5 Billion Dollars.[14]

Generally speaking, commercial space services' providers (such as the Earth observation or satellite communications sectors) and components and sub-systems manufacturers seem to have a greater chance of penetrating foreign markets, especially compared to launch services' providers, or satellites and payloads manufacturers. In spite of the remarkable achievements of some Israeli companies and the good reputation of their products (for example within the European space industry and European Space Agencies, including ESA) there is an urgent need to extend the sales volume. Currently, the global space market's annual volume is estimated at 250 billion Dollars and Israel could carve out at least 5% for itself. Eshed said:

> But, if the government does not insert some money here, the industries that work on space platforms will not have projects, will need to shut down their production lines and we will

[13] About 290 Million Euro.

[14] About 1.8 Billion Euro.

potentially lose all of the knowledge and expertise we have cultivated over the past 30 years.[15]

In general, the major part of the Israeli Space Industry is owned by the Israeli Defence Industry. Mr. Josef Ackerman, the President of ELBIT systems[16] used to say: "In order to keep the Israeli Defence Industry "alive", 75 % of the Israeli Defence sales should be directed for export. The challenge is here: most of the target markets are satisfying 90 % of their needs by their own domestic industries. In order to penetrate to the remaining 10 % of this market and more, you should act as a domestic player for any practical purpose. This is the reason why ELBIT is defined as a "multi-domestic" company that owns facilities in the major target markets territories, acting as a genuine local entity."[17] This attitude has been adopted by other Israeli companies as well. The situation is even more challenging in the Space Industry, where one might think of four possible ways to cope with this problem:

- **Extended sales volume in the domestic market:** in Israel this option is currently almost blocked.
- **Direct exports:** an option that has up to now penetrated a relatively small portion of its potential markets, and which requires massive investments in business development efforts and strong government support, especially for "turn-key" programmes.
- **Indirect export:** this practice (the "multi-domestic" attitude or sales of subsystems to foreign prime contractors) shows better business results in respect to direct exports, and is mainly relevant to the export of components, sub-systems and rarely payloads.
- **Engaging in international cooperation in space with friendly states.**

Parts of the Israeli Space Industry have a successful track record in their business activities in Europe, mainly in sub-systems and services. Here are some examples: Rafael (propulsion systems), Accubeat (atomic clocks), BAE SYSTEMS – ROKAR (Space borne GPS receivers, guidance systems and space electronics), Ramon Chips (unique VLSI/ASIC solutions for space applications), Ricor (Cryogenic cooling and Vacuum systems), SCD (Detectors), Orbit Systems (Tracking, Telemetry & Communication), ImageSat International (Earth observation services), Spacecom (communications and broadcasting services) and others. However, these companies are far from reaching their optimal business potential in Europe, while at the same time larger companies (like IAI or ELBIT SYSTEMS) have even greater difficulties in getting involved in the European market.

[15] Katz, Yaakov. "Security and Defense: Reaching for the Stars." 28 Sept. 2011. The Jerusalem Post 19 Nov. 2011 http://www.jpost.com/Features/FrontLines/Article.aspx?id=239796.

[16] The biggest nongovernmental Defence Industry in Israel.

[17] It is generally stated by most ELBIT Systems senior representatives during introductory presentations. The author.

8.3 The Israeli and the European Motivation for Cooperating in Space

8.3.1 General Common Motivating Reasons

Cooperation between Israel and Europe in a broader sense, through ESA, the EU, or bilaterally with their respective Member States, can benefit all partners and serve their respective needs, interests and aspirations. Cooperation in space activities stems from a number of considerations:

- **Financial burden sharing:** an indigenous space programme is a costly project that a small state has trouble sustaining independently, creating the need to reduce the financial burden that its space programme entails; Israel is no exception to this rule.
- **Successful legacies in space:** both parties have a long and successful legacy in space with lot of complementary fields of interest. While Europe is one of the world's leaders in all aspects of space activities, Israel had to limit itself to developing national capabilities mainly related to building affordable small satellites for remote sensing and communications purposes, an area in which Israel is today acknowledged as a world leader.
- **Space cooperation as a platform for foreign affairs and diplomatic ties:** further improvement in European-Israeli relations would be highly beneficial to both parties, especially under the current geopolitical circumstances they face. Europe is currently exposed to increasing security and economic concerns, (such as illegal immigration, terrorism, piracy, involvement in remote conflicts such as Afghanistan, Libya, etc.), as well as to environmental challenges. Consequently, it would serve both parties' best interest to achieve a rapid response capability in case of emergency events. This would provide an opportunity to join forces, accelerate activities and upgrade the cooperation level in all relevant channels, for the joint achievement of regional stability.
- **Geographic proximity:** the relatively short geographical distance between the parties makes on-going activities quite easy to manage from a logistics' point of view, shortening transportation times of raw materials and subsystems.

8.3.2 The Israeli Space Community's Interests and Expectations

Cooperation in space-related activities would allow Israel to manifest its relative strengths, create work-culture synergies between Israeli institutions and industries and their European counterparts (e.g. through the utilisation of each other's facilities, the standardisation of procedures and the conduct of common R&D projects). By becoming in the long run a full member of ESA (with a guaranteed budget from the State of Israel), the Israeli Space Industry will "upgrade its legitimacy" among its

European partners, increase its chances of obtaining more significant market shares, and provide new business opportunities. ESA membership may also open new career opportunities for Israeli space community personnel who are able and willing to work in Europe.

8.3.3 The European Interests

The highly focused Israeli technological solutions in space systems and the operational experience gained could be a very valuable factor that may contribute to closer cooperation between Europe and Israel. Some reasons that make Israel an attractive partner in space activities for Europe are:

- **The Israeli ethos: a** national space programme fits well with Israel's ethos of being a knowledge-based society, rooted in innovation and entrepreneurship. Technology and innovation have been the primary catalysts of the Israeli economy, since the early days of the country's existence. The acknowledgment that Israel will always remain quantitatively disadvantaged compared to its neighbours, pushed Israeli decision-makers to base Israel's national security and economic development concepts on the principle of "quality over quantity", and to attribute the foremost priority to science and technology.
- **Complementary needs and capabilities:** meeting some of the European operational needs in space would require improving its current technological capabilities in the miniaturisation, availability and affordability of space systems, sub-systems and payloads, especially with regard to communications and propulsion components. Israel specialises in developing lightweight, high-resolution, high precision small satellites for both national security and scientific purposes, and it has unique miniaturisation expertise in these areas. This fact stems, among other reasons, from its limited launch capabilities and its need for affordable, cost-effective space systems. Furthermore, Israel is known as one of the world's leading countries in technological innovation, thanks to the entrepreneurial character of many Israelis, and enjoys a highly skilled scientific and technological labour pool both in academia and industry. Furthermore, this workforce is traditionally characterised by a friendly, open and informal corporate culture that is prone to shortening development cycles and lowering their costs.[18] All of the above factors would offer Europe the opportunity to benefit from cooperating with Israel.

[18] Dvir, D., and A. Tishler. "The Changing Role of the Defence Industry in Israel's Industrial and Technological Development." Defense Analysis 16.1 (2001): 36; see also Battilega, J., R. Greenwalt, D. Beachley, D. Beck, E. Driver, and B. Jackson. "Transformation in Global Defense Markets & Industries: Implication from the Future of Warfare – the Case of Israel." 10 Jan. 2010 http://fas.org/irp/nic/battilega/israel.pdf.

8.4 Main Israeli Civil Space Cooperation Agreements at a Glance

Up until now, ISA has signed cooperation agreements with the following entities: CNES (France), DLR (Germany), NLR (The Netherlands), NASA (U.S.), CSA (Canada) ISRO (India), NSAU (Ukraine) and Roscosmos (Russia). Furthermore, agreements with AER (Brazil) and KARI (South Korea) are in process. The following examples reflect Israel's attempts to increase its participation in international cooperation in space activities:

8.4.1 VENμS (Vegetation and Environment New Micro Spacecraft)

VENμS is a joint Israeli-French initiative to develop, build and launch a new satellite for environmental scientific and technological mission purposes. The related Memorandum of Understanding (MoU) between ISA and CNES was signed in 2005. The scientific objective of VENμS is to provide data for scientific studies that focus on how to monitor, analyse and model land surfaces, given the influence of environmental factors and human activities. To accomplish this objective, VENμS will take high resolution and super-spectral images of predefined sites of interest all around the world every 2 days. The technological objective of VENμS is to demonstrate related technologies and test their hardware, including electrical propulsion systems in order to display an accurate autonomous orbit control capability. The VENμS platform is jointly built by IAI and Rafael, while the Israeli company ELBIT SYSTEMS – ELECTRO-OPTICS (ELOP) was selected by CNES to develop the satellite's optical instrument. The scientific missions of the project have been defined and will be conducted by the Ben-Gurion University in the Negev and CESBIO (Centre d'Etudes Spatiales de la BIOsphère) in France.

8.4.2 MUSAR (Mapping Using Synthetic Aperture RADAR) Mission (Pending for Selection)

In early 2009, cooperation was initiated between NASA and ISA to study the feasibility of building a SAR satellite based on Israeli technologies in order to explore and map the surface of Venus, as part of the Discovery Programme competition. Venus is covered by clouds, which block electro-optic imagery. Therefore, the project requires the use of a synthetic aperture RADAR (SAR) that can obtain imagery through clouds and darkness, while the satellite itself has to be relatively light in weight. Israel's strength in SAR manufacturing and operation, as well as in miniaturisation capabilities were of primary importance for this project. Participating entities include: IAI (through its ELTA and MBT subsidiaries) and the Ben-Gurion University from the Israeli side; Northrop Grumman Space Technology, the LPI-Lunar and Planetary Institute and the NASA Ames Research Centre from the U.S. side. The MUSAR mission and related R&D are all about creating high-value scientific returns at a low cost, based on concepts and spacecraft technologies that already have a proven operational reliability. However, the mission is now "frozen" by NASA due to recent priority changes in the agency's plans, and some attempts have been made to convert this initiative into an alternative application.

8.4.3 The "SHALOM" Programme

In 2009, ISA signed an MoU with the Italian Space Agency (ASI) for the development of a hyper spectral scientific mission. This project involves: the development and launch of two satellites carrying complementary payloads based on hyper spectral imaging instruments; the development and operation of a hyper spectral thermal airborne imaging demonstrator; and the eventual development of relevant applications, including the joint exploitation of the mission's data. The final Agreement on this programme was signed in July 2010 and the study phase [was expected to start][started] by the end of 2011 (Fig. 8.2).

8.5 Current and Near-Future Factors Affecting Cooperation Between Europe and Israel in Space, from an Israeli View Point

While celebrating the signature of the ESA-ISA agreement, we are witnessing the emergence of dramatic events that could influence global stability, such as the financial crisis in the U.S. and similar challenges that the European economy is now facing. At the same time, we are also experiencing major geopolitical developments, like the unprecedented wave of revolutions and riots in the Arab world ("The Arab Spring"),[19] particularly in Egypt, Libya, Morocco, Tunisia, Yemen, and recently Syria. Even Europe is not spared from social unrest, as the August 2011 riots in the United Kingdom, or previous riots in France, have demonstrated. In Israel, the most prominent social protest ever recorded took place only recently, calling for a change in national priorities under the slogan "the people call for/demand social justice." This spontaneous social movement asks for socio-economical reforms such as tax

[19] While experiencing the bloodshed in Syria, the attempt at lynching in the Israeli embassy in Cairo without any control by the law-enforcement forces, it seems that the "Spring" may change into "Dark Winter."

Fig. 8.2 Ofeq satellite launch with Shavit launcher

reductions and the even distribution of the country's resources that are a result of the good condition of the Israeli economy. All of the above developments are occurring in addition to the traditional security and international challenges[20] that the Israeli Government has to cope with. In all likelihood, these factors will lead to one of the most volatile periods of global history since World War 2, and this is a possibility that cannot be ignored. In order to evaluate the prospects of ESA-ISA cooperation, this section briefly reflects on the main supporting and adverse factors for such collaboration from an Israeli perspective:

8.5.1 Internal Harmful Factors: Weaknesses

- **The lack of a coherent governmental policy on space activities.** Up to now, space activities were mainly dedicated to national security purposes, while civil (scientific and commercial) applications drew limited attention in domestic Israeli public dialogue. The participation of Ilan Ramon in the STS 107 mission and the tragic loss of the crew is probably the most common fact in the minds of

[20] Such as the planned September declaration of a Palestinian State, the Iranian "agenda", the tension with Turkey as well as the latest terror attacks from Gaza strip and Sinai including rockets launches directed against the south of Israel while writing this article.

the Israeli public regarding Israel's civil space activity. Dealing with "bizarre space issues" is out of the scope of most politicians that look for "more popular" subjects with broader societal influence and more immediate results. In order to put the space agenda on the table, there is a need for the Israeli President to personally intervene in its favour.

- **The low level of government investment in space, especially in civil programmes.** Dealing simultaneously with a great number of geopolitical and security concerns does not allow for space related issues to be acknowledged in the government work-plan, except for national security space applications that draw their resources from the defence budget.
- **The slow implementation of the recommendations of the committee responsible for formulating the national space programme of Israel, and overcoming related administrative obstacles**. An often repeated argument against increasing the government space budget in Israel is the existence of private funding that could be easily made available for immediate investments in space activities. However, current practice shows that space is still considered as a "risky investment" that is out of the scope of most private investors. Consequently, this entire public discussion could be highly counterproductive, at a time when Israeli decision makers are not available to enforce its implementation.
- **The relative dependence of the Israeli space industry on the government decision making process, which prevents the completion of space business cycles in a reasonable timeframe.** The major Israeli space industry business activities are projects based on government-to-government agreements. In this case, the decision making process and the associated administrative work spreads over a long period that is unacceptable by private business standards. In other cases, such as direct sales to foreign customers, government procedures for export approval are often longer than the deadline to respond to international tenders.
- **The fluid and versatile current geopolitical and social environment, which could deflect the Israeli government's attention and funding from space activities to other, more pressing policy priorities**. The current geopolitical situation in the Middle East might increase the probability of this scenario.

8.5.2 Internal Helpful Factors: Strengths

- **Israel's long-standing space manufacturing heritage,** including in the production of turn-key satellites, payloads and launch systems; the presence of already existing infrastructure that can support future development projects.
- **The presence of successful commercial space companies** in manufacturing, satellite applications and finance, which constitute an excellent business base with further growth potential.

- **The significant academic and research infrastructure,** building on the human capital of an innovative, motivated, and internationally networked student and researcher base.
- **The existence of a strong cooperative spirit,** as is manifested in cooperation agreements with leading space powers, as well as in commercial contracts. Undeniably, the space community of Israel is eager to collaborate internationally and it identifies Europe as a highly valued strategic partner in space activities.
- **The country's vibrant and innovative aerospace and defence sector,** with exceptional technological knowhow in unmanned systems, rockets, remote-sensing instruments and telecommunications equipment, which is relevant to developing civil and commercial space applications as well.

8.5.3 *External Harmful Factors: Threats*

- **The strict export controls** that prevent sales of certain high technology products. However, civil, commercial and scientific space activities based on government-to-government agreements could be relatively immune to this problem.
- **The highly volatile geopolitical environment,** which could reduce commercial space activities and exports. The uncertain political situation in the Middle East, in conjunction with the economic crisis in Europe, could be the cause of a trend change in the current "warming relations" climate.
- **The global economic crisis** in general may downgrade the priority attached to investments in space and decrease the incentive for industries to cooperate, while adopting at the same time a more aggressive competitive stance. If economic conditions were to worsen even further, ESA and ISA could face mounting pressure to reduce the scope of the recently signed cooperation agreement.

8.5.4 *External Helpful Factors: Opportunities*

- **The currently difficult economic situation in Europe** could create a greater incentive to produce scale economies by sharing the burden of developing space technologies. This could create opportunities for Israel to offer lower-cost alternative space products and services.
- **The emerging change in the security requirements of Europe** could make Israel an attractive partner in developing space security concepts and technologies, thanks to its long operational experience in this field. This fact could lead for example to establishing a regional joint coordination centre.
- **The possibility for Israeli companies to build upon their personnel's extensive language skills** in English, French, German, Italian, Spanish, Russian and other European languages, in order to facilitate their communication with their European colleagues and bridge any existing cultural gaps (Fig. 8.3).

Fig. 8.3 The signature of ESA – ISA cooperation agreement, Tel-Aviv, 30 Jan. 2011 (Source: ESA). Seating (left to right): ESA's Director General Jean-Jacques Dordain and the Director General of the Israel Space Agency (ISA), Dr. Zvi Kaplan. Standing (left to right): Karl Bergquist, Director for international relations – ESA, Menachem Greenblum, Director General of the Ministry of Science and Technology, Prof. Daniel Hershkowitz, the Israeli Minister of Science and Technology and Prof. (Gen. Ret) Issac Ben-Israel, the Chairman of ISA

8.6 Conclusion

The cooperation agreement signed by ESA and ISA at the beginning of 2011 is probably one of the symbols of the warming of relations between Europe and the State of Israel. This move is perceived by the Israeli space community as a very meaningful and historical opportunity for moving towards full ESA membership.

This cooperation agreement should be used as a valuable lever for convincing decision makers and administration officials to support its realisation. Such a development would be in line with the recommendations of the special presidential committee, taking into account Israel's need to strengthen its qualitative advantage by investing in scientific and technological "growth engines", such as space. Experience so far has shown that every new position in the space sector can create up to four additional jobs in the broader economy, and this calculation is valid for the Israeli, as well as the European market.[21]

[21] Josef Weiss, Corporate Vice President and General Manager of the Systems, Missiles and Space Group of Israel Aerospace Industries Ltd, in his presentation during a Space Work-Shop held in Tel Aviv University, Nov. 2009.

For all of the aforementioned reasons, this collaboration has an inherently strong chance of success. Nevertheless, as the current situation (both in Europe and Israel) is quite unpredictable, several supporting measures should be taken in order to hedge against the risk factors described above. Such measures should include: raising government support and creating a positive public opinion stance towards space activities in Israel, especially by adopting the appropriate legislative measures; fostering the development of an influential and fully authorised space agency and providing it with the required resources; and fully implementing the space programme outlined in the presidential committee's report. Only close coordination between ESA and ISA has the ability to lead the build-up of mutual trust between the two parties. Inspired by the leadership of the two agencies, the current capabilities and infrastructure of the Israeli space industry could be quickly mobilised to provide additional civilian and dual use space capabilities for Europe.

At the end of the day, we should be aware that beyond the positive political atmosphere, this agreement is the result of efforts invested by a handful of very dedicated people. Their close cooperation created excellent interpersonal relations and mutual trust on both sides of the Mediterranean Sea. The Israeli "space legacy" has so far been the work of "a few crazy guys" with a common vision, which became a success story against all odds. Today, we should use this "revolutionary atmosphere" to accelerate the process of forging stronger ties between the new partners, and thus demonstrate the leadership necessary to making this cooperation a success story and giving hope to the younger generation, interested in extending the boundaries to new horizons (Ofeq = horizon in Hebrew).

8.7 Epilogue

On 6 December 2011, the Minister of Science and Technology and the Minister of Finance agreed on implementing the committee's recommendations including the re-establishment of ISA with an initial budget of 165 Million Israeli Shekels (about 35 million euros) spread over 2012 and 2013.[22] At the end of January 2012 the Israeli Government appointed Mr. Menachem Kidron as the new Director General of ISA. Mr. Kidron served as head of the Manor and Technologies divisions at Rafael.[23] In his inauguration speech, Mr. Kidron stated that the Israeli Space Agency will invest mainly in research and development of miniaturization technologies and will initiate and financially support international cooperation with market leaders. Mr. Kidron further stressed the priority ISA gives to

[22] Singer, Zvi. "Huge Budget Space Agency Will Leverage the Global Space in Israel to the Top." 29 Jan. 2012. Megafon News 13. Feb. 2012 http://megafon-news.co.il/asys/archives/16103.

[23] Ronen, Gil. "Israel's Space Budget Lifts Off: From NIS 10M to 165M." 29 Jan 2012. Israel National News 13 Feb. 2012 http://www.israelnationalnews.com/News/News.aspx/152231.

strengthening ties with ESA, and mentioned the Israeli intention of joining ESA as a non European member state, which is anticipated to be a challenging goal.[24]

Acknowledgements This chapter is dedicated to the so few, owed by so many: the pioneers and the first leaders of the Israeli "space odyssey", the very small but very qualitative and dedicated Israeli Space Community members. I'd like to take this opportunity to thank Dr. Deganit Paikowsky, Academic Director and Fellow at the Yuval Ne'eman Workshop for Science, Technology and Security, Tel Aviv University. Last but not least, I would like to thank my supporting and loving family and especially my wife Orna, for her encouragement and patience.

[24] Menachem Kidron, Head of ISA, The 7th Annual International Ilan Ramon Space Conference, Jan. 2012 http://www.hayadan.org.il/future-of-isa-130112/.

Chapter 9
Recent Development of Japan's Space Policy: The Quasi-Zenith Satellite Programme and the Space Policy Making Process

Hideaki Shiroyama

9.1 Introduction: The Quasi-Zenith Satellite Programme as an Experiment of the Space Policy Making Process

The first Quasi-Zenith satellite "MICHIBIKI" was launched on 11 September 2010. The Quasi-Zenith Satellites System uses multiple satellites that have the same orbital period as geostationary satellites with some orbital inclinations (their orbits are known as "Quasi-Zenith Orbits"). These satellites are placed in multiple orbital planes, so that one satellite always appears near the zenith above the region of Japan. The system makes it possible to provide high accuracy satellite positioning services covering close to 100% of Japan, including urban, canyon and mountain terrain.[1] The image of the Quasi-Zenith Satellite System using three satellites is shown in Fig. 9.1 below.

In the history of Japan's space policy, satellites were to respond to fixed social needs, such as communications and broadcasting. By contrast, the quasi-zenith satellite programme was an experimental challenge to meet new social needs explored by emerging actors. In the meantime, the last decade during which the programme was discussed coincides with a period of change in the Japanese space policy making process as well, triggered by the reorganisation of government ministries etc.

Following the launch of MICHIBIKI, discussion surrounding the quasi-zenith satellite programme is once again taking place on the issues of how and for what practical application Japan should advance the programme. The purpose of this article is to examine the past Japanese space policy making process concerning this

[1] "The Quasi-Zenith Satellites System." Japan Aerospace Exploration Agency http://qzss.jaxa.jp/index_e.html.

H. Shiroyama (✉)
Policy Alternatives Research Institute and The University of Tokyo, Tokyo, Japan
e-mail: siroyama@j.u-tokyo.ac.jp

P. Hulsroj et al. (eds.), *Yearbook on Space Policy 2010/2011*,
Yearbook on Space Policy, DOI 10.1007/978-3-7091-1363-9_9,

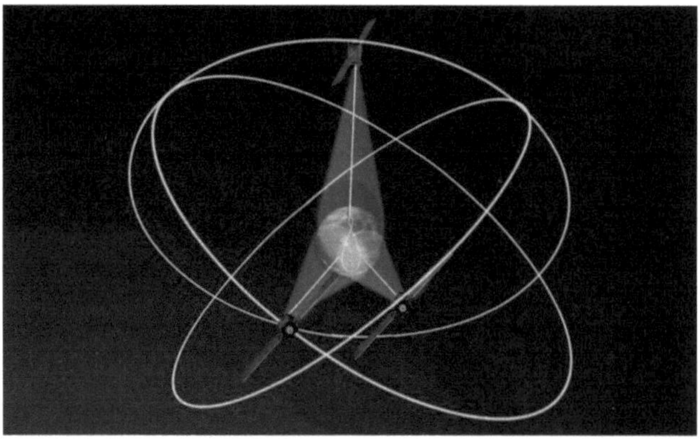

Fig. 9.1 The Quasi-Zenith Satellite System using three satellites (Source: JAXA)

project, summarise the current discussion about it, and finally to extract issues related to the Japanese space policy making process in general.

First, the historical background of the Quasi-Zenith Satellite Programme will be examined, from its initiation to the present. The programme has a history of about 10 years, during which its content and position experienced some changes. This report discusses the programme by dividing it into three periods. The first is from 2001 to 2002. This was the period when the private sector proposed the promotion of the programme and the Council for Science and Technology Policy decided to introduce it. The second period is from 2003 to 2004, during which research and development on the satellite begun. However, discussions on the programme and especially on its operational framework were tangled and its detailed plan not determined. The third period is from 2005 to 2010, when the first satellite "MICHIBIKI" was launched. This was a period when the concept of the programme was re-examined, following the private sector's decision in February 2006 to abandon the commercialisation of the satellite's communication/broadcasting service. As a result, the "Basic Policy for Promotion of the Quasi-Zenith Satellite System" was released by the Conference for Promotion of Positioning/Geographic Information System in March 2006, which stipulated the programme's policy that is still in effect today.

Next, this chapter will also look into the international dimension of the programme, its recent development status and the space policy context linked to its development after the change of government in autumn 2009 [when] discussions about the future of the Quasi-Zenith Satellite Programme resumed in the Strategic Headquarters for Space Policy. Consequently, further discussions on the framework of the space policy making process in Japan resumed as well, in conjunction with the programme's implementation. Based on the structure described above, the last part of this article will attempt to extract issues concerning the status quo of the space policy making process in Japan.

9.2 The Historical Background of the Quasi-Zenith Satellite Programme: Its Position in Japanese Domestic Policy and Its Changing Process

9.2.1 The First Period (2001–2002): The Proposal to Introduce the Quasi-Zenith Satellite Programme

Research on the satellite orbit of the Quasi-Zenith Satellite System was initiated in the 1970s at the former Radio Research Laboratory of the Ministry of Posts and Telecommunications (currently, the National Institute of Information and Communications Technology [NICT] of the Ministry of Internal Affairs and Communications). In the 1990s research and patent application in private companies begun as well. Research on satellite positioning technology associated with the Quasi-Zenith Satellite mission started in the former Communications Research Laboratory (currently NICT) and the former National Space Development Agency of Japan (currently JAXA) in the 1990s. Concrete action towards the programme's realisation commenced in response to a proposal from the private sector urging its development. In July 2001, the Japan Business Federation (Keidanren) prepared a document entitled "Grand Strategy Towards Expansion of Space Utilisation Frontier", which recommended the inauguration of the development of the quasi-satellite system. In January 2002, the "Study Group for New Satellite Business" was established by private companies and it started conducting feasibility studies on the construction of the system.

On 21 February 2002, Keidanren presented its recommendations at the fourth conference of the Special Committee for Space Development and Utilisation of the Council for Science and Technology Policy.[2] In reply to the recommendation, the Council set up its policy "to promote the development and deployment of the Quasi-Zenith Satellite System under the public/private partnership" in the "baseline for the approach to the space development and utilisation in the future."[3] Meanwhile, in May 2002, the "Council for Quasi-Zenith Satellite System Development and Utilisation" was founded. In August 2002, four ministries responsible for R&D (the Ministry of Education, Culture, Sports, Science and Technology, the Ministry of Internal Affairs and Communications, the Ministry of Economy, Trade and Industry, and the Ministry of Land, Infrastructure, Transport and Tourism) requested an approximate budget to undertake R&D on the satellite.

---

[2] Taniguchi, Ichiro. "Japan's Space Development and Utilization for Industrialization." Council for Science and Technology Policy http://www8.cao.go.jp/cstp/tyousakai/cosmo/haihu04/siryou4-4.pdf.

[3] "Baseline for the Approach to the Space Development and Utilization in the Future." 19 June 2002. Council for Science and Technology Policy http://www8.cao.go.jp/cstp/output/iken020619_5.pdf.

Following the approximate budget request, it was decided that the Council for Science and Technology Policy would evaluate the appropriateness of the quasi-zenith satellite programme as a new large-scale development project. In the "Prioritization of Approximate Budget Request Concerning Science and Technology in FY2003" (the so-called SABC ranking) released in October 2002, all the budgetary items related to the Quasi-Zenith Satellite submitted from the R&D ministries were ranked as "S."[4] After that, the Council worked to conclude the evaluation. On 27 November 2002, the "Report (Draft) on Evaluation Result of 'quasi-zenith satellite system'" was prepared and the Council approved the report as drafted on 25 December 2002. In the document, the Council judged that "it is appropriate to advance the programme" for the purpose of creating new business opportunities, utilising a wide range of fields, and potentially expanding its utilisation abroad. Simultaneously, the Council expressed some concerns about the programme, namely on the roles to be shared between the public and private sector, its commercialisation by the private sector, its rather optimistic market demand forecast, and so on.[5]

In the private sector during this period, the Advanced Space Business Corporation (ASBC) was founded by a joint investment of 59 companies, including Mitsubishi Electric Co., Hitachi, Ltd., ITOCHU Co., NEC TOSHIBA Space Systems, Ltd., and TOYOTA MOTOR CO. The objective of ASBC was first to plan and promote business opportunities created by the Quasi-Zenith Satellite System, and second to undertake all necessary R&D associated with the system.

9.2.2 The Second Period (2003–2004): Start of R&D and Postponement of a Decision on the Operational Framework

During the second period in the programme's development, discussions on how to advance the Quasi-Zenith Satellite project began. In August 2003, the sharing of roles and costs between the public and private sector was agreed at the Council for Quasi-Zenith Satellite System Development and Utilisation. More precisely, both sectors confirmed that the government would be responsible for the technological development and in-orbit demonstration of the satellites, while the private sector would assume responsibility for the system's commercialisation. At that time, the

[4] "Prioritization of Approximate Budget Request Concerning Science and Technology in FY 2003." 18 Oct. 2002. Council for Science and Technology Policy http://www8.cao.go.jp/cstp/budget/h15ordergaiyo.pdf.

[5] "Draft Report on Evaluation Result of 'Quasi-Zenith Satellite System." 22 Oct. 2002. Council for Science and Technology Policy http://www8.cao.go.jp/cstp/tyousakai/hyouka/haihu16/siryo3.pdf.

total programme cost was estimated at 170 billion yen, and both sectors agreed on the government paying 90 billion yen and the private sector 80 billion.[6]

Following this, the Council for Science and Technology Policy initiated detailed studies on how to advance the programme. In order to fulfil its role in establishing a positioning policy for the system, the government created the Working Group on Positioning Field, under the Special Committee for Space Development and Utilisation of the Council. Following three rounds of review and study, the group prepared a report entitled "Interim Conclusion (Draft) on Satellite Positioning System in Japan" in December 2003.[7] In the following year, the report title was revised to "Interim Study Result (Draft)" and its purpose was also changed.[8] With regard to the creation of an organisational scheme suitable for the deployment and operation of the system, as well as to the satellite concept itself, the document simply mentioned that "these issues should be continuously reviewed", without addressing concretely the issue of creating an institutional design.

Furthermore, the concept and adequate framework to promote the quasi-zenith satellite programme were continuously discussed in the Council for Science and Technology Policy, but no specific conclusion was formulated. The 23rd Special Committee for Space Development and Utilisation of the Council was held in June 2004. According to the chairperson's judgment, "as for deployment and operation of the complement system for positioning, concrete approach and concept is still under coordination on the government side." In September 2004, the "Basic Strategy for Space Development and Utilisation of Japan – Outline Draft" was concluded during the plenary meeting of the Council for Science and Technology Policy.[9] This document defined the immediate goal in the following words: "the government is to steadily proceed with research, development, and demonstration for high-risk areas including positioning complement and augmentation." However, creating a concrete institutional design was put off. The document said that "how the government involves in deployment and operation will be decided before the completion of the demonstration." As stated above, although detailed studies and discussions were took place on the framework to promote the Quasi-Zenith Satellite Programme during this period, there was no clear conclusion to them.

---

[6] "History and Current Status of QZS." 7 Sept. 2010. Strategic Headquarters for Space Policy http://www.kantei.go.jp/jp/singi/utyuu/QZS/dai1/siryou2.pdf.

[7] "Interim Draft Conclusion on Satellite Positioning System in Japan." 4 Dec. 2003. Special Committee for Space Development and Utilization of the Council for Science and Technology Policy http://www8.cao.go.jp/cstp/tyousakai/cosmo/haihu16/siryo16-2.pdf.

[8] "Interim Study Results (Draft) on Satellite Positioning System in Japan." 16 Jan. 2004. Special Committee for Space Development and Utilization of the Council for Science and Technology Policy http://www8.cao.go.jp/cstp/tyousakai/cosmo/haihu17/siryo17-4-1.pdf.

[9] "Basic Strategy for Space Development and Utilization of Japan – Outline Draft-." 9 Sept. 2004. Council for Science and Technology Policy http://www8.cao.go.jp/cstp/siryo/haihu39/siryo1-2.pdf.

9.2.3 The Third Period (2005–2010): Withdrawal of the Private Sector and the Government Approach on Geospatial Information Policy

During the third period, the private sector gave up its commercialisation plans for communication and broadcasting services, while the government reconsidered the concept of the Quasi-Zenith Satellite Programme. At the same time, satellite positioning came to be discussed in connection with the policy of the geographic information system (GIS), and consequently R&D for the Quasi-Zenith Satellite was discussed and promoted in this context.

In early 2005, the Conference for the Promotion of Positioning/Geographic Information System was established to enhance collaboration between geospatial information policy and satellite positioning policy. The issues addressed by the conference were the following: establishment of positioning/spatial information; effective establishment of a geospatial information system and promotion of its mutual utilisation; deployment and operation of the quasi-zenith satellite system; and the creation of a framework responsible for these. In December 2005, the "Proper Way to Proceed with the Quasi-Zenith Satellite System in the Future (Draft)" was prepared by the conference.[10] In the document, the conference proposed that the business plan should be re-examined since the risks related to the programme's commercial services by the private sector were high.

In February 2006 ASBC decided to abandon the commercialisation of communication and broadcasting services by the private sector.[11] The main reason for this decision was the rapid growth of terrestrial communication and broadcasting services' infrastructure that made it difficult for the private sector alone to implement the private communication and broadcasting business model using the quasi satellite system. Following this decision, ASBC dissolved itself in August 2007. Because of the private sector's decision on the commercialisation issue, the Conference for Promotion of Positioning/Geographic Information System reviewed again the Quasi-Zenith Satellite Programme and substantially changed its plan. In March 2006 the Conference set up the "Basic Policy for Promotion of the Quasi-Zenith Satellite System" that indicated a new policy.[12] Based on the general importance of satellite positioning in the quasi-zenith satellite programme, the conference determined that the government should establish the system on its own initiative.

---

[10] "Proper Way to Proceed with the Quasi-Zenith Satellite System in the Future." 21 Dec. 2005. Conference for Promotion of Positioning/Geographic Information System. http://www.cas.go.jp/jp/seisaku/sokuitiri/171221/doc2-2.pdf.

[11] "Promotion of QZS Plan – Feasibility of Commercialization and Proposal." Feb. 2006. http://www.cas.go.jp/jp/seisaku/sokuitiri/180331/sankou4.pdf.

[12] "Basic Policy for Promotion of the Quasi-Zenith Satellite System." 31 Mar. 2006. Conference for Promotion of Positioning/Geographic Information System http://www.cas.go.jp/jp/seisaku/sokuitiri/180331/siryou2.pdf.

More specifically, the programme was envisaged to advance in two phases. In the first phase, a technology and utilisation demonstration would be made using a single satellite. In the second phase, the system demonstration would be made by a constellation of three satellites. It was also decided that, in the first phase, the Ministry of Education, Culture, Sports, Science and Technology would play a leading role, in cooperation with the Ministry of Internal Affairs and Communications, the Ministry of Economy, Trade and Industry, and the Ministry of Land, Infrastructure, Transport and Tourism. At the same time, JAXA would be responsible for the maintenance and operation of the satellite. In the private sector, the Satellite Positioning Research and Application Centre (SPAC) was founded in January 2007, in order to expand the use of geospatial information in the industrial and commercial areas.

The policy released in March 2006 is in fact valid even today. In April 2008 the "Basic Plan for Promotion of Geospatial Information Utilisation" was approved by the Japanese Cabinet. Policy regarding the respective roles of the public and private sector in the programme was determined anew. According to the new policy, the private sector is required to assess the system's commercialisation prospects based on the phase one technology/utilisation demonstration, and then decide to fund the programme or not, according to its business content, scale etc. In accordance with the process described above, the quasi-zenith satellite programme reached the step of launching the first satellite MICHIBIKI in September 2010. JAXA will conduct the phase one technology/utilisation demonstration after the launch.

9.3 The International Position of the Quasi-Zenith Satellite Programme

9.3.1 The Programme's Relevance to Japan-U.S. Relations

In September 1998, the governments of Japan and the United States agreed on the "Japan-United States Joint Statement on Cooperation in the Use of the Global Positioning System" (hereinafter referred to as the "joint statement").[13] In this document, both governments confirmed the following basic principles on the use of GPS: "the U.S. provides the GPS Standard Positioning Service for peaceful civil, commercial, and scientific use on a continuous, worldwide basis, free of direct user fees"; and Japan agreed that "Japan intends to work closely with the United States to promote broad and effective use of the GPS Standard Positioning Service as a worldwide positioning, navigation, and timing standard."

In accordance with the joint statement, a Japan-U.S. plenary meeting was held in February 2001. It has been held another eight times since (as of June 2011), playing

[13] "Joint Statement by the Government of the United States of America and the Government of Japan on Cooperation in the Use of the Global Positioning System." 22 Sept. 1998. The White House, Office for the Press Secretary http://www.pnt.gov/public/docs/1998/japan.shtml.

an important role as one of the Japan-U.S. collaborations in the space science and technology domain. This framework for Japan-U.S. cooperation on GPS use is also a significantly important element for the Quasi-Zenith Satellite Programme with the aim of complementing and augmenting GPS.

When Japan initiated its original positioning satellite system at the end of the 1990s, there was some concern about it in the U.S. At that time, there was a move in Japan seeking an autonomous space development and utilisation policy, such as the decision to introduce the Information Gathering Satellite programme, triggered by the launch of the Taepodong rocket by North Korea. The main premise of the quasi-zenith satellite, however, was that it would be used in combination with GPS and, in the first place, it was impossible to carry out independent, single positioning with the system consisting of three satellites. The Quasi-Zenith Satellite was accepted as a significant programme from a U.S. policy perspective as well, since it complements GPS in the Asia and Oceania region and can provide users in the region with GPS signals.

From a global viewpoint, there has also been a big change in the international environment regarding the GNSS (Global Navigation Satellite System). In addition to the U.S.'s GPS and Europe's Galileo, Russia has worked on the construction of the GLONASS, its own positioning satellite system dating back to the cold war era. Meanwhile, China is rapidly creating its own positioning satellite system called COMPASS. Consequently, having a positioning satellite system is no longer an exclusive property of the U.S. Under these circumstances, it is likely that the U.S. will change its policy to encourage the interoperability of GPS with positioning satellite systems built by third countries and will improve cooperation with them.[14]

9.3.2 *The Programme Within the Asia-Pacific Region Geopolitical Context*

The international expansion of the Quasi-Zenith Satellite Programme in the Asia-Pacific Region was often mentioned during its development process over the past 10 years, which is considered to be of international "significance."[15] Nonetheless, we cannot say that a detailed "strategy" to expand the system has been fully considered.

The possibility that the Quasi-Zenith Satellite could be utilised in the SE Asia and Oceania region was recognised from the beginning, because of its orbital characteristics. In fact, in the policy document released in December 2002 by the Council for Science and Technology Policy describing the evaluation result of the

[14] "U.S. Space-Based Positioning, Navigation, and Timing Policy." 15 Dec. 2004. The White House http://www.whitehouse.gov/files/documents/ostp/Issues/FactSheetSPACE-BASEDPOSITIONINGNAVIGATIONTIMING.pdf.

[15] "About QZS System." 25 Dec. 2002. Council for Science and Technology Policy http://www8.cao.go.jp/cstp/output/iken021225_3.pdf.

satellite programme, the possibility of expanding its service to the Asia and Oceania region was pointed out in terms of "significance in international relations." In recent years, the word "space diplomacy" is frequently and positively cited. The promotion of space diplomacy is one of the important objectives mentioned in the Basic Plan for Space Policy formulated in June 2009, and international cooperation and Japan's contribution to the Asia-Pacific region is also stated as a major aim.

However, Japan has not yet fully considered its actual international strategy necessary to meet this objective. The cause of this seems to be the lack on the part of the Japanese government of a concrete concept for the operational use of the quasi-zenith satellite. Even today, after the launch of the first technology demonstration satellite, policy planning on the follow-on satellites is still under discussion. Under these circumstances, a concrete and comprehensive policy toward the expansion of the system's use in the Asia-Pacific region has not yet been considered and implemented as a national policy.

9.4 The Space Policy Making Process and the Quasi-Zenith Satellite Programme After the Change in the Japanese Government

9.4.1 Changes in the Space Policy Making Process

In 2008, the Aerospace Basic Act entered into force, and the Strategic Headquarters for Space Policy was established within the Cabinet. However, the Aerospace Basic Act (2008) and the Basic Plan for Space Policy (2009) did not include any changes to the quasi-zenith satellite system programme, as set out in the Basic Policy for Promotion of the Quasi-Zenith Satellite System formulated in March 2006.

After the change in the Japanese government in September 2009, Minister Maehara, who had become the minister in charge of space policy for the new Hatoyama administration, established the "Conference of Experts on the Future Space Policy", and promoted the reconsideration of the country's policies on the subject. Furthermore, a "Taskforce Meeting of Experts on the Future State of Space Policy" was organized in June 2010, in order for the widest possible variety of space policy experts to have inputs to the decision making process on the government's New Growth Strategy. Participants in this meeting included not only leading government officials, but also experts from the Conference of Experts and various other organisations, who took a leading role in the taskforce's deliberations. The meeting was held on 2 and 3 April 2010 at the University of Tokyo, and was attended by nearly 100 people. It was a novel and interesting initiative to gain a wide input on the policy making process, led by the private sector in the space policy field.

With this input from the private sector, the Strategic Headquarters put together the "Priority Policy in the Space Field" on 25 May 2010, indicating that "in order to realise the growth of Japan through space policy when entering an era of full use of space, urgent task is to convert to a policy that is driven by use (demand side)."[16] Furthermore, on 18 June 2010 the government decided on a "New Growth Strategy" aimed to "revive a healthy Japan." After that, the Headquarters invited professor Yamakawa of Kyoto University to the Secretary General and started its new activities.

9.4.2 Prioritisation in the Quasi-Zenith Satellite Programme

Regarding the Quasi-Zenith Satellite System, the "Promotion of Space Policy for the Present" decided on 27 August 2010 adopted the following policy: "In view of the fact that, in September of this year, the first satellite "MICHIBIKI" is scheduled to be launched, it is necessary for us, as the government, to start the necessary planning and review to decide a policy in preparation for future satellites."[17]

Also, at a meeting at the Strategic Headquarters on 27 August 2010, it was decided to establish a project team within the Headquarters at parliamentary secretary level, in order for the government to carry out coordinated planning and review activities of the promotion and development of the future Quasi-Zenith Satellites as one body, under the general coordination of the Cabinet Secretariat. In December 2010, the Special Committee for Space Development Strategy (second term) was established, under the Strategic Headquarters for Space Policy.[18] Regarding the deployment of a Japanese satellite positioning system, the "Working Group for the Study of Quasi-Zenith Satellite System Development and Utilisation" consisting of experts with professional knowledge was established under the Special Committee for Space Development Strategy of the Headquarters. The working group was supposed to carry out expert research and study related to the formulation of business plans, including the use of such a system. The following were listed as issues to be studied by the group:

a. risk analysis and countermeasures when depending on positioning satellites of other countries such as GPS;
b. policy evaluation of public use and private sector's services, etc. of the positioning satellite system of Japan;
c. appropriate international expansion;
d. appropriate architecture of the positioning satellite system of Japan;
e. appropriate cost sharing and promotion organization structure of the positioning satellite system of Japan;
f. other issues related to the concept and business planning for the positioning satellite system of Japan.

[16] "Priority Policy in the Space Field." 25 May 2010. Strategic Headquarter for Space Policy http://www.kantei.go.jp/jp/singi/utyuu/honbu/dai4/siryou2.pdf.

[17] "Promotion of Space Policy for the Present." 27 Aug. 2010. Strategic Headquarter for Space Policy http://www.kantei.go.jp/jp/singi/utyuu/honbu/dai5/siryou1.pdf.

[18] "The 10th Meeting of the Special Committee for Space Development Strategy." 20 Dec. 2010. Strategic Headquarter for Space Policy http://www.kantei.go.jp/jp/singi/utyuu/senmon/dai10/gijisidai.html.

The Working Group's met six times from December 2010 to April 2011. On 22 April 2011, the study results were released in the "Interim Report – Basic Concept of Study on Business Plan for Positioning Satellite System in Japan."[19] This report regarded the positioning satellite system as "indispensable infrastructure for the state and the people" and redefined the practical reasons for building the quasi-zenith satellite system as follows: first the "enhancement of economic society" through the improvement of business efficiency, the creation of new industries and the optimisation/enhancement of administration were mentioned. Second, in relation to the great East Japan earthquake, the report acknowledged its importance in "ensuring the safety and security of the people" through the enhancement of services related to public safety and the improvement of disaster response capabilities by providing safety confirmation and information during disasters. Third, the "guarantee of independence from and contribution to the positioning satellite infrastructure in the world" was highlighted, signifying an active part in international coordination efforts, (e.g. agreements on international standardisation), the avoidance of complete dependence on other countries and the contribution to important social infrastructures. Finally, the report confirmed once again the importance of the system's "contribution and expansion to the Asia-Pacific Region" through the provision of high-accuracy services, and especially disaster prevention and response. This interim report was submitted to the Special Committee for Space Development Strategies on 25 April 2011.

At the same time, discussions were held regarding the specific role of the Cabinet Office in Japan's space policy (with possible options discussed including a coordination promotion function, a comprehensive budget request function, or an implementation function), as well as on the administrative jurisdiction of JAXA (between the Ministry of Education, Culture, Sports, Science and Technology, and the Cabinet Office). Finally, on 15 May 2011, the second Task Force meeting was held again under the leadership of experts outside the government.

9.4.3 The Future Direction Currently Envisaged: Connecting the Quasi-Zenith Satellite Programme with Discussion About Organisational Framework

In response to developing discussions on the subject, the Special Committee for Space Development Strategy announced the "Policy for Prioritization and Optimization of Measures for Strategic Promotion of Space Development and Utilisation (Draft)" on 8 August 2011, shedding light on Japan's future space policy strategies.[20]

---

[19] "Interim Report – Basic Concept of Study on Business Plan for Positioning Satellite System in Japan." 22 Apr. 2011. Working Group of Special Committee for Space Development Strategy http://www.kantei.go.jp/jp/singi/utyuu/QZSkaihatsu/dai6/siryou2.pdf.

[20] "Policy for Prioritization and Optimization of Measures for Strategic Promotion of Space Development and Utilization." 8 Aug. 2011. Special Committee for Space Development Strategy http://www.kantei.go.jp/jp/singi/utyuu/senmon/dai18/siryou1.pdf.

Keeping in mind the country's severe financial situation and the impact of the great East Japan earthquake in March 2011, this document adopted the view that policies should not aim "to try to please everybody," but they should promote the prioritization and optimization of space policy. In this document, the Quasi-Zenith Satellite Programme was mentioned as the country's most important programme in terms of space policy. Furthermore, the document indicated that "deployment of the Quasi-Zenith Satellite System should be carried out as soon as possible, and full development and deployment of the system should be undertaken from FY2012 as the most important theme." This view was based on the recognition of the significance of security in its broad sense, in the aftermath of the great East Japan earthquake.

However, the announcement of this direction was not necessarily based on the concrete needs of the ministries responsible for the demand of the satellite services, such as the Ministry of Land, Infrastructure, Transport and Tourism in charge of disaster management. Also, in terms of the international position of the quasi-zenith satellite, if Japan intends to avoid its complete dependence on other countries for satellite positioning services, to promote the system's deployment in Asia, and to make a contribution to global satellite positioning services, it should construct a system of no less than seven satellites, and thus be able to perform continuous positioning over the SE Asia region. Nonetheless, no definite decision on the system's final architecture has been made yet.

This report also indicated that the "establishment of organizational framework should be advanced in order for the Cabinet Office to promote the quasi-zenith satellite system as an agent responsible for its development, deployment, and operation." The main idea behind the choice of the "Space Agency" plan would be to delegate to it the implementation of the system. On the other hand, opinions vary on whether it would be appropriate to give this implementation function to the Cabinet Office. As for the entire space policy making process, there are also unresolved issues, such as how to divide research and development on the system from its implementation and policy aspects, as well as how to integrate space policy with relevant science and technology policies. Regarding the latter, for instance, the Democratic Party's Science/Technology and Innovation Study Committee proposed to construct a framework that would concentrate governance on space, ocean, IT, intellectual property and other relevant issues in the reformed Council for Science and Technology Policy.

9.5 Epilogue: Issues in the Space Policy Making Process in Japan

Based on the description of the policy making process in the past 10 years regarding the Quasi-Zenith Satellite Programme, some general observations on the wider space policy making process in Japan are possible.

First, the issue of creating a suitable coordination mechanism arises. The time of the Quasi-Zenith Satellite Programme's start coincided with the reform period in

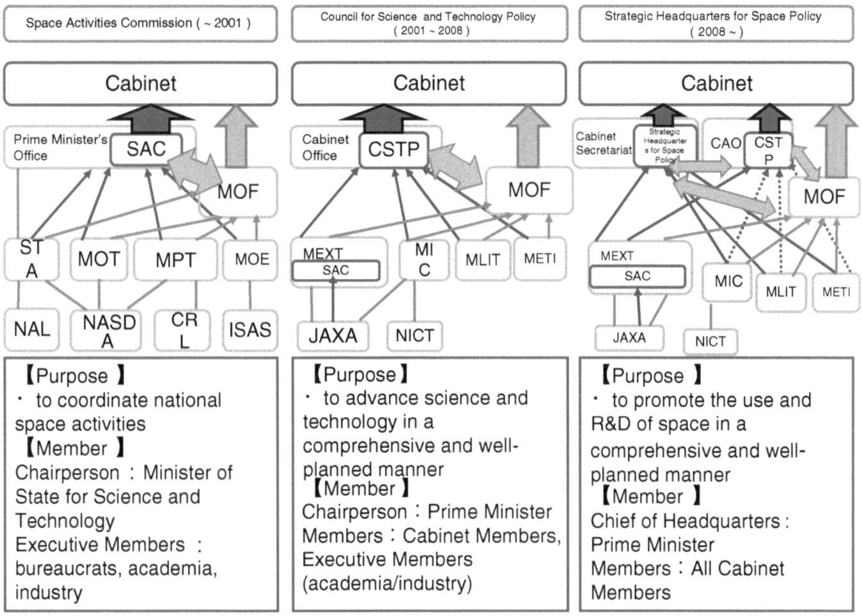

Fig. 9.2 Transformation of structure of Japan's space policy making process (graph prepared by Mr. Shoichi Fujimoto and Mr. Kosuke Sugawara). *CAO* Cabinet Office, *CRL* Communications Research Laboratory (→NICT), *CSTP* Council for Science and Technology Policy, *ISAS* Institute of Space and Aeronautical Science (→JAXA), *JAXA* Japan Aerospace Exploration Agency, *METI* Ministry of Economy, Trade and Industry, *MEXT* Ministry of Education, Culture, Sports, Science and Technology, *MIC* Ministry of Internal affairs and Communications, *MLIT* Ministry of Land, Infrastructure, Transport and Tourism, *MOE* Ministry of Education, Culture and Sports (→MEXT), *MOF* Ministry of Finance, *MOT* Ministry of Transport (→MLIT), *MPT* Ministry of Posts and Telecommunications (→MIC/Japan Post Holdings), *NAL* National Aerospace Laboratory (→JAXA), *NASDA* National Space Development Agency of Japan (→JAXA), *NICT* National Institute of Information and Communications Technology, *SAC* Space Activities Commission, *STA* Science and Technology Agency (→MEXT)

space administration, which was itself triggered by the reorganisation of the government ministries. As Fig. 9.2 shows, the main coordinating body for space policy making changed from the Space Development Commission to the Council for Science and Technology Policy, then to the Strategic Headquarters for Space Policy.

The role of the Space Development Commission, which operated within the Prime Minister's Office with the mandate to promote overall space development, was limited to space development under the jurisdiction of the Ministry of Education, Culture, Sports, Science and Technology, following the reorganisation of the ministries in 2001. At the same time, the Council for Science and Technology Policy was established in the Cabinet Office. This Council became responsible for reviewing overall space development in Japan, regarding science and technology policy.

Table 9.1 Allocation of Quasi-Zenith Satellite System R&D budget among four Ministries (Source: Japan's Strategic Headquarters for Space Policy)

	2003	2004	2005	2006	2007	2008	2009	2009*	2010	Total
MEXT	271	331	331	236	331	744	933	372	813	4,363
MIC	150	255	240	157	135	120	153		106	1,318
METI	190	211	237	219	188	126	86		40	1,296
MLIT	40	54	59	58	53	54	39		39	397
Total	651	851	867	670	707	1,045	1,211	372	999	7,372

2009*: supplementary budget
Denomination: 100,000
Rate: 1EURO = 99.63YEN (2011.12.31)

The Council took up an important role during the initial phase of the quasi-zenith satellite programme, for example by giving it the S (top-priority) rank in the FY2003 approximate budget request for science and technology. However, inter-ministry coordination was not easy. Furthermore, the role and main activity of the Council is only to "study and review" basic policies on science/technology, allocate budget and human resources, and "evaluate" R&D results.

On the other hand, the Strategic Headquarters for Space Policy was established by the Aerospace Basic Act. The Act states that the Headquarters has the task to (a) create and promote the Basic Plan for Space Policy, and (b) study and review plans for other important policies, as well as promote the implementation and general coordination of these plans. However, the Headquarters, contrary to the former Space Development Commission, does not have any clearly-stipulated function regarding cost estimates and, consequently several relevant ministries have to separately carry out the necessary coordination with the fiscal authority.

Second, there is the issue of integrating the utilisation and R&D aspects of space programmes. When the Quasi-Zenith Satellite started, the Council for Science and Technology Policy highly praised the significance of research and development, but there was a lack of concrete planning and commitment from the government's side regarding the future operation and utilisation of the system. This situation created additional investment risks that eventually led to the withdrawal of the private sector from the programme. In order to coordinate R&D and utilisation, the operating model of a public/private partnership was presumed concerning the Quasi-Zenith Satellite Programme. However, the exact roles and responsibility-sharing among government ministries responsible for the system's R&D and utilisation, other relevant space organisations, and the private sector, was not clear.

In the meantime, a suitable mechanism is required within the government to collate the needs on the demand side and not on the supply side of space technology, and then to achieve harmonisation between them. On the supply side, there were no less than four Ministries involved in conducting R&D for the programme. These were the Ministry of Internal Affairs and Communications, the Ministry of Education, Culture, Sports, Science and Technology, the Ministry of Economy, Trade and Industry, and the Ministry of Land, Infrastructure, Transport and Tourism. The budget of the quasi-zenith satellite programme for each ministry is shown in the following Table 9.1.

In recent discussions on the creation of a Space Agency, the focus has been on integration on the supply side between the Ministry of Education, Culture, Sports, Science and Technology, and the Ministry of Economy, Trade and Industry. Consequently, there is not enough discussion on how to incorporate the ministries on the demand side of the space policy formation process, namely the Ministry of Land, Infrastructure, Transport and Tourism (control section as consumer), the Ministry of Defence, the Ministry of Agriculture, Forestry and Fisheries, the Ministry of the Environment, and the National Police Agency.

In the Quasi-Zenith Satellite System example, calculations on future services' demand were too optimistic, leading to final withdrawal of the private sector. In the Council for Science and Technology Policy, criticism was raised already from the beginning of the programme on the business feasibility and the demand predictions stated by the private sector. The fact that the private sector and the government ministries that responded to this criticism were basically on the supply side, seems to have hindered an objective analysis and evaluation from a demand side point of view.

Third, there is an issue of cooperation in various space related policy fields, including diplomacy. In the context of the quasi-zenith satellite programme, the regional SE Asia implications are now emphasised, while in the beginning of the programme its overseas use was not thought to be important. On the other side, cooperation with the U.S. was stressed from the beginning, but the as the programme enters its utilisation phase, this partnership is still insufficient. Satellite systems inherently allow for their global use. Consequently, if there had been a linkage of the project with diplomatic and international trade policy objectives from its very beginning, more solid foundations for meeting the needs of diplomacy in the broad sense might have been successfully built. One should acknowledge, however, that substantial initial investments would have been required, in order to fully exploit the system's technological, operational and policy potential from its beginning.

Acknowledgments This article is based on the discussion at the Space Governance Study Group (2010–2011) at the Graduate School of Public Policy at the University of Tokyo and is supported by Research Assistants, Mr. Yuichiro Nagai, Mr. Shoichi Fujimoto, and Mr. Kosuke Sugawara.

Chapter 10
What Future for GEOSS? Results from the November 2010 GEO Ministerial Meeting

Catherine Doldirina

10.1 Introduction

In November 2010, the Geographic Earth Observation (GEO) group that is responsible for the set up and implementation of the Geographic Earth Observation System of Systems (GEOSS) held its Ministerial Summit in Beijing, China, that resulted in the adoption of a series of important documents that form the basis for the actions aimed at making GEOSS operational. They include the GEOSS Data Sharing Action Plan,[1] the Mid-Term Evaluation of GEOSS Implementation,[2] and the Report on Progress[3] regarding the implementation of GEOSS. In the Declaration of 5 November 2010, GEO stressed the importance of the adoption of the GEOSS Data Sharing Implementation Guidelines and the Action Plan for the success of GEOSS and improved access to global Earth observation (EO) data and resources.[4] Adherence to the data sharing principles embodied in them by all actors involved in GEOSS is seen as quintessential for the successful development and operation of the system of systems. The

[1] "GEOSS Data Sharing Action Plan." 3–4 Nov. 2010. Group on Earth Observations 8 Feb. 2012 http://www.earthobservations.org/documents/geo_vii/07_GEOSS%20Data%20Sharing%20Action%20Plan%20Rev2.pdf.

[2] "Mid-Term Evaluation of GOSS Implementation." 3–4 Nov. 2010. Group on Earth Observations 8 Feb. 2012 http://www.earthobservations.org/documents/geo_vii/06_Mid-Term%20Evaluation%20of%20GEOSS%20Implementation%20Rev1.pdf.

[3] "Report on Progress Beijing Ministerial Summit: Observe, Share, Inform." 5 Nov. 2010. Group on Earth Observations 8 Feb. 2012 http://www.earthobservations.org/documents/ministerial/beijing/MS2_The%20GEO%20Report%20on%20Progress.pdf.

[4] "The GEO Beijing Declaration: Observe, Share, Inform." 5 Nov. 2010. Group on Earth Observation 8 Feb. 2012 http://www.earthobservations.org/documents/ministerial/beijing/MS1_The%20GEO%20Beijing%20Declaration.pdf.

C. Doldirina (✉)
Institute of Air and Space Law, McGill University and Space Generation Advisory Council, Milan, Italy
e-mail: kdoldyrina@yahoo.com

P. Hulsroj et al. (eds.), *Yearbook on Space Policy 2010/2011*,
Yearbook on Space Policy, DOI 10.1007/978-3-7091-1363-9_10,

Declaration furthermore urges GEOSS participants to adopt more open domestic data policies that would extend free and unrestricted use of EO data beyond their participation in GEOSS, as well as enable more effective and efficient implementation of its Data Sharing Principles. The endorsement of GEOSS Data Sharing Principles brings the goal of setting up GEOSS much closer to realisation, enabling "open data exchange across different legal traditions and jurisdictions and reducing institutional, legal, and cultural impediments to data sharing".

The research undertaken supports the assertion that the free and unrestricted sharing encouraged by the Principles should become the rule with regard to the use of data that are integrated into GEOSS, from which there should be no derogations. Such an approach is based on the finding of GEO that all other benefits that GEOSS may bring will be limited without a framework enabling access to and use of data.[5] Furthermore, its recognition confirms the suitability of this regime of data use for both data generators and users, whether public or private,[6] particularly if the unjustified tightening and stretching of intellectual property protection is terminated and more consideration is given to the notion of sharing. This stand is reasonable, particularly when taking into account that the extensive use and exchange of EO data and information increases their respective value and contributes to the development of the market for geographic information products.

The necessity of sharing valuable assets, no matter what the source of their origin is, needs to be realised and – more importantly – articulated, since it enables better enforcement of the principle of free and unrestricted access to and use of EO data and information. The latter, no doubt, contributes to the achievement of the common good of prosperity and better life within a society. The analysis of the goals and aims of GEOSS supports the argument that the protection of the proprietary interests of data owners should not be considered as the only or primary factor that shapes the protection granted to data, particular since the use of EO data contributes to the greater good of the society at large. While popularising and enforcing the proper implementation of GEOSS, emphasis should be placed on the interdependence between the necessity to share EO data on the one hand, and the value and success of GEOSS activities for society as a whole on the other.

This chapter focuses on the analysis of the material outcomes (the Declaration and other documents) of the Beijing Ministerial Summit, with the aim of assessing their value and contribution to the development of GEOSS, the related challenges, and the way forward to enable the success of this system of systems. This chapter does not represent an attempt at a comprehensive overview of GEOSS, it rather concentrates on the organisational and institutional aspects of its implementation and their respective strengths and weaknesses, as well as discussing the best strategy for the system's further development. To provide more context to the discussion of these issues, some of the structural aspects of GEOSS' organisation are outlined, followed by an overview of its Data Sharing Principles and their implementation so far.

---

[5] GEOSS Data Sharing Action Plan, 7.

[6] Ibid: 9.

10.2 GEOSS

Some of the most important goals set for the successful operations of GEOSS include the provision of cross-cutting data and information products[7] and services that support decision making, the promotion of "green" economic growth, the management of natural ecosystems and resources, the improvement of food security, and effective response to disasters, etc.[8] The popularity of GEOSS is growing, and the increasing number of participants (over 130 members and participating organisations) show that the initiative has found substantial support across the globe. The progress of its organisation and implementation is also reflected in the action "GEOSS Data Collection of Open Resources for Everyone" initiated at the Beijing Summit, under which more countries and organisations are contributing their data and information into the system of systems.[9] This action can substantially increase the availability of openly accessible data and information as it multiplies their sources, for example with data from the China-Brazil Earth Resources Satellite (CBERS), the Landsat archives, or rainfall data from China's Earth-based 150 RADAR sites.

The technical side of GEOSS and its goals relates to linking traditional information systems and overcoming problems related to access to and integration of disparate datasets. Its policy side encompasses the resolution of issues of open data exchange across different legal traditions and jurisdictions, as well as of institutional, legal, and cultural impediments to data sharing. Both sides of GEOSS and its goals are equally important. However, it is probably easier to agree on technical standards than on the legal principles governing the process of integrating data into the system, as well as accessing and disseminating them. This chapter does not concentrate on the technical side of GEOSS activities, but it rather highlights the achievements, concerns and future development of the exchange of GEOSS data and information on a transparent basis and equal to all conditions.

The GEOSS Common Infrastructure (GCI) was presented to the GEO community as an operational system for the first time at the Beijing Ministerial Summit of 2010.[10] It provides a sustained entry point to GEOSS data and resources via a dedicated web portal. As of mid-2010, over 300 data bases, information services and other resources had been registered in the GCI. According to information from the GEOSS website,[11] GCI consists of web-based portals, clearinghouses that

[7] The list of products is available in the 2009–2011 Work Plan Progress Report, 2–3. "2009–2011 Work Plan Progress Report." 16–17 Nov. 2011. Group on Earth Observations 8 Feb. 2012 http://www.earthobservations.org/documents/geo-viii/05_2009-2011%20Work%20Plan%20Progress%20Report.pdf.

[8] As outlined in the Report on Progress Beijing Ministerial Summit: Observe, Share, Inform.

[9] Including Brazil, China, Japan, South Africa, Spain, UK, USA, ESA, EUMETSAT. Individual contributions are highlighted in the 2009–2011 Work Plan Progress Report, 13–14.

[10] More information see in the 2009–2011 Work Plan Progress Report, 6–7.

[11] "Data Management: Strategic Target." Group on Earth Observations 8 Feb. 2012 http://www.earthobservations.org/geoss_ta_da_tar.shtml.

Fig. 10.1 The 2010 GEO Ministerial Summit held in Beijing, China (Source: GEO)

enable search of data, information and services, as well as registries and capabilities that support access to GEOSS components, standards, and best practices., In order to address the goal of linking traditional information systems and fostering comprehensive, coordinated, and sustained observations of the Earth system, the infrastructure introduces and incorporates necessary framework and operational interfaces. The resources of the GCI are offered by GEO Members and Participating Organizations, which should act in a way to ensure its sustained operation. GCI should be maintained in such a way as to ensure interoperability and therefore effective access to, exchange of and use of data, metadata and products across all GEOSS components.

GEOSS should improve access to data that will increase the engagement and knowledge of users through its GEO Portal, community web portals, and the satellite-based GEONETCast dissemination system.[12] GEO has already managed to achieve important successes in the operational setting up of GEOSS, on the technical (all the portals and GEONETCast), as well as the policy side. A recent development in this regards has been the declarations made by the U.S., China and Brazil who would share data from their satellite systems with GEOSS according to the principle of full and open access to data and information. At the same time, more countries and organisations are preparing to make similar steps. The GEO Portal in particular, will be the single gateway "for users seeking data, imagery and analytical software packages relevant to all parts of the globe". For users with limited or no access to the Internet, similar information will be made available via the network of telecommunication satellites GEONETCast. The GEO Portal will link them in near real time to satellite-based data dissemination systems and thereby to needed data and information (satellite EO, air-borne and in-situ data, metadata and products) that are pooled into the system of systems (Fig. 10.1).

GEOSS is designed to become a global network of data, information and other geographically referenced content from multiple providers – "an extraordinary range of information". This system of systems aims at offering decision-support tools to a wide variety of users, by linking together national and international EO satellites and systems, as well as other sources of information. It is set up to promote common technical standards, in order to achieve the interoperability and coherence

[12] The 2009–2011 Work Plan Progress Report, 10–11.

of data generated from different sources. It will operate through the GEO Portal as the single gateway "for users seeking data, imagery and analytical software packages relevant to all parts of the globe". For users with limited or no access to the Internet, similar information will be made available via the network of telecommunication satellites GEONETCast. GEO Portal will provide a near real time link to satellite-based data dissemination systems (satellite EO, air-borne and in-situ data, metadata and products) that are pooled into the system of systems.[13] When fully operational, the system should be capable of providing users with reliable, up-to-date and easy-to-use information that is indispensable for planning, decision-making, and managing various situations. The data and information will be provided for use under nine different themes: disasters, health, energy, climate, agriculture, ecosystems, biodiversity, water and weather.

10.3 Access to Data

The ultimate goal of establishing GEOSS is to "open data exchange across different legal traditions and jurisdictions and reducing institutional, legal, and cultural impediments to data sharing."[14] This principle is necessary to ensure that the processes of making decisions and taking actions for the benefit of mankind are made on the basis of "coordinated, comprehensive and sustained Earth observations and information"[15] that are available to those who need them. This is one of the most important aims and features of GEOSS that cannot be discarded, but to the contrary should be reflected in and supported by the regulatory framework in which it operates. The common, or in other words communal or societal good of making people information rich (one of the central goals that GEOSS pursues) is emphasised in its Data Sharing Principles, the widest possible implementation of which is crucial to GEOSS' overall success.

The 2007 GEO Capetown Ministerial Summit agreed on the three overarching principles that govern the exchange and use of data within GEOSS. The first is the full and open basis of sharing data, metadata, and products integrated within GEOSS. Full implementation of this general rule may be limited by provisions of "relevant international instruments and national policies and legislation". The second is the principle of "minimum time delay and at minimum cost" at which GEOSS data and information should be made available. The third stipulates that the provision of GEOSS data for research and education purposes should be carried out

[13] "GEONETCast." Group on Earth Observations 1 Feb. 2011 http://www.earthobservations.org/geonetcast.shtml.

[14] "Report on Progress Beijing Summit: Observe, Share, Inform." 5 Nov. 2010. Group on Earth Observations 29 Aug. 2011, 38 http://www.earthobservations.org/documents/geo_vii/geo7_report_on_progress.pdf.

[15] "White Paper on the GEOSS Data Sharing Principles." 27 Sept. 2008. Group on Earth Observations 1 Feb. 2011, lines 14–16 http://www.earthobservations.org/documents/dsp/Draft%20White%20Paper%20for%20GEOSS%20Data%20Sharing%20Policies_27Sept08.pdf.

"free of charge or at no more than cost of reproduction". The Implementation Guidelines for the GEOSS Data Sharing Principles provide the interpretation of the principles that corresponds to the vision and goals of GEOSS. The members and participating organisations are advised to follow them when they set up their relationships with GEO and contribute data to GEOSS. This makes the fulfilment of otherwise non-binding GEOSS Data Sharing Principles more realistic and coherent. The third paragraph of the GEO Beijing Declaration expresses the intention to "maximize the number of documented datasets made available on the basis of full and open access".

Apart from ensuring that access to EO data and information integrated in GEOSS should be free and open, this principle brings forward the necessity of enabling the user of data to extract the most value from it by granting him the essential freedom to re-use and re-disseminate it.[16] The possibility of imposing restrictions on these activities should be kept to a minimum. Furthermore, the principle of open and unrestricted access to GEOSS data should also be applicable to data (when integrated into the system) from private[17] and mixed[18] sources. This decision reflects the finding that the principle of sharing should be applicable to the utilisation and distribution of EO data notwithstanding its source, because it makes its use more beneficial for all stakeholders and thereby contributes to the achievement of the common good. The drafting history shows that the principle of free and open access to GEOSS data is based on the premise that the shared data and information represent a public good.[19] This choice is made because GEOSS helps to disseminate EO data as widely as possible, thereby maximising societal benefits from its use.[20]

However, the voluntary nature of GEOSS' implementation was found by the GEO Executive Committee to be potentially detrimental to the sustainability of GEOSS. Therefore, additional measures and actions need to be carried out in order to ensure that the adoption of these principles and their undisrupted enforcement is secured. The Data Sharing Implementation Plan, also adopted in the course of the Beijing Ministerial Summit, restates and elaborates upon the GEOSS 10-Year Implementation Plan by stating that "the societal benefits of Earth observations cannot be achieved without data sharing". The concrete steps that need to be pursued by the GEO, its members and participating organisations are outlined in

---

[16] "Implementation Guidelines for the GEOSS Data Sharing Principles." 17–18 Nov. 2009. Group on Earth Observations 29 Aug. 2011, Section 2 http://www.earthobservations.org/documents/geo_vi/07_Implementation%20Guidelines%20for%20the%20GEOS%20Data%20Sharing%20Principles%20Rev2.pdf.

[17] Section 2.2 of the Implementation Guidelines for the GEOSS Data Sharing Principles, supra note 673; White Paper on the GEOSS Data Sharing Principles, supra note 659, lines 645 ff.

[18] White Paper on the GEOSS Data Sharing Principles, lines 606 ff.

[19] Ibid: supra note 659, lines 539–543. See also Section 4 of Implementation Guidelines for the GEOSS Data Sharing Principles, supra note 673, where they also refer to the concept of the public good with regard to GEOSS data pricing policy.

[20] White Paper on the GEOSS Data Sharing Principles, lines 545–550.

this document that provides the practical aspects of achieving planned results. The GEO Secretariat believes that its members and participating organisations need to be familiarised with examples of success stories regarding data access and the use, re-use, sharing and dissemination of information on a free and unrestricted basis. This will enable them to "demonstrate benefits and entice data contributors to voluntarily and enthusiastically adhere to the Data Sharing Principles".[21] One of the ways to show these benefits, which was recognised by GEO and declared as necessary for implementation, is to monitor and document the use and impact of resources made available through full and open access and widely distribute the results that they help to achieve.

10.4 Challenges

According to the Beijing Declaration, the success of GEOSS and improved access to global EO data and resources relies on the adoption and enforcement of the GEOSS Data Sharing Implementation Guidelines and of the Action Plan. Adherence to the data sharing principles by all actors involved in GEOSS is quintessential for the successful development and operation of the system of systems, because only such an approach may result in the satisfaction of the truly global scale needs that GEOSS is aiming to achieve as an invaluable information resource.[22] As highlighted above, data should be accessible with minimal time delay and with as few restrictions as possible, on a non-discriminatory basis, at a minimum cost of no more than the cost of reproduction and distribution. For this purpose, more open domestic data policies should be adopted to effectively implement the Data Sharing Principles.

Some of the implementation concerns that were identified in the course of the GEOSS mid-term evaluation[23] include: the voluntary nature of GEOSS (no legal mandate) that nonetheless was beneficial at the stage of engaging partners; the lack of sufficient resources to sustain efforts into the future (both human and financial, e.g. the lack of a formal budget); the inadequate dissemination of progress evidence demonstrating the value-added results unique to the implementation of GEOSS and the positive return on investment; the absence of an explicit end-to-end process that utilises existing resources in support of the overall GEOSS vision and goals; and the lack of a comprehensive gap analysis regarding the implementation approach adopted or the relevant EO operational requirements.[24] These issues need to be addressed, explained and resolved, in order for GEOSS to successfully move

[21] GEOSS Data Sharing Action Plan, 16.

[22] Ibid: 11.

[23] The GEOSS Mid-Term Evaluation Report. June, 2010. It is suggested by some experts that GEO should organize and carry out the transition from "voluntary acceptance" into a priority "commitment to action".

[24] GEO is aware of these concerns or hurdles and has attempted to address them or to outline the methodology for addressing them in the GEOSS Mid-Term Evaluation Report from 3 to 4 Nov. 2010.

forward, and this is the most challenging task for GEO following the successful decisions made at the Beijing Ministerial Summit of 2010.

Other challenges arise from the actual milestones of implementing GEOSS and making it fully operational. GEO needs to sustain comprehensive and coordinated EO networks, as well as GEOSS components and information infrastructure that require resources and commitment. Furthermore, GEOSS infrastructure needs to accommodate the need for timely, global and open data sharing across borders and disciplines, within the framework of national policies and international obligations. For this purpose, interoperability amongst observational, modelling, data assimilation and prediction systems needs to be achieved and sustained. In addition, GEO needs to foster research and development activities and ensure the coherent planning of future observation and information systems – goals that require actions on all levels involving the provision, exchange and use of GEOSS data and information. One of the challenges in achieving this concerns pooling national, regional and global investments in scientific and technological advances and innovative approaches for upgrading and expanding EO. Since GEOSS is a global system, it will face the issue of the different capacity levels of its members, participants and users. Capacity building is hence another challenge that the GEO will need to resolve for the successful implementation of GEOSS, as well as for the wider adherence to Data Sharing Principles.[25] At the same time, the GEO needs to develop mechanisms to ensure that the members who have contributed more resources, technology or other types of input into the development of GEOSS do not suffer from the free-riding of smaller or less developed members. Nevertheless, this concern may not yet represent a significant problem today, particularly due to the very active involvement of countries like Brazil and China and their engagement in helping developing countries (particularly in Africa) to participate in GEOSS too.

The GEO Beijing Report on Progress lays down some of the key actions and activities that need to be undertaken to meet these challenges and make GEOSS a true success,[26] while it is reinforced and updated by the details included in the new GEO 2012–2015 Working Plan.[27] For instance, policy-makers and managers need to make use of GEOSS to support the decision-making process.[28] Environmental

[25] "GEOSS Strategic Targets." 17–18 Nov. 2009. Group on Earth Observation 8 Feb. 2012 http://www.earthobservations.org/documents/geo_vi/12_GEOSS%20Strategic%20Targets%20Rev1.pdf.

[26] To achieve this, primarily and above all, "strong, high-level support from governments and leading organizations is vital for maintaining the momentum generated by GEO and ensuring that GEOSS becomes recognized as a vital infrastructure that serves the global public good. Building the capacity of users to exploit GEOSS is also essential."

[27] "GEO 2012–2015 Work Plan." 13 Dec. 2011. Group on Earth Observations. 8 Feb. 2012 http://www.earthobservations.org/documents/work%20plan/GEO%202012-2015%20Work%20Plan_Rev1.pdf.

[28] "The true value of GEOSS is its ability to support decision–making. As a key user group, senior policymakers can help to ensure that GEOSS addresses the UN's Millennium Development Goals and other priority issues facing the global community." This message needs to be clearly articulated and transmitted to all those that can potentially benefit from using GEOSS.

experts have to be in a position to trust GEOSS and its Common Infrastructure and treat it as a unique and essential tool for accessing EO data and information. On an operational level, GEO needs to attract both public and private resources in order to support capacity-building, strengthen national EO programs, and maintain the GEOSS shared architectural and information infrastructure components. Additional efforts should be made to emphasise the collective spirit that unites all those participating in GEOSS and promote its success. In more practical terms, it is essential for GEO to establish an effective governance structure for the post-2015 period.

10.5 Ways Forward

Form-wise, GEO needs to think strategically about the desirability of and potential for its evolution into an independent organisation under the United Nations system, in order to facilitate its relationships with its member states and participating organisations, as well as with the "outside" world in general. Operation-wise, the GEO community should demonstrate that its activities have the potential to, or in fact do, lead to new in-demand applications[29] and additional jobs, while fostering a more open competition environment, as opposed to the old model of data protection. The accomplishment of this step (or at the very least the initiation of a process that could lead to it) will bring forward the realisation of the necessity to fully implement Data Sharing Principles as a prerequisite for ensuring free and open exchange of data, metadata and products made available through the GEOSS.[30] The 2012–2015 GEO Working Plan, adopted in December 2011, points out that one of the most important milestones to achieve during this period is the development of "flexible national and international policy frameworks to ensure that a more open data environment is implemented."[31] The Data Sharing Task Force leads the execution of this task.

The Implementation Strategy adopted specifically for the GEOSS Data Sharing Principles emphasises that evidence from pilot projects launched or carried out recently is necessary to draw the attention of member states and prove to them that thinking about society at large pays off for its individual members as well. This will reinforce the notion that the common or public good aspect of GEOSS, which is the only capable way of truly unfolding its benefits, requires data sharing, as opposed to protecting the interests of data generators alone. Furthermore, the promotion of

[29] As per findings within the GEOSS Mid-Term Evaluation Report, 14.

[30] The concern is important and cannot be put aside. For instance, 29% of key informants and 19% of survey respondents also recognized the Data Sharing Principles as one of the most important challenges to overcome in GEOSS implementation. The challenge was to create a discussion forum and develop guidelines agreeable to all parties, in light of the divergence between national laws and policies.

[31] GEO 2012–2015 Work Plan, 16.

GEOSS and the process of making it operational and developing it further should engage the users of GEOSS data and services,[32] and focus on the users' operational requirements and the development of end-products tailored to them. This is particularly true when looking at the GEOSS goal of acting as a decision-making support on a global scale. Improving the working relations between individual GEOSS users and data providers will contribute to the gradual adoption of balanced domestic data access policies that equally respect the interests of data generators and data users through increased access and use opportunities. Seeing and experiencing the concrete benefits that the GEOSS operation can deliver will be the best way towards implementing the principle of full and open access data, as the only approach capable of reconciling individual and societal well-being.

A practical strategy to promote the principle of free and unrestricted access to GEOSS data and information necessitates outreach, particularly to users at the grassroots level who may contribute to the popularisation of GEOSS and to lobbying for the adoption of the principle by their respective organisations or governments. In this direction GEO should assess the possibility of cooperation with other organisations that promote access to and wide use of EO data, or space activities in general. One of the options could be a targeted partnership with, for example, the Space Generation Advisory Council whose network of students and young professionals encompasses all the continents. Its members could help spread the message about the goals and aims of GEOSS, as well as highlight the benefits of its use to the relevant communities.

The limited analysis undertaken above does not allow making appropriate suggestions for all areas where action needs to be taken in order to facilitate GEOSS operational use and development. Nevertheless, some thoughts regarding the most suitable strategy for the popularisation of the GEOSS Data Sharing Principles need to be articulated. One of the most important issues that GEO needs to formulate and promote is the establishment of a conceptual link between the successful implementation of the GEOSS Data Sharing Principles and the promotion of overall societal prosperity. The societal benefits that can be derived from the use of data integrated by GEOSS should be assessed and compared to the benefits of introducing and enforcing restrictions on access that aim to protect the proprietary interests of the few. Taking into account that GEOSS will ideally integrate data generated by publicly owned sources, as well as by private satellite operators or other providers, respect for the greater societal benefits should be decisive when adopting and enforcing regimes governing the distribution and use of EO data. The nature and the goals of GEOSS should be presented as evidence of the fact that "private" and "communal" interests are not necessarily antagonistic. Society protects private or proprietary interests because it decides that their existence will be beneficial for the society in general and contributes to the common good for which society exists. In the case of data, retaining very strict conditions of

[32] Mid-Term Evaluation of GEOSS Implementation, Recommendation 6, 10.

access and use will in the long run also undermine interests of their owners, as complete exclusion is unnatural to handling data.

Consequently, sharing EO data could be instrumental to achieving an information rich society, or (in the terms of the documents setting up GEOSS) at least to providing decision-making resources and tools.[33] At the same time, because of its importance in the overall process of acquiring and using knowledge, EO data should itself be regarded as a common good. In addition, and in abstraction from the situations that involve handling of data and information, sharing constitutes an element of healthy relations among individuals that leads to the improvement of society. It is the nature of humans to share things amongst themselves, so that providing the possibility to do so through the adoption and implementation of the GEOSS Data Sharing Principles unmistakably has the potential to contribute to this process. For the nations, sharing is also important and sometimes in fact stipulated. Since a large part of EO data integrated into GEOSS will be generated by satellites, principles of international space law should be applicable and relevant for this type of activity. Since "use of outer space ... shall be carried out for the benefit and in the interests of all countries",[34] then if unrestricted and free access to and sharing of data is beneficial for all participants, then it is this interest that should be the basis of adhering to GEOSS Data Sharing Principles. The common good itself should also be presented and marketed as a key criterion for shaping the data policies of GEO members (at least within the framework of relationships with the system of systems) so that they correspond to the main requirements set by the GEOSS Data Sharing Principles.

The complexity of GEOSS and its all-encompassing ambitions regarding the range of uses, applications and fields of activities for which the integrated data can be accessed, analysed and disseminated require us to understand how important its benefits can be to societies worldwide. In some cases, an accurate assessment of the needs of the user community might not be readily calculable. In others, however, quantifying the benefits of using EO data when responding to disasters against the disadvantages of their absence can be expressed in precise monetary values that could be used as evidence to promote GEOSS' effectiveness. Both tangible and less measurable benefits should be highlighted when arguing in favour of adopting the GEOSS Data Sharing Principles as a global standard for using EO data; especially as they speak in favour of the ability of GEOSS to contribute to the achievement of the common good.[35]

---

[33] For the author the latter is narrower than the former.

[34] Article 1 Treaty on Principles Governing the Activities of States in the Exploration and Use of Outer Space, including the Moon and Other Celestial Bodies. 27 January 1967. 610 U.N.T.S. 205.

[35] Case studies of various GEOSS Societal Benefit Areas represent one of the best logical ways to highlight how use of GEOSS EO data and information services based on them will benefit its users worldwide once GEOSS becomes fully operational. Focus on the practical achievements that GEOSS contributes, enables assessment of the societal value or benefits of each of the GEOSS nine areas. Adopting this approach will help in advocating wider access to data integrated in GEOSS and the increased possibility of its use and further dissemination.

10.6 Conclusion

Societies all around the world have a vested interest in secure access to EO data, information and related services, because its use contributes to the sustainable development, along with other benefits, of society at large, as well as of its individual members who are GEOSS users. Providing resources and tools for informed decision-making and other activities that require access to and use of EO data and information requires recognising the principle of sharing as part of the legal regime governing EO activities and the relationship of GEOSS with its participants. The plea to nation states, relevant agencies and other involved organisations to open up access to EO data, particularly for the purposes of the successful operations of GEOSS, can reach its aim if it is focused on the realisation and acceptance of the necessity of making decisions that accommodate the interests not only of the actual or potential owners of information, but also of its users. (NB: sentence too long) This is particularly true because EO data and information are of little value if they are not used, and the benefits of their use cannot be achieved without data sharing. The rules governing the distribution and use of data protected in this way, particularly within projects or initiatives comparable to GEOSS, should incorporate the interests of the users. This is especially true as such data and information can be of great value to society as a whole.

GEOSS is a unique initiative that has a tremendous potential to increase the use of EO data and information for the benefit of the entire society. This is a noble aim that sometimes seems unrealistic for those involved in EO activities. Therefore, this should be precisely the line of action of GEO: to convince its members and participating organisations that the adoption and enforcement of the principle of free and open access to EO data and information is more practical than its extensive protection. Even more so since it enables the extensive use of these resources and allows for more actors to engage in EO activities, especially in the field of the analysis and use of geographic information. The success cases in GEOSS operations that can be witnessed already today can be utilised to provide evidence of the viability of this approach. In addition, the rhetoric on the long-term benefits of open and free access to EO data and information needs to be highlighted, both for public and private actors potentially involved with GEOSS, in contrast to the short-term achievements that other regimes for treating geographic information may demonstrate.

Chapter 11
Digital Divide and the Global Crisis

Julia Neumann

11.1 Introduction

Discrepancies in access to telecommunications have been perceived since the late 1970s and gained attention when two Commissions set up by UNESCO and the ITU respectively presented their reports: the "McBride-Commission"[1] in 1980 and the "Maitland-Commission"[2] in 1984. The latter observed that two-thirds of the world population had no access to telephone services, and that Tokyo had more telephones than the whole of the African continent.[3] In the age of digitisation, these discrepancies are usually referred to as the "digital divide."

In the last decade of the previous century, the digital divide and the underlying problems were on the agenda of several UN organisations and received considerable public attention. In particular, the digital divide is addressed in the UN Millennium Development Goals of 2000: Goal 8 aims at developing "a global partnership for development", and its Target 8.F foresees that the benefits of

This article is a summary of the findings of the author's doctoral thesis "Bridging the Digital Dive – Eine Analyse völkerrechtlicher Ansprüche und internationaler Solidarstrukturen zur Überbrückung des digitalen Grabens mittels Satellitenkommunikation", to be published by Wolters Kluwer in 2012 (in German).

Rechtsanwältin (Attorney-at-Law) with White & Case LLP, Berlin. The content of this paper reflects the personal views of the author and is no statement on behalf of White & Case whatsoever.

[1] United Nations Educational, Scientific and Cultural Organization, ed. Many Voices, One World. Oxford: Rowman & Littlefield, 1980.

[2] Cf. Res. No. 20 of the Plenipotentiary Conference of the International Telecommunication Union, Nairobi, 1982.

[3] International Telecommunication Union. The Missing Link. Geneva: ITU, 1984: 13.

J. Neumann (✉)
White & Case LLP, Berlin, Germany
e-mail: jneumann@berlin.whitecase.com

P. Hulsroj et al. (eds.), *Yearbook on Space Policy 2010/2011*,
Yearbook on Space Policy, DOI 10.1007/978-3-7091-1363-9_11,

new technologies, especially related to information and communications, shall be made available in cooperation with the private sector.

Moreover, 21 December 2011 marked the 10-year anniversary of UN General Assembly Resolution 56/183, endorsing the holding of the World Summit on the Information Society (WSIS). The Summit was held in two phases, in Geneva from 10 to 12 December 2003 and in Tunis from 16 to 18 November 2005. Together, the two sessions of the WSIS produced four outcome documents: the Geneva Declaration of Principles and the Geneva Plan of Action, as well as the Tunis Agenda and the Tunis Plan of Action. None of them is in itself a legally binding document of public international law. Albeit on a non-binding level, WSIS has led to a follow-up process, by which progress in bridging the digital divide shall be observed and furthered.[4]

It is worthwhile taking a closer look at the legal regime that might provide tools for bridging the digital divide, specifically by asking whether there are legal claims of developing countries with regard to satellite communications. This question is particularly interesting against the background of the global financial crisis, which might render developed states more reluctant to commit to international cooperation in the form of binding legal provisions.

11.2 The Digital Divide: What Is It?

The term "digital divide" refers to the gap between those who have access to the latest information technologies and those who do not.[5] A digital divide can exist either between different countries (industrial nations/developing countries), or between regions of a single country (urban/rural areas). This article is only concerned with the international dimension of the divide, however.

While there is consensus that the term digital divide refers to differences in access to hardware, some authors give it a broader meaning. The concept can be extended to discrepancies in the ability to use information and communication technologies (ICTs), in the actual use of ICTs, and in the effective use of ICTs. However, the differences in hardware access will be the sole focus of the following analysis.

[4] "WSIS Follow-up." United Nations Conference on Trade and Development 19 Oct. 2011. http://www.unctad.org/templates/Page.asp?intItemID=4239.

[5] Compaine, Benjamin M. The Digital Divide: Facing a Crisis or Creating a Myth? Cambridge and London: MIT Press, 20011: xi.

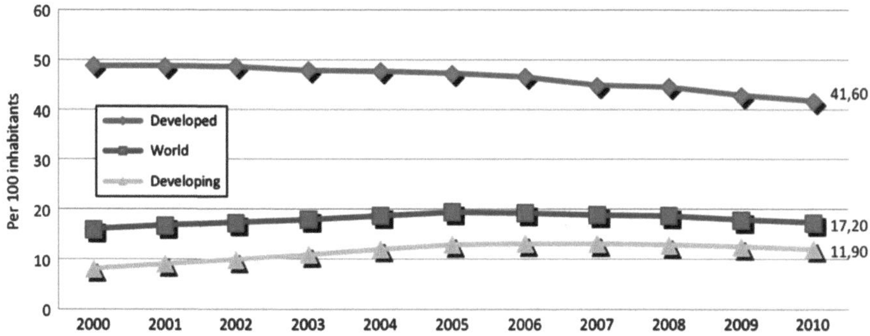

Fig. 11.1 Fixed telephone lines per 100 inhabitants, 2000–2010 (Source: ITU)

11.3 Digital Divide: Facts and Figures

Many regions of the world still have only very limited telecommunications services' penetration,[6] with major differences between developing and developed countries: in 2010, developing countries had only 11.9 fixed telephone lines per 100 inhabitants, whereas developed countries had 41.6.[7] Per 100 inhabitants, developing countries counted 70.1 mobile phone users as opposed to 114.2 mobile phone users in developed countries.[8] With regard to internet use, developing countries had 21.1 and developed countries 68.8 internet users per 100 inhabitants.[9] There is only very limited fixed internet access in developing countries and, if available, it is still very slow and/or expensive. Fast (broadband) connections are a rarity and mobile broadband penetration in most developing countries is still insignificant, while it increases rapidly in developed countries (Figs. 11.1, 11.2, and 11.3).[10]

Whereas the divide is diminishing with regard to certain technologies, such as mobile telephony, it persists with regard to other technologies. Viewed in total and not sectoral figures, the digital divide has been reduced only marginally[11] and it continues to exist. The disparities are such that they will not vanish in the

[6] Malanczuk, Peter. "Telecommunications, International Regulation." Encyclopaedia of Public International Law (EPIL) Vol. IV. Ed. Rudolf Bernhardt. Amsterdam: North Holland, 2000: 791, 796.

[7] "World Telecommunication/ICT Indicators Database." International Telecommunication Union. 19 Oct. 2011. http://www.itu.int/ITU-D/ict/statistics/.

[8] Ibid.

[9] Ibid.

[10] International Telecommunication Union. ICT Development Index 2009. Geneva: ITU, 2009: 1.

[11] Ibid: 45.

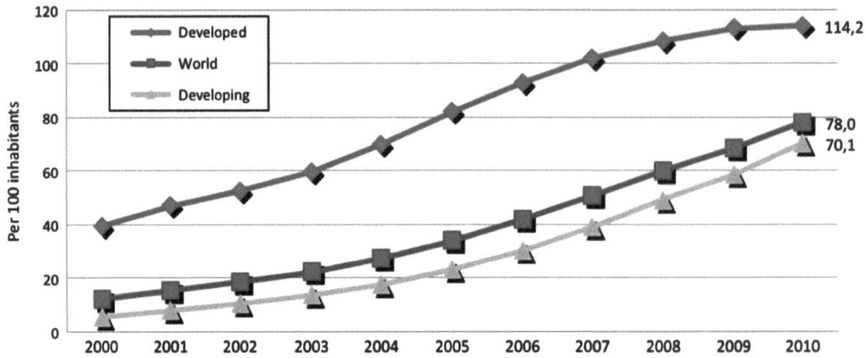

Fig. 11.2 Mobile cellular subscriptions per 100 inhabitants, 2000–2010 (Source: ITU)

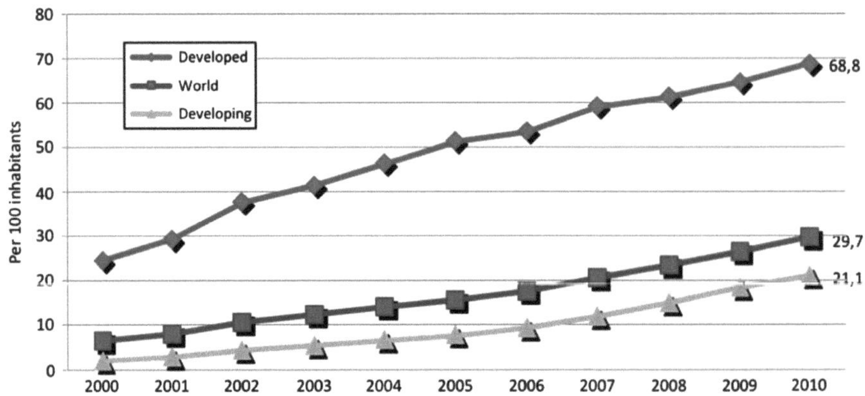

Fig. 11.3 Internet users per 100 inhabitants, 2000–2010 (Source: ITU)

foreseeable future without additional efforts. Moreover, the ICT sector is characterized by continuous technological progress. New developments and patents (e.g. LTE,[12] WiMAX[13]) are announced almost daily. With every new technological development, the problem of unequal access becomes even greater.[14]

[12] Long Term Evolution (LTE) is a technology supposed to replace UMTS, which is still classified as third generation.

[13] Worldwide Interoperability for Microwave Access (WiMAX) is a new standard for regional networks.

[14] Kelly, Tim. "Twenty Years of Measuring the Missing Link." Maitland+20. Ed. Gerald Milward-Oliver. Bradford on Avon: Anima Centre, 2005: 23, 27.

11.4 Digital Divide: Why Is It a Problem?

The effects of the digital divide pose problems from different perspectives:

For developing countries, the digital divide has economic as well as socio-political implications. ICTs can be used to promote economic growth[15] and telecommunications infrastructure and services are widely acknowledged as indispensable tools for the socio-economic and cultural development of a country, offering unprecedented opportunities for the betterment of the citizen's quality of life.[16] Studies show that a 10% increase in high-speed internet connections results in economic growth of 1.3% in developing countries, yet many people live in rural areas without access to such services, or they are too poor to afford it.[17]

For developed countries, the digital divide is a matter of indirect interest, in the sense that their export potential is proportionally dependent on the increasing prosperity of developing countries.[18] Whereas this interdependence might lead to more competition and less tax income for developed countries, the latter's industry can nevertheless benefit from the growing prosperity of developing countries. Specifically with regard to telecommunications, it is generally recognised that network externalities rise with a growing number of participants.[19] Whereas these interests might not render action on behalf of developed countries necessary, it can reduce their freedom of action, especially in an increasingly interdependent world.

Bridging the digital divide is also *in the interest of the international community as such.* This term does not describe the interests of a specific group of states, or of the sum of individual states, but the genuine interest of the entire international community.[20] An international community interest exists where states face

[15] International Telecommunication Union. Revised Draft of the ITU Strategic Plan for 2008–2011. Doc. PP-06/28 of 27 June 2006: para. 2.1.7.

[16] Jakhu, Ram S. "Safeguarding the Concept of Public Service in View of Increasing Commercialization and Privatization of Space Activities, with Particular Attention to the Global Public Interest & the Needs of Developing Countries." 5 Sing. J. Int'l & Comp. L.: 71.

[17] "Goal 8: Develop a Global Partnership for Development by 2015." World Bank 19 Oct. 2011 http://www.worldbank.org/mdgs/global_partnership.html.

[18] Bleckmann, Albert. "Souveränitätsprinzip." AVR 23 (1985): 450 (473); Hailbronner, Kai. "Foreign Investment Protection." Reforming the International Economic Order. Eds. Thomas Oppermann, and Ernst-Ulrich Petersmann. Berlin: Duncker & Humblot, 1987: 99, also points to the opening up of new markets by foreign investment.

[19] Röller, Lars-Hendrik, and Leonard Waverman. "Telecommunications Infrastructure." Discussion Paper No. 2, March 2000: 14; Yu, Peter K. "Terrorism and the Global Digital Divide." 11 Feb. 2002. FindLaw 19 Oct. 2011 http://writ.news.findlaw.com/commentary/20020211_yu.html.

[20] Tomuschat, Christian. "Obligations Arising for States without or against their Will." RdC 241 (1993): 195, 222.

Simma, Bruno. "Community Interest." RdC 250 (1994): 217, 244; Delbrück, Jost. "Laws in the Public Interest." Liber amicorum Günther Jaenicke. Eds. Volkmar Götz, Peter Selmer, and Rüdiger Wolfrum. Berlin: Springer, 1998. 17 et seq.; Dahm, Georg, Delbrück, Jost, and Rüdiger Wolfrum. Völkerrecht. Vol. I/3. Berlin: de Gruyter, 200:. 778.

problematic situations on a global scale that they cannot solve by themselves,[21] e.g. underdevelopment. The international community's interest in bridging the digital divide can be particularly perceived with regard to the correlation of the lack of access to ICTs and other major problems concerning the international community, such as underdevelopment. To that end, the final WSIS outcome documents emphasize the growing importance of ICTs as a development enabler, and as a tool for achieving the internationally agreed development goals and objectives, including the Millennium Development Goals (MDGs).[22] ICTs' role in promoting peace, security and stability, in enhancing democracy, social cohesion, good governance and the rule of law, at national, regional and international levels, has also been underscored by the ITU.[23] These problems, which individual states cannot solve themselves, could thus be addressed in a joint effort by the international community to bridge the digital divide. Especially viewed in connection with development and peace-keeping, bridging the digital divide by promoting access to telecommunications can thus be perceived to be in the interest of the international community as a whole.

11.5 Addressing the Problem in Times of Global Crisis

The digital divide can be bridged by various technological means, yet satellites offer some specific advantages making them extremely apt for this purpose. The major advantages of satellites are their broad geographical coverage, the avoidance of expensive terrestrial infrastructures, such as cable, and the practically unlimited number of transmission recipients, coupled with the possibility of building up global networks.[24] Moreover, satellites are the only means to connect states that are not linked by optical fibre networks to the international communications network ("lifeline connectivity").[25] As an integral part of the global "information highways", satellite communications complement the linking of distant regions

[21] Simma, Bruno. "Community Interest." RdC 250 (1994): 217, 235, also points to the common nature of the problematic situations.

[22] Tunis Commitment, WSIS-05/TUNIS/DOC/7-E of 18 November 2005 http://www.itu.int/wsis/docs2/tunis/off/7.html, and Tunis Agenda for the Information Society, WSIS-05/TUNIS/DOC/6 (Rev.1)-E of 18 Nov. 2005 http://www.itu.int/wsis/docs2/tunis/off/6rev1.html.

[23] International Telecommunication Union. Revised Draft of the ITU Strategic Plan for 2008–2011. Doc. PP-06/28 of 27 June 2006, para. 2.1.7.

[24] Von Kries, Wulf, Schmidt-Tedd, Bernhard, and Kai-Uwe Schrogl. Grundzüge des Raumfahrtrechts. München: Beck, 2002. 155; Hudson, Heather E. "From Missing Links to Digital Divides." Maitland+20. Ed. Gerald Milward-Oliver. Bradford on Avon: Anima Centre, 2005. 37, 41.

[25] Baumann, Ingo. Das internationale Recht der Satellitenkommunikation. Frankfurt/Main: Lang, 2005. 548.

through submarine cables,[26] and alongside optical fibre constitute the backbone of the information society.[27] Satellites are thus perfectly well suited for bridging the digital divide.

11.6 What's of Interest for Developing Countries?

In order to bridge the digital divide by using satellite communications, developing countries can either create their own infrastructure, or participate in and benefit from existing infrastructures.

With respect to the creation of their *own infrastructure,* developing countries may be specifically interested in *access to the most favourable orbit/spectrum resources,* as well as in the transfer of technology. In terms of orbital positions for telecommunications, the Geostationary Orbit (GEO) is usually preferred since it enables global coverage with only a few satellites of a considerably long life span (of approximately 15 years). Located at an altitude of roughly 36,000 km above the equator, satellites revolve at the same rate as the Earth, so that they appear stationary when viewed from the ground.[28] For an almost complete global coverage, three satellites are sufficient.[29] Due to the physical characteristics of radio waves, only certain frequencies are suitable for communications via satellite.[30] Developing countries may also have an interest in *technology transfers*, so as to take an active part in the development and operation of satellite systems and to participate in their industrial development.[31] However, it should be kept in mind that intellectual property rights (IPRs)[32] in technologies are usually vested with subjects of private law. Transfer of intellectual property comprises the transfer of technical literature, the education of local workers, the delegation of technicians, engineers and managers, the supply of industrial goods and/or industrial plants, and the provision of patents and know-how licenses, technical assistance, and direct

---

[26] Malanczuk, Peter. "Telecommunications." IV. Ed. Rudolph Bernhardt. Amsterdam: North Holland, 2000: 791, 795.

[27] Baumann, Ingo. Das internationale Recht der Satellitenkommunikation. Frankfurt/Main: Lang, 2005. 547.

[28] Paterna, Mischa. Globalisierung der Telekommunikationsmärkte – Internationalisierung der Netzbetreiber am Beispiel der Deutsche Telekom AG. Doctoral Disertation, St. Gallen, Hallstadt, 1996: 59.

[29] Wolfrum, Rüdiger. "Geostationäre Umlaufbahn." Handbuch des Weltraumrechts. Ed. Karl-Heinz Böckstiegel. Cologne: Carl Heymanns Verlag, 1991: 351.

[30] Smith, M. L. "International Regulation of Satellite Communication." Utrecht Studies in Air and Space Law 7 (1990): 7.

[31] Cf. Matte, Nicolas Mateesco. "Aerospace Law: Telecommunications Satellites." RdC 1980: 123, 215.

[32] For a definition of "intellectual property", see Art. 2 (viii) of the Convention Establishing the World Intellectual Property Organization of 14 July 1967, as amended on 28 September 1979.

investments (industrial cooperation, joint ventures, subsidiaries).[33] The transfer of know-how is particularly important, and is defined as "the sum total of specialist knowledge, operational experience, and the technical production of a manufacturer."[34] Irrespective of the failure of negotiations on an international Code of Conduct on the Transfer of Technology in 1985, technology transfer to developing countries has been a recurrent theme in the multilateral discussions of recent years.[35]

With a view to *participating in existing infrastructures*, developing countries will have a strong interest in being provided satellite communications services at good quality and at no or little cost ("global universal service").

11.7 Legal Claims of Developing Countries in Public International Law?

Public international law[36] has the potential to play an important role in bridging the digital divide, since it regulates the behaviour between sovereign states. It can serve as a tool for bridging the digital divide particularly by enabling developing countries to make legal claims concerning satellite telecommunications. Such claims can result from all the sources of public international law listed in Art. 38 (1) of the Statute of the International Court of Justice, and especially from international agreements and rules of customary international law.[37]

11.7.1 What Is a "Claim" in Public International Law?

Any investigation into legal claims by developing countries raises the question of what constitutes a legal claim in public international law. There has been very little legal analysis on this topic so far.[38] An analysis of the theories on subjective rights

---

[33] Schmidt-Diemitz, R. "Geistiges Eigentum und entwicklungspolitischer Wissenstransfer."GRURInt 1988: 287, 290.

[34] Stumpf, H. The Know-How Contract in Germany, Japan and the United States. Deventer: Kluwer Law and Taxation Publishers, 1984: 10.

[35] Roffe, P. and T. Tesfachew. "Revisting the Technology Transfer Debate: Lessons for the New WTO Working Group." Bridges 5.2 (2002): 5.

[36] Particularly international telecommunications law, international economic law (esp. within the WTO), and space law.

[37] Since a detailed analysis would go beyond the scope of this paper, focus will be on the main aspects.

[38] Cf. Bleckmann, Albert. Allgemeine Staats- und Völkerrechtslehre: Vom Kompetenz- zum Kooperationsvölkerrecht. Köln: Heymann, 1995: 579, who has dealt intensively with the "subjective right" in public international law.

and the transposition of their criteria to public international law shows that the concept of a legal claim as a subjective right does in fact exist in public international law. Its elements can be summarised as follows: (1) there has to be a legal norm of public international law, (2) this legal norm must name a person/entity that is entitled and one that is obliged, and (3) the norm must contain an obligation automatically correlating with a right.[39]

11.7.2 Access to the Orbit/Spectrum Resource

For developing countries that are members of the ITU, access to the orbit/spectrum resource is to a certain degree ensured by Art. 44(2) of the ITU Constitution,[40] but only within the limits of allotment plans and allocation of current spectrum and orbital positions. Under general international law, and particularly under the principles of sovereignty and territorial jurisdiction, the regulation and control of national telecommunications belongs to the *domaine réservé* of states.[41] Therefore, each nation state is responsible for the assignment[42] of frequencies to certain users, while the ITU itself is not in a position to assign the frequencies.[43] International coordination is necessary to avoid conflicts and enable the most efficient use of these resources.[44] The international recognition of a nationally assigned frequency is vital for its unobstructed use, free from harmful interference from other users.

Pursuant to Art. 44 (2) of the ITU-Constitution (CS), "Member States shall bear in mind that radio frequencies and any associated orbits, including the geostationary orbit, are limited natural resources", which "must be used rationally, efficiently and economically [...] so that countries or groups of countries may have equitable access to those orbits and frequencies, taking into account the special needs of the developing countries and the geographical situation of particular countries." Art. 44 (2) emphasises, on the one hand, that the frequency/orbit spectrum is physically limited. On the other hand, Art. 44(2) underscores the importance of the rational,

[39] A detailed analysis can be found in Neumann, Julia. Bridging the Digital Divide. Cologne: WoltersKluwer, forthcoming 2012.

[40] Constitution of the International Telecommunication Union, done in Geneva on December 1992 as amended in Kyoto 1994, Minneapolis 1998 and Marrakesh 2002.

[41] Malanczuk, Peter. "Telecommunications." Encyclopeadia of Public International Law IV. Ed. Rudolf Bernhardt. Amsterdam: North Holland, 2000. 791, 792; Jakhu, Ram S. and Virginia R. Serrano. "International Regulation of Radio Frequencies for Space Services." Project 2001, Legal Framework for Commercial Satellite Telecommunications. Cologne: DLR, 2000: 71.

[42] "assignment of a radio frequency (channel) is the "authorization given by an administration for a radio station to use a radio frequency or radio frequency channel under specified conditions", Art. S.18 of the ITU Radio Regulations.

[43] Baumann, Ingo. Das internationale Recht der Satellitenkommunikation. Frankfurt/Main: Lang, 2005: 284.

[44] Ibid: 276.

efficient and economic use of the orbit/spectrum resource. The ITU radio frequencies' distribution process tries to take both objectives into account by encompassing the allocation[45] of radio frequencies to certain services, and the allotment[46] of radio frequencies to different countries (a priori planning).[47]

A nationally assigned frequency is internationally recognized on the basis of "first come, first served", meaning that the user that first notified ITU of the use of a specific frequency will be legally protected from interference. This obviously favours the technologically advanced countries and has thus been challenged by developing countries.[48] The system of a priori planning, with WARC 1977s allotment plan for regions 1 and 3 (Europe, Asia, Australia), and with the WARC 1983s allotment plan for region 2 (America), indicated a trend towards an international administration of frequencies and orbital positions.[49] Nevertheless, to date "first come, first served" is still the dominant principle.[50] The problem that becomes evident from the discussions on the two systems is that of a fair balance between "efficient and economic use" and "equitable access" within the meaning of Art. 44 (2) CS.

The interpretation of the meaning of efficient and economic use of the orbit/ spectrum resource is left to the discretion of each ITU member. However, member states have impliedly agreed to the objective of "equitable access." Generally, the concept of equity relates to principles of justice and fairness. In the context of Art. 44 (2) CS, the meaning of the term "equitable access" is further clarified when considering two other provisions contained in Art. 44(2), i.e. that countries may have equitable access only in conformity with the Radio Regulations, and that the special needs of the developing countries and the geographical situation of particular countries must be taken into account while making use of the orbit/spectrum resource.[51] Against this background, developing countries have a right to access these orbital positions and use the associated frequencies that have been assigned to them in the allotment plans. Their notifications to the Table of Frequency Allocations of the ITU Radio Regulations are also internationally protected. Any "equitable access",

---

[45] "Allocation" of a frequency band is the "entry in the Table of Frequency Allocations of a given frequency band for the purpose of its use by one or more terrestrial or space radiocommunications services [...] under specified conditions", Art. S1.16 of the ITU Radio Regulations.

[46] "Allotment" is the "entry of a designated frequency channel in an agreed plan, adopted by a competent conference, for use by one or more administrations for a terrestrial or a space radiocommunication service in one or more identified countries or geographical areas and under specified conditions", Art. S1.17 of the ITU Radio Regulations.

[47] Jakhu, Ram S. and Virginia R. Serrano. "International Regulation." Project 2001: Legal Framework for Commercial Satellite Telecommunications. Cologne: DLR, 2000: 73.

[48] Malanczuk, Peter. "Telecommunications." EPIL IV. Ed. Bernhardt, Rudolf. Amsterdam: North Holland, 2000: 791, 793.

[49] Hobe, Stephan. Die rechtlichen Rahmenbedingungen der wirtschaftlichen Nutzung des Weltraums. Berlin: Duncker & Humblot, 1992: 177.

[50] Baumann, Ingo. Das internationale Recht der Satellitenkommunikation. Frankfurt/Main: Lang, 2005: 274.

[51] Jakhu, Ram S. and Virginia R. Serrano. "International Regulation." Project 2001: Legal Framework for Commercial Satellite Telecommunications. Cologne: DLR, 2000: 72.

however, depends on the design of the Radio Regulations. Developing countries are thus dependant on further international cooperation in this field.

Article I of the 1967 Outer Space Treaty (OST) has been operationalised regarding the use of the orbit spectrum resource for satellite communications by the ITU legal framework.[52] Therefore, it cannot go further than Art. 44(2) of the ITU Constitution and it does not provide developing countries with an effective legal claim to access the GEO and the associated frequencies.[53] Article I (1) does, however, contain a legal claim on the part of developing countries that their interests be considered in the use of outer space by developed countries, by providing that "...the exploration and use of outer space [...] shall be carried out for the benefit and in the interests of all countries, irrespective of their degree of economic or scientific development, and shall be the province of all mankind." Therefore, space faring states are, in principle, obliged to take into consideration the interests of developing countries and to enable them to participate in their outer space activities in one way or the other.[54] However, such legal claims of developing countries are not effective, since it is up to the developed states to decide on the form of cooperation. Accordingly, spacefaring nations interpret their obligation under Art. I (1) OST in a broad way and consider their own uses of outer space as being beneficial also for other states.[55]

11.7.3 Technology Transfer

Developing states that are parties to the Agreement on Trade-Related Aspects of Intellectual Property Rights (TRIPS) have a legal claim to the "promotion" of technology transfer by virtue of Art. 66(2)[56] TRIPS, but such claim will be difficult – if not impossible – to enforce. Art. 66 (2) TRIPS contains a general obligation on developed country Members to promote technology transfer to least-developed countries (LLDCs). To this end, they are under a duty to provide incentives to enterprises and institutions in their territory to increasingly engage in the transfer of

---

[52] Hobe, Stephan. Die rechtlichen Rahmenbedingungen der wirtschaftlichen Nutzung des Weltraums. Berlin: Duncker & Humblot, 1992: 181.

[53] The principles of the OST are generally viewed as part of customary international law, thus binding also states that are not members to the Treaty, Jakhu, Ram S. "Safeguarding the Concept of Public Service." 5 Sing. J. Int'l & Comp. L.: 71, 91.

[54] Matte, Nicolas Mateesco. "Aerospace Law." RdC 1980: 123, 147.

[55] Hobe, Stephan. "Article I." Cologne Commentary on Space Law (CoCoSL), Vol. 1. Eds. Hobe, Stephan, Schmidt-Tedd, Bernhard, and Kai-Uwe Schrogl. Cologne: Heymann, 2009. 42; Hobe, Stephan. "Outer Space as the Province of Mankind – An Assessment of 40 Years of Development." Proc. 50th Coll. IISL, 2008: 448.

[56] Art. 66 (2) TRIPS: "Developed country Members shall provide incentives to enterprises and institutions in their territories for the purpose of promoting and encouraging technology transfer to least-developed country Members [...]."

technology to developing countries. Whereas the precise scope of this duty suggests considerable discretion on the part of developed countries, it is clear that the obligation does exist and that it must be discharged.[57] This is emphasised by the general objectives of TRIPS as contained in Articles 7 and 8, which stipulate that the protection of IPRs has to contribute to the promotion of technical innovation and the transfer and dissemination of technology, in a manner conducive to social and economic welfare and to the balance of rights and obligations.[58] The (obligatory) annual notification by industrialised countries members of measures taken with respect to their obligation under Art. 66 (2) TRIPS[59] show that this obligation is taken seriously. Developed countries are thus obliged to *promote* the transfer of technology to developing countries.[60] However, this does not entail their right to be transferred any *specific* kind of technology.

Moreover, the increasing participation of developing country members in global trade shall be facilitated through negotiated specific commitments according to Art. IV (1) of the General Agreement on Trade in Services (GATS), which Member States are thus required to negotiate under Parts III and IV of GATS. However, Art. IV (1) GATS does not oblige industrialised countries to make specific concessions to developing countries. The provision has thus been correctly referred to as "a statement of good intentions"[61] concerning the recognition of the interests of developing countries.[62]

Art. IV (2) GATS obliges developed members to establish contact points within 2 years of the date of entry into force of the WTO Agreement, so as to facilitate the access of developing members' service suppliers to information related to their respective markets. In this respect, information concerning the improvement of professional qualifications and the availability of service technology is of specific interest to developing countries.[63] Nevertheless, the formulation of Art. IV (2)

---

57 United Nations Conference on Trade and Development-International Centre for Trade and Sustainable Development, ed. Resource Book on TRIPS and Development. New York: Cambridge Univ. Press, 2005: 725, 730.

58 Ibid: 725, 730.

59 Para. 11.2 of the Decision on implementation-related issues and concerns of 14 November 2001, WT/MIN/(01)/17. See, e.g., IP/C/W/431/Add.3 of 7 January 2001, and IP/C/W/431/Add.3/Suppl.1 of 14 Apr. 2001.

60 Other TRIPS provisions are concerned with technical cooperation to improve the protection of intellectual property, not with project-oriented transfer of technology, cf. Werner, Walter. "Grundlagen zum Technologietransfer in der WTO." RIW 2006: 187, 188.

61 United Nations Conference on Trade and Development. The Outcome of the Uruguay Round: An Assessment. United Nations, 1994: 156.

62 Koehler, Matthias. Das Allgemeine Übereinkommen über den Handel mit Dienstleistungen. Berlin: Duncker & Humblot, 1999: 140.

63 Ibid: 141.

GATS indicates that the right to obtain specific information concerning technologies is not included.

Para. 6 (c) of the GATS Annex on Telecommunications[64] provides that "[...] Members shall make available, where practicable, to developing countries information with respect to telecommunications services and developments in telecommunications and information technology to assist in strengthening their domestic telecommunications services sector." The wording "where practicable" indicates that members retain a wide discretion regarding the provision of such information. Consequently, any specific rights of developing countries to obtain such information cannot be derived from Para. 6 (c) of the Annex. The same applies to Para. 6 (d) of the Annex, which stipulates that "Members shall give special consideration to opportunities for the least-developed countries to encourage foreign suppliers of telecommunications services to assist in the transfer of technology [...]." Once again, no specific developing countries' rights to technology transfers are contained in the provision, despite the fact that members are called upon to promote such transfers. In conclusion, Para. 6 of the Annex does not stipulate a specific legal claim of developing countries regarding the transfer of technology.

11.7.4 Provision of Continued Telecommunications Services

The provision of continued telecommunications services was of great concern to developing countries in the course of INTELSAT's and INMARSAT's privatisation. For INTELSAT, countries qualified as "lifeline countries" were granted entry into Lifeline Connectivity Obligation (LCO) commitments.[65] Lifeline Connectivity Obligation (LCO) is the obligation assumed by Intelsat Ltd. to provide continued telecommunications services to the LCO customer.[66] The LCO commitment guarantees that until 2013, Intelsat Ltd. may not raise its rates to lifeline users above pre-privatization levels.[67] LCOs are obligations stemming from agreements in private law between the private company Intelsat Ltd. and the respective states. ITSO, the remaining international satellite organisation, is not a party to these agreements (and does not even have the facilities to provide such services), but instead is responsible for ensuring that the private company fulfils its obligations contained in these contracts.[68] As a result, claims in public international law can only be raised against ITSO in relation to its obligation to ensure Intelsat Ltd.'s

[64] Hereinafter referred to as the Annex.

[65] Katkin, Kenneth. "Communication Breakdown?" 38 Vand. J. Transnat'l L.: 1323, 1361. The equivalent for INMARSAT were so-called "Public Services Agreements" (PSAs).

[66] See Art. I (h) ITSO-Agreement.

[67] Katkin, Kenneth. "Communication Breakdown?" 38 Vand. J. Transnat'l L.: 1323, 1361.

[68] See Art. III ITSO-Agreement.

conformity with its LCO commitments. Consequently, there is no valid claim against ITSO to provide such services.

11.7.5 International Cooperation

What becomes clear is that most of the provisions analysed stipulate a general claim of developing countries regarding cooperation by industrial states, but they are all very vague as to how this cooperation is to be achieved in practice. Nevertheless, basically all of the aforementioned are expressions of the striving towards international cooperation in the field of satellite telecommunications. In this respect, they may be viewed in the spirit of the purposes of the United Nations as laid down in Art. I of the UN Charter, and the UN Friendly Relations Declaration of 1970,[69] which reaffirms that the maintenance of international peace and security and the development of friendly relations and cooperation among nations constitute the fundamental purposes of the United Nations.[70]

A concretization of this general obligation to cooperate cannot be deducted from the multitude of UNGA Resolutions in the field of (satellite) telecommunications. UNGA Resolutions are not legally binding, but they may reflect an *opinio iuris*, which in connection with state practice is a constituent element for the creation of customary international law.[71] However, an analysis of the relevant resolutions[72] regarding satellite communications and the bridging of the digital divide has failed to identify such an *opinio iuris*. While the documents examined contain palpable approaches to bridging the digital divide based on international cooperation, it is difficult to derive specific corresponding rights. Instead, it becomes clear that states are reluctant to commit to specific obligations of cooperation, especially without due consideration.

[69] Declaration on Principles of International Law Concerning Friendly Relations and Co-Operation among States in Accordance with the Charter of the United Nations. UNGA Res. 2625 (XXV) of 24 Oct. 1970. New York: United Nations.

[70] Principle (d) of the Friendly Relations Declaration.

[71] Ipsen, Knut, Menzel, Eberhard and Volker Epping. Völkerrecht. Munich: Beck, 2010: Para. 32 marginal note 47.

[72] E.g. Art. 17 of the Charter of Economic Rights and Duties of States, UNGA Res. 3281 (XXIX) of 12 Dec. 1974; G8 Okinawa Charter on Global Information Society of 2000; Paras. 1, 17 of the Geneva Declaration of Principles of 12 Dec. 2003, WSIS-03/GENEVA/DOC/4; Geneva Plan of Action of 12 Dec. 2003, WSIS-03/GENEVA/DOC/5; Para. 7 of the Tunis Commitment of 18 Nov. 2005, WSIS-05/TUNIS/DOC/7; Paras. 83 et seq. of the Tunis Agenda for the Information Society of 18 Nov. 2005, WSIS-05/TUNIS/DOC/6(Rev.1).

11.7.6 The Solidarity Concept

There is, however, a solidarity concept of public international law that applies specifically to bridging the digital divide through satellite communications.[73] This concept can be deduced from all the legal provisions and non-legally binding documents in this field. Whereas such a solidarity concept does not establish legal claims in terms of establishing specific obligations on the part of developed countries, it nevertheless has an impact on the legal claims founded on the aforementioned provisions. Indeed, the application of the solidarity concept can create "legitimate expectations"[74] on the part of developing countries that their interests regarding the use of satellite communications for bridging the digital divide will be taken into account. The wide discretion of developed states is thus reduced, since developing countries may expect industrial states to make use of their discretion so as to bridge the digital divide.

11.8 Possible Solutions

"Positive" norms of public international law would help bridge the digital divide. *De lege ferenda*, such norms could be shaped in different ways. In terms of access to the orbit/spectrum resource, one suggestion voiced by some authors is the establishment of an international space authority tasked with the resource management. However, to date this option seems to lack political consensus.[75] States are simply not yet willing to accept the curtailing of their sovereign power in their *domaine ré servée*. Another suggestion is to make certain frequencies and orbital positions for public services exempt from the principle of "first come, first served", by introducing a fee or an auction process for orbital positions and the related frequencies.[76] However, such a fee would have to be sufficiently high to discourage

[73] Neumann, Julia. Bridging the Digital Divide. Cologne: WoltersKluwer, forthcoming 2012.

[74] Verwey, W. D. "Least Developed Countries." Essays in International Law in Honour of Judge Manfred Lachs. Ed. Jerzy Makarczyk. The Hague: Nijhoff, 1984: 531, 548.

[75] Lyall, Francis. "Deriving more 'Common Benefit' from Space Telecommunications." Proc. 48th IISL Coll. 2005, 2006: 462; "A World Communications Commission?" Proc. 37th IISL Coll. 1993, 1994: 42.

[76] Lyall, Francis. "Deriving more 'Common Benefit' from Space Telecommunications." Proc. 48th IISL Coll. 2005, 2006: 463 et seq.; "Expanding Global Communications Services." Proc. UNISPACE III, 1999 (A. CONF-184/7): 77 et seq.; "Rational, Efficient and Economic Use of Space". Liber amicorum Karl-Heinz Böckstiegel. Eds. Benkö, Marietta and Walter Kröll. Cologne: Heymann, 2001: 386, 389.

developed countries from obtaining the frequencies and orbital positions concerned, and yet affordable to developing states. This would require an adjustment mechanism to guarantee that the fee would not weigh too much on the developing countries' economies. Moreover, an auctioning process[77] would also raise problems, since it is highly questionable that the interests of developing countries could be taken into account on the basis of the purely market-based approach that an auction implies.[78]

In any case, developed countries seem rather more willing to let developing countries participate in existing infrastructures than to help them build up their own infrastructure. This perception may be explained by the dual character of satellite communications, since they can not only be used for civil, but also for military purposes. At the same time, developing countries show a greater interest in participating in existing infrastructures than in building up their own. Also, developing countries would stand to benefit the most from the creation of an internationally owned satellite system.[79] A first step toward this direction would be to prolong the LCOs and Public Services Agreements (PSAs) within the frameworks of ITSO and IMSO. This, however, requires the willingness of the now private enterprises Intelsat and Inmarsat to cooperate with the states and international organisations.

Therefore, it seems necessary to safeguard the provision of basic satellite communications services to developing countries through the creation of a legal framework for a "global universal service" that lasts beyond 2013 (when the LCO and PSA commitments of Intelsat and Inmarsat expire). For instance, industrial states could undertake the obligation to assign frequencies only in connection with universal service obligations. Under their licenses, private satellite operators would be obliged to provide services also to remote areas worldwide. Since universal service does not mean "free of charge", they would only need to be provided on the basis of "affordable cost." One could also think of a gradual approach whereby in a first step, the services would be provided free of charge, and only in a second step they would be provided at an affordable price. The implementation of such a model may encounter some problems, but it nevertheless seems a viable solution. The already existing "Digital Solidarity Fund"[80] could be used as a funding tool for such a global universal service, provided that it acquires a more legally binding design. It would be particularly important to establish mandatory contributions on its participating users, in order to guarantee planning reliability and sufficient financial means.

[77] Frieden, Rob. "Shared Global Radiocommunication Resources." 24 U. Pa. Int'l Econ. L. (2003): 289.

[78] This is also considered problematic by Lyall, Francis. "Deriving more 'Common Benefit' from Space Telecommunications." Proc. 48th IISL Coll. 2005, 2006: 461, 464.

[79] Cf. Jakhu, Ram S. "Global Public Interest in Outer Space." J. Space L. 32 (2006): 31, 75.

[80] Global Digital Solidarity Fund. 14 Jan. 2012 http://www.dsf-fsn.org.

11.9 How Realistic Is This in Times of Global Crisis?

Even before the global financial crisis, states were reluctant to lay down specific binding international legal obligations. In times of the global (financial) crisis, states may seem to focus on the immediate problems and to be less inclined to commit, in a legally binding form, to solving such problems that seem more distant to the actual financial crisis.

However, developed states also seem to see the bigger picture even in times of global crisis.[81] States are becoming increasingly aware of their interdependence within the context of today's globalized world, which underpins the need to cooperate in bridging the digital divide. Bridging the digital divide, however, requires strong political will, both at the international and the national level, so as to eventually lead to a legally binding solution to the problem.[82]

Especially in times of global crisis, international cooperation is necessary to address the crisis and prevent the underlying problems from intensifying. As the former chief economist of the World Bank has stated, the two major problems of our times are climate change and worldwide poverty.[83] With regard to these problems, states can only succeed or fail together. Considering the potential of telecommunications for overcoming poverty, bridging the digital divide must be one of the main objectives of the international community. This necessitates a holistic approach, as focussing on technology alone will not be sufficient. What is clear, however, is that satellite communications can contribute significantly to solving the problems connected with the digital divide. Therefore, a more legally binding approach to bridging the digital divide is inevitable in the long term.

---

[81] For instance, the former German Federal Minister for Economic Cooperation and Development announced in 2009 that Germany would not reduce its development expenditure abroad since it was aware that this could result in a humanitarian catastrophe. 14 Jan. 2012.
http://www.welt.de/politik/article3111426/Wieczorek-Zeul-verteidigt-Entwicklungshilfe.html.

[82] Cf. Jakhu, Ram. "Global Public Interest". La Revue Québécoise de Droit International 18 (2005): 173.

[83] Stern, Nicolas H. The Economics of Climate Change. Cambridge: Cambridge Univ. Press, 2007.

Chapter 12
Space Debris: Legal Framework, Issues Arising and New ISO Guidelines in 2010/2011

Joanne Wheeler

12.1 Introduction

We are all conscious about being environmentally friendly on Earth, but the phrase "out of sight, out of mind" seems rather more appropriate when it comes to outer space.

The issue of space debris and the environment of outer space was highlighted earlier this year with the uncontrolled re-entry of the 6-t decommissioned US Upper Atmosphere Research Satellite (**UARS**)[1] and the defunct German 2.4 t Rontgensatellit, or **ROSAT**. The re-entry of ROSAT was deemed to be of greater concern. The issue was the probable mass of surviving material due to the mirror system made of reinforced carbon composite material on the ROSAT x-ray telescope satellite. ROSAT has been, in effect, space debris since 1999, when it was shut down.

There is a legacy of over 50 years of spacecraft, so these re-entry events could become more common. In fact, it seems that high levels of solar activity in 2012 may damage more defunct spacecraft and send them heading towards Earth.[2] On the 24 September 2011, NASA's UARS fell back to Earth. According to NASA,[3] 26 components of

---

Partner, CMS Cameron McKenna LLP, Mitre House, 160, Aldersgate Street, London, EC1A 4DD, UK specialising in the regulatory and commercial aspects of space and satellite law. Expert representative for the UK Government at the Legal Subcommittee of the Committee on the Peaceful Uses of Outer Space.

[1] "UARS Entered the Atmosphere Over the Pacific." 24 Sept. 2011. EarthSky 15 Mar. 2011 http://earthsky.org/space/where-will-nasas-uars-satellite-land.

[2] Bloxman, Andy. "Second Giant Chunk of Space Junk Heading for Earth." 24 Spt. 2011. The Telegraph 15 Mar. 2012 http://www.telegraph.co.uk/science/space/8785971/Second-giant-chunk-of-space-junk-heading-for-Earth.

[3] "UARS Re-Entry Overview." NASA 15 Mar. 2012 http://www.nasa.gov/mission_pages/uars/index.html.

J. Wheeler (✉)
CMS Cameron McKenna LLP, Mitre House, 160, Aldersgate Street, London, UK
e-mail: Joanne.WHEELER@cms-cmck.com

P. Hulsroj et al. (eds.), *Yearbook on Space Policy 2010/2011*,
Yearbook on Space Policy, DOI 10.1007/978-3-7091-1363-9_12,

UARS, weighing a total of 1,200 pounds, could have survived the re-entry to reach the surface of the Earth. The re-entries this year of NASA's UARS satellite and Germany's ROSAT have illustrated that space debris is a growing issue for us on Earth, just as much as for vulnerable and valuable space assets.

On 10 February 2009 the first collision, or "intersection" as more technically known, of two intact spacecraft occurred. Iridium 33, a US communications satellite, and Cosmos 251, a decommissioned Russian satellite collided as the two objects passed over northern Siberia.[4] This intersection caused two distinct clouds of debris to extend through a substantial part of low Earth orbit. Four accidental collisions in outer space had occurred prior to this event, but the Iridium/Cosmos collision was the first to involve two intact satellites. It also probably contributed more to the public awareness of the issue of space debris than any other collision and stimulated more media attention. Another recent incident that dramatically increased the amount of catalogued fragmentation debris in outer space (and public interest) was the destruction of an orbiting weather satellite by China with a missile in 2007.[5]

The issue of space debris has been the focus of precise scientific and technical study for many years. The risks of space debris and measures to mitigate it, however, have not perhaps achieved the legal recognition they deserve, in order to encourage the drafting of an adequate international framework to deal with the complex legal issues raised by this growing hazard. Accentuating this, there are difficulties inherent in the negotiation and drafting of any future legal regulation; an international-level playing field is required to avoid the creation of uncompetitive advantages or disadvantages.[6]

There is an internationally recognised need to deal with the issue of space debris. As Donald Kessler, retired head of NASA's orbital debris programme, has recently stated:

> The longer you wait to do this the more expensive it's going to be. Given the economy, we'll probably end up putting it off, but that's really not very wise. This scenario of increasing space debris will play out even if we don't put anything else in orbit.[7]

In this important year when awareness of space debris has been heightened due to the re-entry of UARS and ROSAT, and when two important international standards on space debris have been published, it seems topical to write about this phenomenon.

This chapter summarises and discusses briefly the extent to which some of the issues raised by space debris are dealt with in the current international, European, US and UK

---

[4] "Satellite Collision Leaves Significant Debris Clouds." NASA Orbital Debris Quarterly News 13.2 (2009): 1.

[5] This incident has lead to several predicted "close approaches" (predicted conjunctions of space debris and active satellites), as identified by the United States Joint Space Operations Center (JSpOC).

[6] Jasentuliyana, N. "Space Debris and International Law," Journal of Space Law 26–27 (1998–1999): 139–162.

[7] Donald Kessler, chair of the committee appointed by the US National Research Council to review NASA's meteoroid and orbital debris programmes. Sample, Ian. "Space Junk Rising Exponentially." 1 Sept. 2011. The Guardian 15 Mar. 2012 http://www.guardian.co.uk/science/2011/sep/01/space-junk-rising-exponentially.

legal framework. It continues by analysing some difficulties inherent in the legal regulation of space debris and comments briefly on measures now being considered to mitigate the problem more directly, in order to ensure safe and sustainable access to space for future generations. In particular, such measures include the following ISO International Standards published this year:

(a) ISO 26872: Space systems – Disposal of satellites operating at geosynchronous altitude, published by the British Standards Institute (BSI) on 15 September 2010; and
(b) ISO 24113: Space systems – Space debris mitigation requirements, published by the BSI on 31 October 2011, some of the prescribed requirements of which are also mentioned.

12.2 The Current Legal Framework

12.2.1 The Law of Outer Space

Human activities in outer space are governed by a branch of international law known as space law. This comprises United Nations (UN) treaties, principles and declarations, strengthened by other rules originating from different legal entities such as sovereign states, intergovernmental organisations and further developed by custom and contracts. There are also various non-binding guidelines and codes of conduct at national, regional and international level, which are increasingly adhered to and which cover issues of mitigation of space debris.

12.2.2 Legal Issues of Space Debris

The legal issues of space debris are to some extent dealt with in current international space law, which confirms general obligations without creating clear rules. The Magna Carta of space law is the Outer Space Treaty of 1967 (Outer Space Treaty).[8] The Outer Space Treaty's provisions, however, alone are too generic to deal with the complex problems of space debris. In fact, despite efforts by legal writers to define the concept of "space debris", no internationally agreed definition or description exists.[9] In this chapter, I shall refer to space debris as constituting any man-made object that is either: (a) Earth-orbiting

[8] Treaty on Principles Governing the Activities of States in the Exploration and Use of Outer Space, including the Moon and Other Celestial Bodies 1967, 610 UNTS 205.

[9] The Annex to Resolution 5 of the Buenos Aires International Instrument on the Protection of the Environment from Damage Caused by Space Debris, adopted by the International Law Association took an important step forward towards an agreed definition. It states that space debris constitutes, "man-made objects in outer space, other than active or otherwise useful satellites, where no change can reasonably be expected in these conditions in the foreseeable future." Cf. Bockstiegel, K-H. "The Draft of the International Law Association for a Convention on Space Debris", IISL Colloquium, 95-IISL. 2.03, 69–77.

and is non-functional with no reasonable expectation of assuming or resuming its intended function; or (b) re-entering the Earth atmosphere. This concept covers fragments and component parts of space objects, as well as decommissioned/failed spacecraft and spent upper stages of launchers. Therefore, with this definition, Cosmos 2251 would be defined as space debris.

The Outer Space Treaty does, however, offer indirect minimal guidance at state level with much interpretation left to lawyers and other experts. Article IX provides that State Parties to the Outer Space Treaty:

> shall conduct all their activities in outer space... with due regard to the corresponding interests of all other State Parties to the Treaty.

This provision can be interpreted generally to oblige State Parties to avoid the creation of, and to attempt to reduce and even remove, space debris so as to allow all states to participate in the exploration and use of outer space with minimal risk from debris.

Interpretative difficulties are further illustrated by the next sentence of Article IX which explains that the study and exploration of outer space shall be conducted, "*so as to avoid their harmful contamination*", and that States Parties, "*shall adopt appropriate measures for this purpose.*" The Outer Space Treaty does not enlighten us by providing a definition or guidance as to what constitutes "*harmful contamination*", or what such "*appropriate measures*" consist of. While open to various interpretations, these provisions are in practice limited. Space debris is not normally classed as "*harmful contamination*"; the phrase being usually construed as biological or radioactive contamination.

An international consultation process is provided for by Article IX of the Outer Space Treaty, whereby if a State Party believes that an activity or experiment planned by it or its nationals would "*cause potentially harmful interference*" to the activities of another State Party, it shall undertake appropriate consultations before proceeding. A State Party may also request consultations if it has reason to believe that an activity or experiment planned by another State Party would cause potentially harmful interference. One problem with the implementation of these provisions is that it is difficult to describe the existence or creation of space debris as a future "*planned*" activity. In addition, the provisions do not address the issue of existing or completed activities, or the problem of current space debris.

12.2.3 Liability Regime for Damage Caused by Space Debris

Other provisions of the UN treaties that may be relevant to space debris can best be discussed through the use of an example. Suppose a UK based private communications operator launches a communications satellite into low earth orbit. A bolt from a component part of the satellite becomes loose, forming identifiable space debris. This space debris collides with a French satellite causing damage to it.[10] As a State Party to the

[10] The first suspected collision between two catalogued space objects was detected in low earth orbit in 1996. The collision was between the French CERISE microsatellite and a piece of debris which impacted the satellite's stabilisation boom http://www.ph.surrey.ac.uk/satellites/main/introducing_the_satellites6.html.

Outer Space Treaty, pursuant to Article VI the UK is responsible for the activities conducted by its nationals in outer space, whether such activities are carried out by governmental agencies or by non-governmental entities, and for ensuring that such activities are compliant with the Outer Space Treaty. The UK is, therefore, responsible for the actions of its private commercial operators and thus for the consequences if such operators create space debris.

Under the Registration Convention of 1976,[11] each state must hold a register of the objects that it launches into outer space. The UK deals with this obligation through the Outer Space Act 1986, under which the Secretary of State maintains a register of space objects. Article VIII of the Outer Space Treaty provides that each state retains ownership[12] and control over objects launched into space that are registered on its national registry:

> Ownership of objects launched into outer space. . . and of their component parts, is not affected by their presence in outer space.

An interpretation of space debris which would include a bolt as a sub-category of space "*objects. . . and of their component parts*" leads to the conclusion that the principle of permanent ownership also extends to space debris.

The UK may, therefore, be held to be liable for damage caused by the bolt (space debris) under Article VII of the Outer Space Treaty:

> Each State Party [. . .] that launches or procures the launching of an object into outer space [. . .] and each State Party from whose territory or facility an object is launched, is internationally liable for damage to another State Party to the Treaty [. . .] by such object or its component parts on the Earth, in air space or in outer space.

Article VII of the Outer Space Treaty is further elaborated in the Liability Convention of 1972,[13] Article III of which provides that:

[11] Convention on Registration of Objects Launched into Outer Space, London/Moscow/Washington, done 11 November 1974, entered into force in 1976, 1023 UNTS 15. Article II states that, "(W)hen a space object is launched into Earth orbit or beyond, the launching State shall register the space object by means of an entry in an appropriate registry which it shall maintain". By Article VI: "Each State of registry shall furnish to the Secretary-General of the United Nations [. . .] the following information concerning each space object carried on its registry: (a) Name of launching State or States; (b) An appropriate designator of the space object or its registration number; (c) Date and territory or location of launch; (d) Basic orbital parameters, including: (1) Nodal period; (2) Inclination; (3) Apogee; (4) Perigee; (e) General function of the space object".

[12] Treaty on Principles Governing the Activities of States in the Exploration and Use of Outer Space, Including the Moon and other Celestial Bodies, London/Moscow/Washington, done 27 January 1967, entered into force 10 October 1967, 610 UNTS 205, 6 ILM 386 (1967). Article VIII, "A State Party to the Treaty on whose registry an object launched into outer space is carried shall retain jurisdiction and control over such object, and over any personnel thereof, while in outer space or on a celestial body".

[13] Convention on International Liability for Damage Caused by Space Objects, London/Moscow/Washington, entered into force 1 September 1972, 961 UNTS 187.

> (I)n the event of damage being caused elsewhere than on the surface of the Earth to a space object[14] of one launching State or to persons or property on board such a space object by a space object of another launching State, the latter shall be liable only if the damage is due to its fault or the fault of persons for whom it is responsible.[15]

It would be difficult to apply the above statement to particulate debris, the owner of which cannot be readily identified, or to attempt to establish "fault."

12.2.4 Issues Regarding Filing a Claim Under the Liability Convention

Only States that are parties to the Liability Convention can file a claim under this regime. In the example above, for the UK to be liable, France would have to:

(a) Prove that damage was caused to French citizens or to space objects that are registered by France on the registry which it maintains;
(b) Establish that the UK is the "launching State"[16] of the space object (the debris bolt) that caused the damage and therefore has ownership and control over the debris; and
(c) Prove that the damage was caused by the fault (as the damage has occurred in outer space) of the UK or the fault of the private operator for whom the UK is responsible under Article VI of the Outer Space Treaty.

Establishing the elements of a claim under the Liability Convention for damage caused by space debris would be challenging for France. While the first element may be relatively easy to prove, establishing the causality of damage caused by space debris may be difficult given that the origin of the majority of space debris that are not catalogued, or which is particulate, cannot be determined.

Debris is "identified" if it is possible to determine who is responsible for launching it into space (the "launching State"). Identification requires that the launch of the object be registered, as mentioned, and that the space object be continuously tracked in orbit. Earth-based tracking stations can currently monitor only pieces of space debris that are larger than approximately 10 cm. Smaller pieces are not tracked or catalogued and generally cannot be "identified" with certainty for the purpose of determining the launching State, or attributing fault under the provisions of the Liability Convention.

---

[14] The term "space object" is defined in Article I(d) of the Liability Convention as including, "component parts of a space object as well as its launch vehicle and parts thereof."

[15] In accordance with Article II of the Liability Convention however, "A launching State shall be absolutely liable to pay compensation for damage caused by its space object on the surface of the Earth or to aircraft in flight."

[16] More than one State may qualify as the "launching State" of the space object, for example: (a) the State on whose territory the launch site is located; (b) the State where the launch services provider is incorporated; (c) the State of incorporation of the company owning the payload on the spacecraft; or (d) the State where the spacecraft was manufactured. As provided in Article V, (1) of the Liability Convention "Whenever two or more States jointly launch a space object, they shall be jointly and severally liable for any damage caused."

Establishing the "fault" of the responsible launching State, as required in the case of damage caused to the space object of a claimant State in outer space, is even more challenging.[17] The scope of what constitutes "fault" is unclear. The Liability Convention does not define the term and there are several interpretations of "fault" as applied to public international law instruments, under common and civil law theories. However, it is generally the case that fault liability presumes the existence of a standard of care, against which the reasonableness of the defendant's actions can be judged.

In order to establish whether a launching State is at fault for a collision in outer space, there must be an applicable standard of conduct in outer space, and a breach of that standard. In this context, proving the fault of a launching State would require the claimant State to establish that the owner of the space debris that caused the damage did not comply with national or international standards, guidelines and practices for conducting space activities and/or for debris mitigation. This would be to the extent that such standards or practices exist, apply to and are binding on the relevant party. The Liability Convention is silent as to what standard of care applies to space activities. There are currently only voluntary and non-binding national, regional and internationally accepted standards applicable to the operation of space objects and the mitigation of space debris, including the ISO standards. There are no mandatory international standards of conduct regarding space debris mitigation to establish a standard of care against which fault can be assessed.

Given the current absence of binding requirements and standards of care or conduct regarding space debris, in case of a collision in space between two space objects (including space debris), it would be difficult for the claimant State to establish that a launching State's conduct was not in conformity with applicable standards of conduct and care under the Liability Convention. It would thus be challenging to establish a cause of action based on fault or negligence, under international or national law. Scholars have argued that general principles of international law, equity and justice will provide a solution with respect to applicable standards of care. From a practical perspective, however, it remains doubtful that such principles are sufficiently clear (especially in the context of damage caused by space debris) to provide clear guidance. For damage caused by space originating debris on the ground or to aircraft in flight,[18] a cause of action based on strict (or absolute) liability is possible under either international law or national law.

12.2.5 Relevance of the Rescue Agreement

Another space treaty that some might consider relevant is the Rescue Agreement of 1968,[19] Article 5 of which deals with the return of space objects or their component parts to their launching state. However, the opinion of most scholars is that this

[17] Article III of the Liability Convention.

[18] Article II of the Liability Convention.

[19] Agreement on the Rescue of Astronauts, the Return of Astronauts and the Return of Objects Launched into Outer Space, London/Moscow/Washington entered into force 3 December 1968, 672 UNTS 119.

provision deals with space debris etc. found on Earth, rather than in space. It may therefore be of some relevance to the surviving components of UARS or ROSAT which fall to Earth, but not to space debris in orbit.

12.3 Mitigation Measures: Current Guidelines

Any mandatory measures to mitigate space debris are likely to lead to an increase in the cost of space activities. An international solution to space debris, therefore, can only be achieved after the appropriate national or regional regulations are as far as possible harmonised internationally to avoid unfair competition. An international solution must be based on a "level playing field." Legal certainty can only be achieved through binding international law; a treaty dealing precisely with state responsibility, liability and the mitigation of space debris and ratified by all space-faring states. However, as mentioned, there are presently no binding mandatory international standards or guidelines to deal with the issue of space debris, and this is likely to be the case for some time.

The next best solution is voluntary guidelines and standards accepted by the international community or codes of conduct adopted at a regional level. Such voluntary standards can be used as indicators of the expected standard of care of satellite operators with regard to the generation of space debris. Some examples of standards and guidelines in this area are mentioned below.

12.3.1 Inter-agency Space Debris Coordination Committee

The aim of the Inter-Agency Space Debris Coordination Committee (IADC) is to coordinate activities related to space debris on a world-wide basis.[20] The IADC has produced draft Space Debris Mitigation Guidelines (IADC Guidelines)[21] reflecting and consolidating mitigation measures developed by several national and international organisations. These were based on three principles: the prevention of orbital break-ups; the removal of end of mission spacecraft and orbital stages; and the limitation of objects released during normal space operations. Although non-binding, the IADC Guidelines have achieved wide international recognition.

[20] The IADC consists of the Agenzia Spaziale Italiana (ASI), the British National Space Centre (BNSC), Centre National d'Etudes Spatiales (CNES), the China National Space Administration (CNSA), the German Aerospace Centre (DLR), the European Space Agency (ESA), the Indian Space Research Organisation (ISRO), the Japan Aerospace Exploration Agency (JAXA), the National Aeronautics and Space Administration (NASA), the National Space Agency of Ukraine (NSAU) and the Russian Federal Space Agency (ROSCOSMOS). "Inter-Agency Space Debris Coordination Committee." 15 Mar. 2012 www.iadc-online.org.

[21] "IADC Space Debris Mitigation Guidelines, Revision 1." Sept. 2007. Inter-Agency Space Debris Coordination Committee 15 Mar. 2012 http://www.iadc-online.org.

12.3.2 UN Space Debris Mitigation Guidelines

A Working Group on Space Debris was also established by the United Nations (UN) Scientific and Technical Committee of the UN Committee on the Peaceful Uses of Outer Space (COPUOS). Its task was to develop a set of recommended guidelines based on the technical content and basic definitions of the IADC Guidelines, taking the UN treaties and principles of outer space into consideration. The "Space Debris Mitigation Guidelines of the Scientific and Technical Subcommittee of the Committee on the Peaceful Uses of Outer Space" (COPUOS Guidelines) were published in 2007.[22] The seven guidelines represent voluntary measures applicable to mission planning and the operation of newly designed spacecraft and orbital stages and, if possible, to existing ones. The COPUOS Guidelines are to be implemented by UN Member States and international organisations through their own mechanisms. As such, the COPUOS Guidelines are not legally binding, although the rules of particular licensing states are binding on those they license.[23] The work of the Scientific and Technical Committee in producing the COPUOS Guidelines is much respected.

In the 2011 session of the Legal Subcommittee of COPUOS, the delegation of the Czech Republic presented a working paper calling for the Legal Subcommittee to elaborate binding space debris mitigation principles for adoption. Many other delegations were in favour, although the consensus among members of the Legal Subcommittee was that further work was necessary within the Scientific and Technical Subcommittee, before effective binding legal guidelines could be prepared. The topic however remains on the agenda of both the Scientific and Technical Subcommittee, and the Legal Subcommittee of UNCOPUOS.

12.3.3 European Code of Conduct for Space Debris Mitigation

Another example is the European Code of Conduct for Space Debris Mitigation (Code of Conduct), which is consistent with, but expands on, the IADC Guidelines.[24] Although this is a policy statement containing general guidelines and voluntary measures that are

[22] United Nations General Assembly. Space Debris Mitigation Guidelines of the Scientific and Technical Subcommittee of the Committee on the Peaceful Uses of Outer Space. UN Doc. A/RES/62/217 of Dec. 2007. New York: United Nations.

[23] Space debris mitigation measures are required by the UK as part of the licence application to the UK licensing authority (the UK Space Agency). See BS ISO 24113 Space systems – Space debris mitigation requirements and the Database of standards in support of the Outer Space Act 1986, QinetiQ, 24 July 2007. In the US licensing process, space debris mitigation is required by the Federal Aviation Administration (see. 14 Code of Federal Regulations § 400 ff. and the Federal Communications Commission (see "Mitigation of Orbital Debris, Second Report and Order", 2004 19 FCC Rcd 11567).

[24] European Code of Conduct for Space Debris Mitigation 28 June 2004. 15 Mar. 2012 http://www.cnsa.gov.cn/n615708/n676979/n676983/n893604/appendix/2008529151013.pdf.

not legally binding, it does represent a valuable start for space debris mitigation measures in Europe.

The Code of Conduct was drafted by the Network of Centres Space Debris Coordination Group, which has been coordinating space debris work at a European level between national space agencies since 1999.[25] It is intended to be used practically and by a wide range of industry players (space agencies, space industry manufacturers, the operators and launchers of satellite systems and other concerned private or public entities) to assist in the early consideration and application of measures reducing or preventing the build up of space debris, and to encourage space systems to be designed to achieve this end. Time will tell if the Code of Conduct does in fact encourage an improved and consistent understanding of the best practices to minimise space debris in Europe, and it will be interesting to see if and how it develops to include changes in technology and industry accepted guidelines.

12.3.4 ESA Space Debris Mitigation Handbook

The ESA Space Debris Mitigation Handbook 2007[26] (ESA Handbook) was jointly produced by an industrial consortium and ESA, under an ESA contract. It is a non-binding, non-regulatory, self-standing document that provides technical information in support of European debris mitigation standards.[27]

The ESA Handbook intends to provide technical support to projects in the areas of describing the current space debris and meteoroid[28] environment, assessing risk potentials due to debris and meteoroid impacts, analysing the future space debris environment, implementing space debris mitigation measures and planning of debris shielding and collision avoidance concepts. The ESA Handbook along with the European Space Debris Mitigation and Safety Standard[29] jointly define ESA's policies concerning space debris mitigation and collision risk reduction, for any mission under ESA's control.

12.3.5 International Organization for Standardization

The International Organization for Standardization (ISO) is a worldwide federation of national standards bodies. The work of preparing the international standards is normally

---

[25] The members of the Network of Centres Space Debris Coordination Group are ASI, BNSC, CNES, DLR and ESA.

[26] "Update of the ESA Space Debris Mitigation Handbook. Executive Summary." July 2002. European Space Agency 15 Mar. 2012 http://www.esa.int/gsp/completed/execsum00_N06.pdf.

[27] Klinkrad, H. et al., "The ESA Space Debris Mitigation Handbook 2002." Advances in Space Research 34.5 (2004): 1251–1259.

[28] Meteoroids are described in the ESA Handbook as being particles in space which are of natural origin, originating primarily from comets and asteroids.

[29] Alby, F. et al. "The European Space Debris Safety and Mitigation Standard." Advances in Space Research, 34.5 (2004): 1260–1263.

carried out through ISO technical committees. The ISO is responsible for the transformation of debris mitigation guidelines (for example, as published by the IADC, the UN and also the International Telecommunication Union (ITU)[30]) into a set of high-level debris mitigation requirements, methods and processes and a series of lower level implementation standards to ensure that they are in turn incorporated into engineering practice at a national level.

12.3.5.1 BS ISO 24113 Space Systems: Space Debris Mitigation Requirements

Perhaps one of the most important of the ISO standards is ISO 24113 on Space Systems – Space debris mitigation requirements. ISO 24113 sets out the primary space debris mitigation requirements applicable to all elements of unmanned systems launched into or passing through near-Earth space, including launch vehicle orbital stages, operating spacecraft, and any objects released as part of normal operations or disposal actions. Its requirements aim to reduce the growth of space debris by ensuring that spacecraft and launch vehicle orbital stages are designed, operated and disposed of in a manner that prevents them from generating debris throughout their orbital lifetime. It does not however cover launch phase safety.

ISO 24113 is classed as a "top-level standard" in a series of standards aimed at addressing space debris mitigation. Section 5 describes the low Earth orbit (LEO) and geostationary Earth orbit (GEO) regions as protected regions with regard to space debris. Both regions have been defined by the IADC.[31]

Section 6 sets out a series of technical requirements covering the following.

(a) Avoiding the intentional release of space debris into Earth orbit during normal operations. For example, space debris released into Earth orbit as part of normal operations shall, in general, remain outside the GEO protected region, while the presence of space debris in the LEO protected region is to be limited to a maximum of 25 years after release.
(b) Avoiding break-ups in Earth orbit. For example, during the disposal phase a spacecraft or launch vehicle orbital stage shall permanently deplete or make safe all remaining on-board sources of stored energy in a controlled sequence.
(c) Removing a spacecraft or launch vehicle orbital stage from the protected regions, after end of mission. For example, a spacecraft or launch vehicle orbital stage operating in the GEO protected region shall be manoeuvred in a controlled manner during the disposal phase to an orbit that is entirely outside the GEO protected region. In the LEO protected region, it shall limit its post-mission presence to a maximum of 25 years from the end of mission.

---

[30] ITU Recommendation on GEO Disposal, ITU-RS.1003, January 2004.

[31] "IADC Space Debris Mitigation Guidelines, Revision 1." Sept. 2007. Inter-Agency Space Debris Coordination Committee 15 Mar. 2012 http://www.iadc-online.org.

Any provider of a spacecraft or launch vehicle shall prepare a space debris mitigation plan (SDMP). The SDMP is to list the applicable space debris mitigation requirements, include a compliance matrix and provide justification for any non-compliance. As the implementation of space debris mitigation measures can have a significant impact on the design and operation of space systems (and therefore on the mission cost), it is in the interest of applicants for a licence to be aware at an early stage in their mission planning of which debris mitigation measures will apply to their specific space system, and with which the applicant must demonstrate compliance.

Space activities carried out by individuals established in the UK, its Overseas Territories or Crown Dependencies are required to be licensed under the UK Outer Space Act 1986. An application for a licence must be submitted to UK Space Agency along with information, such as technical details of the mission, orbital location, and plans for the disposal of the space object at the end of its life.

During the licensing process, the UK Space Agency and its advisors will assess this information to ensure that the launch, operation and disposal of the space object conform to international treaties and other applicable laws and regulations and that they do not pose an unacceptable risk to lives or property. The assessment also considers applicable international guidelines and standards on space debris mitigation. ISO 24113 would be one such standard that would be considered to be particularly important as part of this assessment prior to the grant of an outer space licence.

12.3.5.2 BS ISO 26872 Space Systems: Disposal of Satellites Operating at Geosynchronous Altitude

ISO 26872 prescribes requirements for planning and executing manoeuvres and operations to remove a satellite from geosynchronous orbit at the end of its mission, and the final disposal of the satellite in an orbit where it will not pose a future hazard for satellites operating in the geosynchronous orbit.

More specifically, the standard specifies requirements for the following.

(a) Planning for disposal of a satellite operating in the geosynchronous orbit to ensure that efficient propellant is reserved for such manoeuvres.
(b) Selecting final disposal (or "graveyard") orbits where the satellite will not re-enter the operational region in the next 100 years and executing the disposal manoeuvre successfully.
(c) Depleting all energy sources on board the satellite before the end of its mission to minimise the possibility of a debris-producing event.

Independent of the other aspects of disposal action, a contingency plan should also be developed to deplete all energy sources and secure the satellite before its final disposal. The objective is to ensure that actions necessary to secure the vehicle are taken before the end of life of the satellite.

The standard sets out techniques for planning and executing the disposal of space hardware that reflect internationally accepted guidelines. It also states that ISO 24113 is "*indispensible for the application of this standard*".

12.4 Further Considerations and Conclusion

> Long-term projections of the space debris environment indicate that even drastic measures, such as an immediate, complete halt of launch and release activities, will not result in a stable environment of man-made space objects.[32]

Many studies have shown that the current debris problem in LEO has reached the critical point where the environment is unstable and collisions will become the most dominant debris-generating mechanism.[33] The current guidelines and codes of conduct prescribe the post-mission disposal of space objects. This will slow down the growth of future debris populations but will not prevent such growth. Therefore, as mentioned, consideration must be given to the actual removal of existing space debris from orbit. However, further research needs to be carried out as to the technical and economic feasibility of such a proposal, which would be much assisted if the concept would be driven by an international body on an international level. This proposal will also require further legal study, in particular because the removal of space debris (from identifiable fragments to defunct satellites) will require the consent of the "launching State", as the continuing owner of the space debris.

The ISO standards published this year, ISO 26872 and ISO 24113, take us closer to the creation of an international level playing field for the mitigation of space debris. I have been privileged to have been part of the discussions leading up to the publication of these measures over the last year. Work is still required in relation to measures for the mitigation of space debris, on-orbit capture and the removal of existing space debris from orbit. Such a multi-disciplinary challenge as space debris requires the on-going collaboration of both governments and the private sector and also of scientists, engineers, industry players, economists and lawyers to achieve a practical solution to mitigate the potential risks. The sustainable future of man's activities in outer space, whether in exploration, communications, observation, broadcasting or navigation, demands it. Meanwhile, the public's attention will be increasingly drawn upwards to the growing problem of space junk, the problems it may cause and what the media write about the consequences of its re-entry, which may be an even greater issue in 2012.

[32] Johnson, N. and Klinkrad, H. "Space Debris Environment Remediation Concepts." Orbital Debris Quarterly News 13.2 (2009).

[33] Liou, J.-C., and Johnson, N.L. "Risks in Space from Orbiting Debris." Science 311.20 (2006).

Part III
Facts and Figures

Chapter 13
Chronology: June 2010–May 2011

Spyros Pagkratis and Blandina Baranes

13.1 Access to Space

All launch dates are calculated using Greenwich Mean Time (GMT), therefore the date at the launch site may differ from the date listed here by 1 day.[1]

Europe	Other countries
Launch log	
June 2010	
26 Ariane 5ECA- Arabsat 5A (C) and COMS 1 (C/M)	01 Long March 3C - Beidou 2 (N)
	01 Rockot - SERVIS 2 (D)
	03 Proton M - BADR-5 (C)
	03 Falcon 9 - Falcon 9 Demo Flight (D)
	08 KSLV I - STSAT 2b (failure, S)
	14 Dnepr M - Prisma Main (D), Picard (S) and Prisma Target (D)
	14 Long March 2D - Shijian 12 (S)
	15 Soyuz - Soyuz ISS 23S (ISS)
	20 Dnepr M - TanDEM X (R)
	21 Shavit - Ofeq 9 (R)
	29 Soyuz - Progress ISS 38P (ISS)

(continued)

[1] "Semi-Annual Launch Report, First Half of 2011." Federal Aviation Administration (FAA) 31 Oct. 2011 http://www.faa.gov/about/office_org/headquarters_offices/ast/reports_studies/; "Commercial Space Transportation: 2010 Year in Review." Federal Aviation Administration (FAA) 31 Oct. 2011 http://www.faa.gov/about/office_org/headquarters_offices/ast/reports_studies/; "Space Launch Report – 2011 Launch Vehicle/Site Statistics." 31 Oct. 2011 http://www.spacelaunchreport.com/log2011.html; "Spaceflight Now | Tracking Station | Worldwide launch log" 31 Oct. 2011 http://spaceflightnow.com/tracking/launchlog.html .

S. Pagkratis (✉) • B. Baranes
ESPI, Vienna, Austria
e-mail: spyros.pagkratis@espi.or.at

P. Hulsroj et al. (eds.), *Yearbook on Space Policy 2010/2011*,
Yearbook on Space Policy, DOI 10.1007/978-3-7091-1363-9_13,

Europe	Other countries
Launch log	
July 2010	
	09 Proton M - EchoStar XV (C) 12 PSLV- Cartosat 2B (R), AISSat-1 (D), TIsat-1 (D), Alsat 2A (R) and StudSat (D)
August 2010	
04 Ariane 5ECA- Nilesat 201 (C) and RASCOM 1R (C)	01 Long March 3A - Beidou 2-IGS 1 (N) 10 Long March 4C - Yaogan 10 (R) 14 Atlas V 531 - Advanced EHF 1 (C) 24 Long March 2D - Tian Hui-1 (R)
September 2010	
	02 Proton M - Glonass M36, M37 and M38 (N) 04 Long March 3B - Chinasat 6A (C) 08 Rockot - Gonets M 2 (C) 10 Soyuz - Progress ISS 39P (ISS) 11 H IIA 202 - QZS-1 (C) 20 Atlas V 501 - NRO L-41 (I)* 21 Long March 2D - Yaogan 11 (R) 26 Minotaur IV - SBSS 1 (I)* 30 Molniya-M - Kosmos 2469 (I)*
October 2010	
28 Ariane 5ECA- BSAT 3B (C) and Eutelsat W3B (failure: fuel leak, C)	01 Long March 3C - Chang'e 2 lunar orbiter (S) 06 Long March 4B - Shijian 6 (S) 07 Soyuz - Soyuz ISS 24S (ISS) 14 Proton M - XM 5 (C) 19 Soyuz 2 1A - Globalstar 2nd Gen. 01, 02, 03, 04, 05 and 06 (C) 27 Soyuz - Progress ISS 40P (ISS) 31 Long March 3C- Beidou 2-G4 (N)
November 2010	
26 Ariane 5ECA- Intelsat 17 and HYLAS 1 (C)	02 Soyuz 2 - Meridian 3 (C)* 05 Delta II 7420–10 - COSMO-SkyMed 4 (R) 05 Long March 4C - Fengyun 3B (M) 14 Proton M - SkyTerra-1 (C) 19 Minotaur IV - Space Test Program Satellite 26 (D) 21 Delta IV Heavy - NRO L-32 (I)* 24 Long March 3A - Shen Tong 1-B (C)
December 2010	
29 Ariane 5ECA- KoreaSat 6 and Hispasat 1E (C)	05 Proton M - Glonass M39, M40 and M41 (failure: overfuelling of upper stage, N)* 08 Falcon 9 - Dragon COTS Demo 1 (D), SMDC ONE (C)*, QbX-1, QbX-2, Perseus 000, 001, 002, 003 (I), and Mayflower & CAERUS (C/D)* 15 Soyuz - Soyuz ISS 25S (ISS) 16 Long March 3A - Beidou 2C/M1 (N) 25 GSLV - INSAT 4D (GSAT 5) (failure: loss of booster actuator control, C) 26 Proton M - KA-SAT (C)

(continued)

Europe	Other countries
Launch log	
January 2011	
	20 Delta IV Heavy - NRO L-49 (I)*
	20 Zenit 3F - Electro-L 1 (M)
	22 H 11B - Kounotori (HTV 2) (ISS)
	28 Soyuz - Progress M-09M (ISS)
February 2011	
16 Ariane 5 ES-ATV- ATV 2 (ISS)	01 Rockot - GEO IK-2 No.II (failure: did not reach planned orbit, S)
	06 Minotaur 1 - RPP (NRO L066) (I)*
	24 Shuttle Discovery - STS 133 (ELC-4, PMM) (ISS)
	26 Soyuz 2 – Glonass K1-1 (N)
March 2011	
	04 Taurus XL - Glory (S), Kysat 1 (D), Hermes (C), and Explorer 1 (Prime) (S) (failures: payload fairing failed to separate)
	05 Atlas V 501 - X-37B OTV 2 (I)*
	11 Delta IV Medium + (4,2) – NRO L-27 (I)*
April 2011	
22 Ariane 5ECA- Yahsat 1A and Intelsat New Dawn (C)	04 Soyuz - Soyuz TMA-21 (ISS)
	10 Long March 3A - Beidou 2-IGS 3 (N)
	14 Atlas V 401- NRO L-34 (I)*
	20 PSLV - Resourcesat 2 (R), Youthsat (S), and X-Sat (R)
	27 Soyuz - Progress M-10M (ISS)
May 2011	
20 Ariane 5ECA- Insat 4G/GSAT-8 and ST 2 (C)	04 Soyuz 2 - Meridian 4 (C)*
	07 Atlas V 401 - SBIRS GEO 1 (EW)*
	16 Shuttle Endeavour - STS 134 (AMS, ELC-3) (ISS)
	20 Proton M - Telstar 14R (C)

C communications, *D* development, *I* intelligence, *ISS* international space station, *M* meteorological, *MF* manned flight, *N* navigation, *R* remote sensing, *S* scientific, *EW* early warning system

13.2 Space Science and Exploration

Europe	Other countries
Earth sciences	
June 29 Gravity field and steady-state Ocean Circulation Explorer (GOCE, ESA) released first detailed map of the geoid[a]	*June 10* Failed launch of research satellite STSAT 2B (KARI, Korea) with solar spectrometer and lidar, due to first stage failure[b]
	June 15 Launch of Shijian 12 (Shanghai Academy of Spaceflight Technology, China)

(continued)

Europe	Other countries
Earth sciences	
	to conduct various earth science experiments[b]
	February 01 Failed launch of geodesy satellite GEO-IK-2 (Russia Space Forces, Russia) due to failure of Breeze KM stage control system[b]
	March 04 Failed launch of Glory (NASA, U.S.), intended to measure Earth's energy balance and Explorer 1 (PRIME), intended to measure radiation levels in the Van Allen belts, due to fairing separation failure[b]
	April 20 Launch of YouthSat (India and Russia) to study the upper atmosphere and measure cosmic rays[b]
Astronomy	
June 15 Launch of Picard (CNES, France) to study solar irradiance, diameter, shape and interior[b]	*August 26* Kepler (NASA) discovered two planets transiting the same star[c]
	November 09 Fermi (NASA) discovered large gamma-ray "bubbles" extending beyond the galactic disk[d]
	January 11 Kepler (NASA) discovered the first rocky exoplanet, Kepler 10-b[e]
	January 26 Hubble (NASA) discovered the most distant object yet seen, a compact galaxy 13.2 billion light-years away[f]
	May 16 Alpha Magnetic Spectrometer (AMS-02) installed on the ISS to detect cosmic rays and search for dark matter and antimatter[g]
Exploration	
July 10 Rosetta (ESA) flew by the asteroid Lutetia[h]	*June 13* Hayabusa (JAXA, Japan) returned to Earth[i] with samples from the asteroid Itokawa[j]
August 02 ESA and NASA selected the scientific instruments for the joint ExoMars Trace Gas Orbiter (TGO) mission[k]	*September 16* Lunar Reconnaissance Orbiter, (LRO, NASA), completed the exploration phase of its mission[l]
	October 01 Launch of Chang'e 2 (CNSA, China), China's second lunar orbiter[b]
	November 04 Extrasolar Planet Observations and characterisation/deep impact eXtended Investigation (EPOXI, NASA) flew by comet Hartley 2[m]
	March 18 MErcury Surface, Space ENvironment, GEochemistry and Ranging (MESSENGER, NASA) became the first spacecraft to orbit Mercury[n]
	May 25 Origins Spectral Interpretation Resource Identification Security Regolith Explorer (OSIRIS-REx, NASA), first NASA asteroid sample return mission, announced[o]

(continued)

Europe	Other countries
Earth sciences	
Manned spaceflight	
June 03 Mars500, a full-length simulated mission to Mars began with six crew in Moscow[p]	*June 15* Launch of Soyuz TMA-19 (Roscosmos, Russia) with three ISS Expedition 25 crew[b]
November 22 ESA's astronaut class of 2009 completed basic training[q]	*June 30, September 10, October 27, January 28, April 27* Launches of 5 Progress (Roscosmos, Russia) cargo transfer vehicles to the ISS[b]
December 15 – May 24 Paolo Nespoli (Italy) participated in ISS Expeditions 26/27 as flight engineer for the ESA MagISStra mission[s]	*October 07* Launch of Soyuz TMA-01M (Roscosmos, Russia) with three ISS Expedition 26 crew[b]
February 14 Mars500 crew (ESA) began surface operations simulations[t]	*December 15* Launch of Soyuz TMA-20 (Roscosmos, Russia) with three ISS Expedition 27 crew[b]
February 16 Launch of ATV-2, Johannes Kepler (European Space Agency), ISS resupply vehicle[b]	*January 22* Launch of HTV-2 (JAXA, Japan) ISS resupply vehicle[b]
February 18 Luca Parmitano (Italy) assigned to ISS Expeditions 36/37, to launch in 2013[u]	*February 24* Launch of STS 133 (NASA, U.S.) on ISS ULF5 using OV-103, Discovery. Final mission of Discovery. Seven crew, payload PMM and ELC4[b]
May 16–June 01 Roberto Vittori (Italy) participated in STS 134 as mission specialist and the last non-U.S. astronaut to fly on the Space Shuttle[v]	*April 04* Launch of Soyuz TMA-21 (Roscomos, Russia) with three ISS Expedition 28 crew[b]
	May 16 Launch of STS 134 (NASA, U.S.) on ISS ULF6 using OV-105, Endeavour. Final mission of Endeavour and shuttle programme. Four crew, payload Alpha Magnetic Spectrometer (AMS) and ELC3[b]
	May 24 NASA formally announced that the Orion crew capsule will serve as the Multi-Purpose Crew Vehicle (MPCV) mandated by Congress[r]

[a] "ESA – GOCE – GOCE giving new insights into Earth's gravity." 29 June 2010. European Space Agency 31 Oct. 2011 http://www.esa.int/SPECIALS/GOCE/SEMY0FOZVAG_0.html

[b] "Semi-Annual Launch Report, First Half of 2011." Federal Aviation Administration (FAA) 31 Oct. 2011 http://www.faa.gov/about/office_org/headquarters_offices/ast/reports_studies/; "Commercial Space Transportation: 2010 Year in Review." Federal Aviation Administration (FAA) 31 Oct. 2011 http://www.faa.gov/about/office_org/headquarters_offices/ast/reports_studies/; "Space Launch Report – 2011 Launch Vehicle/Site Statistics." 31 Oct. 2011http://www.spacelaunchreport.com/log2011.html;"Spaceflight Now | Tracking Station | Worldwide launch log" 31 Oct. 2011 http://spaceflightnow.com/tracking/launchlog.html

[c] "NASA – NASA's Kepler Mission Discovers Two Planets Transiting Same Star." 26 Aug. 2010. NASA 31 Oct. 2011 http://www.nasa.gov/home/hqnews/2010/aug/HQ_10-197_Kepler_Results.html

[d] "NASA – NASA'S Fermi Telescope Discovers Giant Structure In Our Galaxy." 9 Nov. 2010. NASA 31 Oct. 2011 http://www.nasa.gov/home/hqnews/2010/nov/HQ_10-295_FERMI.html

[e]"NASA – NASA's Kepler Mission Discovers Its First Rocky Planet." 10 Jan. 2011. NASA 31 Oct. 2011 http://www.nasa.gov/home/hqnews/2011/jan/HQ_11-007_Kepler_Rocky_Planet.html
[f]"NASA – NASA's Hubble Finds Most Distant Galaxy Candidate Ever Seen in Universe." 26 Jan. 2011. NASA 31 Oct. 2011 http://www.nasa.gov/home/hqnews/2011/jan/HQ_11-025_Hubble_Distant_Galaxy.html
[g]"AMS-02 » Archive » AT HOME, AT LAST." 19 May 2011. AMS-02 31 Oct. 2011 http://www.ams02.org/2011/05/at-home-at-last/
[h]"ESA – Rosetta – Rosetta triumphs at asteroid Lutetia." 10 July 2010. European Space Agency 31 Oct. 2011 http://www.esa.int/SPECIALS/Rosetta/SEM44DZOFBG_0.html
[i]"JAXA | Asteroid Explorer 'HAYABUSA' (MUSES-C) Reentry." 14 June 2010. JAXA 31 Oct. 2011 http://www.jaxa.jp/press/2010/06/20100614_hayabusa_e.html
[j]"JAXA | Identification of origin of particles brought back by Hayabusa." 16 Nov. 2010. JAXA 31 Oct. 2011 http://www.jaxa.jp/press/2010/11/20101116_hayabusa_e.html
[k]"NASA – NASA And ESA'S First Joint Mission To Mars Selects Instruments." 2 Aug. 2010. NASA 31 Oct. 2011 http://www.nasa.gov/home/hqnews/2010/aug/HQ_10-181_Joint_Mars_Mission.html
[l]"NASA – NASA'S Lunar Spacecraft Completes Exploration Mission Phase." 15 Sep. 2010. NASA 31 Oct. 2011 http://www.nasa.gov/home/hqnews/2010/sep/HQ_10-223_LRO_Success.html
[m]"NASA – NASA EPOXI Flyby Reveals New Insights Into Comet Features." 4 Nov. 2010. NASA 31 Oct. 2011 http://www.nasa.gov/home/hqnews/2010/nov/HQ_10-291_EPOXI.html
[n]"NASA – NASA'S MESSENGER Spacecraft Begins Historic Orbit Around Mercury." 17 Mar. 2011. NASA 31 Oct. 2011 http://www.nasa.gov/home/hqnews/2011/mar/HQ_11-079_MESSENGER_Orbits.html
[o]"NASA – NASA Selects OSIRIS-REx as Next New Frontiers Mission." 25 May 2011. NASA 31 Oct. 2011 http://www.nasa.gov/centers/goddard/news/releases/2011/11–037.html
[p]"ESA – Mars500 – Hatch closed: 18-month Mars500 mission has begun." 3 June 2010. European Space Agency 31 Oct. 2011 http://www.esa.int/esaMI/Mars500/SEMLUB5XT9G_0.html
[q]"ESA Portal – Graduation of Europe's new astronauts." 22 Nov. 2010. European Space Agency 31 Oct. 2011 http://www.esa.int/esaCP/SEMRFLIRPGG_index_0.html
[r]Berger, Brian. "NASA Makes It Official: Orion To Be Multi-Purpose Crew Vehicle." Space News 30 May 2011: 5
[s]"ESA – MagISStra – MagISStra mission key data." 20 May 2011. European Space Agency 31 Oct. 2011 http://www.esa.int/esaMI/magisstra/SEMUJ7QOHEG_0.html
[t]"ESA – Mars500 – Walking on 'Mars'." 14 Feb. 2011. European Space Agency 31 Oct. 2011 http://www.esa.int/esaMI/Mars500/SEMRCFOT1KG_0.html
[u]ESA Portal – ESA astronaut Luca Parmitano assigned to 2013 Space Station mission." 18 Feb. 2011. European Space Agency 31 Oct. 2011 http://www.esa.int/esaCP/SEM5J3PT1KG_index_0.html
[v]"NASA – Shuttle Missions – STS-134." 9 June 2011. NASA 31 Oct. 2011 http://www.nasa.gov/mission_pages/shuttle/shuttlemissions/sts134/main/index.html

13.3 Applications

Europe	Other countries
Earth observation	
June 21 Launch of TanDEM-X (DLR, Germany) to gather precise elevation data[a]	*June 26* Launch of COMS 1 (KARI, South Korea), an ocean, meteorological and communication satellite[a]
July 26 Astrium-led consortium selected by the U.K. government to build an Earth observation satellite operations and data processing centre[b]	*July 12* Launch of Cartosat 2B (ISRO, India), a multi-purpose Earth observation satellite and Alsat 2A (Algerian National Space Technology Centre, Algeria)[a]
	August 09, September 22 Launches of Yaogan (China) general purpose Earth observation satellites[a]
	August 14 Decommissioning of the Ice, Cloud, and Land Elevation Satellite (ICESat) after 7 years of operations[c]
	August 24 Launch of Tian Hui-1 (China Aerospace Science and Technology Corp.) mapping satellite[a]
	November 04 Launch of Fengyun 3B (China Meteorological Administration) weather satellite[a]
	November 06 Launch of COSMO-SkyMed 4 (ASI, Italy) with X-band radar synthetic aperture radar for civil and military applications[a]
	January 20 Launch of Electro-L 1 (Roshydromet, Russia) weather satellite[a]
	February 24 Funding for the Climate Absolute Radiance and Refractivity Observatory (CLARREO) and Deformation, Ecosystem Structure and Dynamics of Ice (DESDynI) terminated by the U.S. Administration[d]
	April 20 Launch of Resourcesat 2 (ISRO, India)[a]
Intelligence and early warning	
	June 22 Launch of Ofeq 9 (Israeli Defence Ministry) reconnaissance satellite[a]
	September 21, November 21, January 20, February 06, March 11, April 15 Classified launches (L-41, 32, 49, 66, 27 and 34, National Reconnaissance Office, NRO, U.S.)[a]
	September 26 Launch of first Space-Based Surveillance System (SBSS, USAF) satellite[a]

(continued)

Europe	Other countries
Earth observation	
	September 30 Launch of a military reconnaissance satellite (Russian Space Forces)[a]
	March 16 The Space Tracking and Surveillance System (U.S.) tracked a ballistic missile through all phases of flight for the first time[e]
	May 07 Launch of Space-Based Infra-Red System (SBIRS) GEO 1 (USAF) for missile early-warning detection[a]
Navigation	
March 02 European Geostationary Navigation Overlay Service (EGNOS) Safety-of-Life signal made available to aviation. The first time that space-based navigation signals were used for vertical guidance of civil aircraft during landing[f]	*June 02, July 31, October 31, December 17 and April 09* Launches of Beidou (CAST, China) satellites[a]
May 20 Svalbard Galileo ground station inaugurated[g]	*September 02* Launch of Glonass M36, M37 and M38 (Ministry of Defence, Russia) satellites[a]
	September 11 Launch of H 11A 202 (JAXA, Japan) to enhance GPS signals for use in rural and mountainous areas[a]
	December 05 Failed launch of Glonass M39, M40 and M41 (Ministry of Defence, Russia) satellites due to Block DM upper stage failure[a]
	February 26 Launch of first next-generation Glonass K1-1 (Russia Space Forces, Russia) satellite[a]
Telecommunications/broadcasting	
July 09 Comsat Bw-2, a military communications satellite, declared operational in orbit[h]	*June 03* Launch of BADR-5 (Arabsat, Saudi Arabia), to provide direct broadcasting to the Middle East[a]
October 28 Launch of W3B (Eutelsat, France) Mission failure due to fuel leak[a]	*June 26* Launch of Arabsat 5A (Arabsat, Saudi Arabia)[a]
November 26 Launch of HYLAS 1 (Avanti Communications, U.K.) to provide broadband services[a]	*July 10* Launch of EchoStar 15 (EchoStar Communications Corp., U.S.) to serve United States direct broadcast customers[a]
December 26 Launch of KA-SAT (Eutelsat, France) to provide broadband Internet services to Europe[a]	*August 04* Launch of Rascom 1R (RascomStar-QAF, Africa) and Nilesat 201 (Nilesat, Egypt) providing direct broadcast to Egypt and the Middle East[a]
December 29 Launch of Hispasat 1E (Hispasat, Spain) to serve Europe, North and South America[a]	*August 14* Launch of Advanced Extremely High Frequency (AEHF) 1 (USAF) to provide highly-secure communications[a]
April 22 Launch of New Dawn (Intelsat, Luxembourg)[a]	*September 04* Launch of Chinasat 6A (SINO – Satellite Communication Co. Ltd., China)[a]
	September 08 Launch of Gonets M 2 (Roscosmos, Russia)[a]

(continued)

Europe	Other countries
Earth observation	
	October 14 Launch of XM 5 (XM Satellite Radio, Inc., U.S.) to provide digital music and entertainment programming for North America[a]
	October 19 Launch of six Globalsat (Globalstar, Inc., U.S.) mobile communications satellites[a]
	October 28 Launch of BSAT 3B (Broadcasting Satellite System Corp., Japan)[a]
	November 02, May 04 Launch of Meridian (Defence Ministry, Russia) for military communications[a]
	November 14 Launch of SkyTerra-1 (SkyTerra Communications, U.S.) for mobile communications in North and Central America[a]
	November 24 Launch of Shen Tong 1-B (Ministry of Defence, China) communications satellite[a]
	November 26 Launch of Intelsat 17 (Orbital Sciences Corp., U.S.)[a]
	December 08 Launch of SMDC ONE (U.S. Army), a demonstration of a triple CubeSat communications satellite[a]
	December 25 Failed launch of GSAT 5P (Insat, India), due to first stage control electronics[a]
	December 29 Launch of KoreaSat 6 (Korea Telecom, South Korea)[a]
	March 04 Launch of Hermes (Colorado Space Grant Consortium, U.S.)[a]
	April 22 Launch of Yahsat 1A (Al Yah Satellite Communications Co., Abu Dhabi)[a]
	May 20 Launch of Telstar 14R (Telesat, Canada), ST 2 (ST 2 Satellite Ventures, Singapore and Taiwan) and Insat 4G/GSAT 8 (ISRO, India)[a]
Technology development	
June 11 Ikaros (JAXA, Japan) finished deploying the first solar sail[i]	*June 02* Launch of SERVIS 2 (Institute for Unmanned Space Experiment Flyer, Japan) to demonstrate commercial off-the-shelf technologies for space applications[a]
June 15 Launch of Prisma (Swedish Space Corp.) to demonstrate formation flying[a]	*June 04* Launch of demonstration Dragon spacecraft (SpaceX, U.S.), and first flight of Falcon 9 launcher[a]
July 12 Launch of AlSSat-1 (Norwegian Defence Research Establishment) and TIsat-1 (University of Applied Sciences of Southern Switzerland)[a]	*July 12* Launch of StudSat (ISRO, India)[a]

(continued)

Europe	Other countries
Earth observation	
	October 06 Launch of two Shijian (CNSA, China) research satellites for technology demonstration[a]
	November 20 Launch of Space Test Program Satellite 26 (USAF), containing multiple payloads[a]
	November 22 First FAA commercial re-entry license issued for the Dragon (SpaceX, U.S.) spacecraft[j]
	December 03 First Boeing X-37 Orbital Test Vehicle (OTV, USAF) lands at Vandenberg AFB[k]
	December 08 Launch of active Dragon spacecraft, Dragon C1 (SpaceX, U.S.), first spacecraft successfully launched and recovered from orbit by a commercial company; Perseus 000, 001, 002 and 003 (Los Alamos National Lab., U.S.); QbX-1 and QbX-2 (National Reconnaissance Office, NRO, U.S.); and Mayflower-CAERUS (Northrop Grumman and University of Southern California, U.S.) to validate nanosatellite applications[a]
	March 04 Launch of Kysat 1 (Kentucky Space, U.S.)[a]
	March 05 Launch of second X-37 OTV (USAF)[a]
	April 20 Launch of X-Sat (Centre for Research in Satellite Technologies, Singapore), Singapore's first national satellite, to demonstrate remote sensing technology[a]
Business	
September 16 ESA signed contract with EADS Astrium to start phase B1 study of the ESA Lunar Lander[l]	*June 15* Rocketplane, Inc. (U.S.) filed for Chap. 7 bankruptcy[m]
January 10 It is reported that Inmarsat (U.K.) secured a $666 million loan from the Export–import Bank (U.S.) to develop three large Ka-band satellites[n]	*June 16* SpaceX announced firm $492 million contract to launch Iridium's second-generation constellation of 72 satellites[o]
	July 20 Hedge fund Harbinger (U.S.) entered into a $7 billion agreement with Nokia Siemens Networks on hybrid networks[p]
	August 30 NASA's Commercial Reusable Suborbital Research Program (CRuSR) awarded approximately $475,000 to Armadillo Aerospace and Masten Space Systems[q]

(continued)

Europe	Other countries
Earth observation	
	August 31 NASA awarded a $0.91 billion contract extension to United Space Alliance, LLC, for Space Shuttle and ISS programmes
	September 14 NASA awarded a 5-year, $1.24 billion contract extension to The Boeing Co. for ISS engineering support
	October 04 Iridium (U.S.) signed a $1.8 billion credit arrangement with Coface (France) to finance its next-generation constellation. Largest commercial space deal[r]
	October 25 TerreStar Networks filed for Chap. 11 bankruptcy[s]
	February 14 EchoStar Corp. (U.S.) won bidding to purchase Hughes Communications (U.S.) for approximately $2 billion[t]

[a]"Semi-Annual Launch Report, First Half of 2011." Federal Aviation Administration (FAA) 31 Oct. 2011 http://www.faa.gov/about/office_org/headquarters_offices/ast/reports_studies/; "Commercial Space Transportation: 2010 Year in Review." Federal Aviation Administration (FAA) 31 Oct. 2011 http://www.faa.gov/about/office_org/headquarters_offices/ast/reports_studies/; "Space Launch Report – 2011 Launch Vehicle/Site Statistics." 31 Oct. 2011 http://www.spacelaunchreport.com/log2011.html; "Spaceflight Now | Tracking Station | Worldwide launch log" 31 Oct. 2011 http://spaceflightnow.com/tracking/launchlog.html

[b]"Astrium-Led Group to Build Satellite Center in Britain." Space News 26 July 2010: 8

[c]"NASA: ICESat." 31 Oct. 2011 http://icesat.gsfc.nasa.gov/icesat/

[d]Brinton, Turner. "Two High-priority Climate Missions Dropped from NASA's Budget Plans." Space News 28 Feb. 2011: 1

[e]Brinton, Turner. "STSS Satellites Demonstrate 'Holy Grail' of Missile Tracking." Space News 28 March 2011: 10

[f]"ESA Portal – EGNOS navigation system begins serving Europe's aircraft." 2 Mar. 2011. European Space Agency 31 Oct. 2011 http://www.esa.int/esaCP/SEM98MUTLKG_index_0.html

[g]"ESA – Navigation – Galileo takes its place on top of the world." 20 May 2011. European Space Agency 31 Oct. 2011 http://www.esa.int/esaNA/SEMZICMSNNG_index_0.html

[h]"German ComsatBw-2 Declared Operational." Space News 19 July 2010: 9

[i]"JAXA | Small Solar Power Sail Demonstrator 'IKAROS' Successful Solar Sail Deployment." 11 June 2011. JAXA 31 Oct. 2011 http://www.jaxa.jp/press/2010/06/20100611_ikaros_e.html

[j]Clark, Stephen. "FAA issues first commercial re-entry license to SpaceX." 22 Nov. 2010. Spaceflight Now 31 Oct. 2011 http://spaceflightnow.com/falcon9/002/101122license/index.html

[k]"Factsheets: X-37B Orbital Test Vehicle." U.S. Air Force 31 Oct. 2011 http://www.af.mil/information/factsheets/factsheet.asp?id=16639

[l]"ESA News – Next step for ESA's first Moon lander." 16 Sep. 2010. European Space Agency 31 Oct. 2011 http://www.esa.int/esaCP/SEMUV2KOXDG_index_0.html

[m]De Selding, Peter B. "Inmarsat Secures Ex-Im Financing for Global Xpress Satellites." Space News 10 Jan. 2011: 7

[n]Berger, Brian. "Rocketplane, Owner File for Chapter 7 Bankruptcy Liquidation." Space News 12 July 2010: 11

[o]De Selding, Peter B. "SpaceX Undercut Competition To Clinch $492M Iridium Deal." Space News 21 June 2010: 5

[p]De Selding, Peter B. "Harbinger Enlists Nokia Siemens as Hybrid Network Partner." Space News 26 July 2010: 5

[q]"NASA Selects Two Firms for Experimental Space Vehicle Test Flights." 30 Aug. 2010. NASA 31 Oct. 2011 http://www.nasa.gov/home/hqnews/2010/aug/HQ_10-203_CRuSR_Awards.html
[r]"Iridium Communications Inc. – Iridium Signs Coface Facility Agreement." Iridium Everywhere 31 Oct. 2011 http://investor.iridium.com/releasedetail.cfm?ReleaseID=514289
[s]De Selding, Peter B. "TerreStar Files for Bankruptcy, Court OKs EchoStar Cash Infusion." Space News, 25 Oct. 2010: 10
[t]De Selding, Peter B. "EchoStar Corp. To Acquire Hughes in $2 Billion Deal." Space News 21 Feb. 2011: 4

13.4 Policy and International Cooperation

Europe	Other countries
General policy	
June 16 ESA and the European Parliament endorsed the idea of free and open access to data from future Sentinel Earth observation satellites[a]	*June 28* U.S. adopted a new space policy[b]
October 21 A new ESA policy was announced that allows EU and ESA nations not participating in the ISS to place experiments there[c]	*June 29* NASA issued a Broad Agency Announcement (BAA) seeking proposals and industry input on heavy-lift system concepts and propulsion technology[d]
November 30 Germany adopted a new space policy with a utilitarian, sustainable, and cooperative approach to space[e]	*July 20* Russia announced an investment of $800 million in a new launch site, the Vostochny cosmodrome[f]
December 06 ASI (Italy) published its strategic vision for 2010–2020 and considered a bilateral agreement with the U.K. on space-related R&D[g, h]	
April 02 European Commission laid out, for the first time, its strategic view on space activities[i]	
January 18 ESA reported that 14 of its 18 Member States have agreed to raise their contributions to the Agency[j]	*July 29* JAXA announced the start of year-round launch capability at Tanegashima, following on from its Basic Plan for Space Policy[k]
	August 31 U.S. President Barack Obama outlined reforms to the U.S. export licensing system[l]
	October 11 U.S. President Barack Obama signed the 2010 NASA Authorisation Act[m]
	February 14 2012 budget for National Oceanic and Atmospheric Administration (NOAA, U.S.) announced of $5.5 billion, most of which was for space programmes[n]
	April 14 The USAF, NASA, and National Reconnaissance Office (NRO) conclude a memorandum to allow commercial rocket

(continued)

<table>
<tr><th>Europe</th><th>Other countries</th></tr>
<tr><td colspan="2">General policy</td></tr>
<tr><td></td><td>developers to launch national security payloads[o]</td></tr>
<tr><td></td><td>May 25 Indian defence ministry released the “Technology Perspective and Capability Roadmap”, calling for anti-satellite weapons[p]</td></tr>
<tr><td colspan="2">General cooperation</td></tr>
<tr><td>September 22 Cooperation agreement signed between Norway and EU to permit Norway to participate in the Galileo programme[q]</td><td>July 21 JAXA and the Asian Development Bank signed a letter of intent to improve cooperation[r]</td></tr>
<tr><td>October 06 ESA reported on cooperation with the International Telecommunications Union (ITU) and national governments to reduce radio interference with the Soil Moisture and Ocean Salinity (SMOS) mission[s]</td><td>October 16 NASA administrator Charles Bolden started a visit to China, despite strong opposition from some Republican lawmakers[t]</td></tr>
<tr><td>November 19 Leaders of all 28 NATO nations endorsed a ballistic missile shield to defend Europe[u]</td><td>November 17 The heads of 27 space agencies agreed on the need for international cooperation[v]</td></tr>
<tr><td>March 17 ESA formally committed to extending Europe’s participation in the ISS until 2020[w]</td><td>May 04 In his testimony before the House Appropriations Committee, the White House science adviser stated that China is viewed as a potential U.S. partner for a human Mars mission[x]</td></tr>
<tr><td colspan="2">July 21 UKSA (U.K.) and Roscosmos signed a Memorandum of Understanding[y]</td></tr>
<tr><td colspan="2">November 20 NATO and Russia were reported to agree to cooperate on a joint ballistic missile threat assessment[z]</td></tr>
<tr><td colspan="2">December 08 DLR (Germany) and NASA concluded a framework agreement for bilateral cooperation[aa]</td></tr>
<tr><td colspan="2">March 29 NASA advised ESA that it cannot fund its rover for the joint ExoMars mission, leading ESA to issue stop-work orders for all ExoMars contracts[ab]; contract payments were later expected to restart in July 2011[ac]</td></tr>
<tr><td colspan="2">Agreements between JAXA, NASA and ESA signed to permit free distribution of data from Japan’s Advanced Land Observing Satellite (ALOS)[ad]</td></tr>
<tr><td colspan="2">Ongoing collaboration between DLR (Germany) and NASA (U.S.) to operate the Stratospheric Observatory For Infrared Astronomy (SOFIA) airborne telescope[ae]</td></tr>
<tr><td colspan="2">Space science</td></tr>
<tr><td>July 05 ESA’s Planck microwave telescope delivers its first all-sky image[af]</td><td>August 13 The National Academy of Sciences (U.S.) released its Decadal Survey of Astronomy and Astrophysics[ag]</td></tr>
<tr><td></td><td>November 10 An independent review found that the James Webb Space Telescope (JWST) had a cost overrun of $1.5 billion[ah]</td></tr>
<tr><td colspan="2">June 14 NASA (U.S.) and DLR (Germany) signed an agreement to extend the Gravity Recovery And Climate Experiment (GRACE) mission by 2 years[ai]</td></tr>
<tr><td colspan="2">Applications</td></tr>
<tr><td>October 11 It is reported that ESA is collaborating with the Defence Advanced</td><td>July 15 The Federal Communications Commission (FCC, U.S.) proposed rules to</td></tr>
</table>

(continued)

Europe	Other countries
General policy	
Research Projects Agency (DARPA, U.S.) to study using Inmarsat-4 satellites for space data relay[aj]	make it easier for terrestrial wireless broadband providers to use radio spectrum reserved for satellites[ak]
November 18 ESA signed a contract with an industrial consortium to build six next-generation meteorological satellites[al]	*November 08* U.S. and Australian defence departments signed an agreement to cooperate on space situational awareness activities[am]
January 18 The European Commission published a report on Galileo which forecasts low revenue and larger-than-expected operating costs,[an] and states that efforts to resolve signal overlaps between it and Chinese satellite navigation systems pose a "major problem for the security of the EU"[ao]	*March 27* Russia and Israel signed a framework agreement that could lead to an Israeli-based Glonass ground station[ap]

[a]De Selding, Peter B. "European Officials Embrace Open Data Policy for GMES Satellites." Space News 5 July 2010: 5

[b]De Selding, Peter B. "New ESA Policy Broadens European Access to the International Space Station." Space News 25 Oct. 2010: 6

[c]Federal Ministry of Economics and Technology. Making Germany's Space Sector Fit for the Future: The Space Strategy of the German Federal Government. Berlin: BMWI, 2010: 9. http://www.bmwi.de/English/Redaktion/Pdf/space-strategy,property=pdf,bereich=bmwi,sprache=en,rwb=true.pdf

[d]Italian Space Agency. Strategic Vision 2010–2020. Rome: ASI, 2010. http://www.asi.it/files/ASI_DVS_2010_2020_ENG_0.pdf

[e]Nativi, Andy and Taverna, Michael A. "Rome, London Seek ways To Compensate for Limited Space Budgets." Aviation Week and Space Technology 26 July 2010: 44

[f]European Commission. Communication from the Commission to the Council, the European Parliament, the European Economic and Social Committee and the Committee of the Regions: Towards a Space Strategy for the European Union that Benefits its Citizens. COM (2011) 152 final of 4 April 2011. Brussels: European Union

[g]De Selding, Peter B. "ESA Budget Rises to $4B as 14 Nations Boost Contributions." Space News 24 Jan. 2011: 5

[h]"Briefing by Senior Administration Officials on the President's National Space Policy Via Teleconference." 28 June 2010. U.S. Department of State 31 Oct. 2011 http://www.state.gov/r/pa/prs/ps/2010/06/143752.htm

[i]"NASA Issues Broad Agency Announcement For Heavy Lift Studies." 29 June 2010. NASA 31 Oct. 2011 http://www.nasa.gov/home/hqnews/2010/jun/HQ_10-147_Heavy_Lift_BAA.html

[j]"Russia to kick off construction of a new spaceport." 20 July 2010. BBC News 31 Oct. 2011 http://www.bbc.co.uk/news/science-environment-10698433

[k]"Deal Permits Year-round Launches from Japan." Space News 9 Aug. 2010: 9

[l]Klamper, Amy. "Obama Outlines Plan To Tackle U.S. Export Control Reforms." Space News 6 Sept. 2010: 10

[m]"Statement by the Press Secretary on S. 3729 | The White House." 11 Oct. 2010. The White House 31 Oct. 2011 http://www.whitehouse.gov/the-press-office/2010/10/11/statement-press-secretary-s-3729

[n]"NOAA – National Oceanic and Atmospheric Administration – NOAA Announces FY 2012 Budget." 31 Oct. 2011 http://www.noaanews.noaa.gov/stories2011/20110214_fy2012budget.html

[o]De Selding, Peter B. "SpaceX Receives Boost in Bid To Loft National Security Satellites." Space News 18 April 2011: 1

[p]"Anti-Satellite Weapons Part Of Indian Technology Vision." Space News 7 June 2010: 8
[q]"Norway Pledges 70M Euros As It Joins Galileo Programme." Space News 27 Sept. 2010: 8
[r]"ESA News – SMOS water mission winning battle with interference." 6 Oct. 2010. European Space Agency 31 Oct. 2011 http://www.esa.int/esaCP/SEMB5USOREG_index_0.html
[s]Brinton, Turner. "NATO To Deploy Missile Shield, Seeks Russian Cooperation." Space News 29 Nov. 2010: 10
[t]De Selding, Peter B. "Renewed Arianespace Subsidies Prompt Protest Threat from ILS." Space News 21 March 2011: 1
[u]"JAXA Agrees to Promote Satellite Use in Asia-Pacific." Space News 2 Aug. 2010: 9
[v]"NASA Chief Says Visit to China Lays Foundation for Future Cooperation." 25 Oct. 2010. Space.com 11 Aug. 2011 http://www.space.com/9391-nasa-chief-visit-china-lays-foundation-future-cooperation.html
[w]Svitak, Amy. "Heads of 27 Space Agencies Agree on Need for Cooperation." Space News 22 Nov. 2010: 14
[x]Svitak, Amy. "China Viewed as Potential U.S. Partner in Future Mars Exploration." Space News 9 May 2011: 5
[y]"News – UK Space Agency and Roscosmos sign Agreement." 14 July 2011. UK Space Agency 31 Oct. 2011 http://www.bis.gov.uk/ukspaceagency/news-and-events/2011/Jul/uk-space-agency-and-roscomos-sign-agreement
[z]Brinton, Turner. "NATO To Deploy Missile Shield, Seeks Russian Cooperation." Space News 29 Nov. 2010: 10
[aa]"ISECG Annual Report 2010." International Space Exploration Coordination Group (ISECG) 31 Oct. 2011 http://www.globalspaceexploration.org/c/document_library/get_file?uuid=927d699a-2126-4205-b182-2fc8409eff88&groupId=10812
[ab]De Selding, Peter B. "ESA Halts ExoMars Orbiter Work To Rethink Red Planet Plans with NASA." Space News 25 Apr. 2011: 1.
[ac]De Selding, Peter B. "ESA Expects by July To Restart Work on 2016 Mars Mission." Space News 30 May 2011: 6
[ad]"Partial Privatization Eyed For Japan's ALOS System." Space News 5 July 2011: 15
[ae]"About SOFIA: The Next Generation Airborne Observatory." 31 Oct. 2011 http://www.sofia.usra.edu/Sofia/sofia.html
[af]"ESA News – Planck unveils the Universe – now and then." 5 July 2010. European Space Agency 31 Oct. 2011 http://www.esa.int/esaCP/SEMF2FRZ5BG_index_0.html
[ag]"BPA: Board on Physics and Astronomy." National Academy of Sciences 31 Oct. 2011 http://sites.nationalacademies.org/bpa/BPA_049810
[ah]Klamper, Amy. "Latest $1.5B in JWST Cost Overruns Imperils Other High-Priority Projects." Space News 15 Nov. 2010: 1
[ai]"United States and Germany Agree to Extend GRACE Mission." Space News 14 June 2010: 13
[aj]De Selding, Peter B. "ESA and DARPA To Study Use of Inmarsat-4 for Space Data Relay." Space News 11 Oct. 2010: 1
[ak]De Selding, Peter B. "European Space Agency Signs Long-contended MTG Contract." Space News 22 Nov. 2010: 5
[al]De Selding, Peter B. "European Commission Report Shines Harsh Light on Galileo." Space News 24 Jan. 2011: 1
[am]De Selding, Peter B. "Europe, China at Impasse on Satellite Navigation Overlap." Space News 24 Jan. 2011: 4
[an]De Selding, Peter B. "FCC Moves To Free Up Spectrum Reserved for Satellites." Space News 19 July 2010: 5
[ao]Brinton, Turner. "U.S., Australia Sign Agreement On Space Situational Awareness." Space News 15 Nov. 2010: 4
[ap]Opall-Rome, Barbara. "Israel Could Host Glonass Ground Station Under Framework Deal with Russia." Space News 4 April 2011: 10

13.5 Country Profiles

Austria

FFG

Population[a]	8.40 million
GDP[b]	286.2 billion euros
Responsibility[c]	The Austrian Space Program is funded by the Federal Ministry for Transport, Innovation and Technology and managed by the Agency for Aeronautics and Space of the Austrian Research Promotion Agency
Activities[c]	In addition to ESA programmes, two main national programmes: the Austrian Space Applications Programme and the Austrian Radionavigation Technology and Integrated Satnav Services and Products Testbed
Budget[d]	2011 ESA contribution of 54.0 million euros
Staff[e]	ALR – 13
Direct employment in Space Manufacturing Industry[f]	320

[a]"Eurostat – Tables, Graphs and Maps Interface (TGM) table: Population." 28 Oct. 2011 http://epp.eurostat.ec.europa.eu/tgm/refreshTableAction.do?tab=table&plugin=1&pcode=tps00001&language=en

[b]"Eurostat – Tables, Graphs and Maps Interface (TGM) table: GDP." 28 Oct. 2011 http://epp.eurostat.ec.europa.eu/tgm/refreshTableAction.do?tab=table&plugin=1&pcode=tec00001&language=en

[c]"Die Österreichische Forschungsförderungsgesellschaft FFG | FFG." 31 Oct. 2011 http://www.ffg.at/

[d]"ESA Budget for 2011." European Space Agency 31 Oct. 2011 http://download.esa.int/docs/DG/ESA_2011_Budget_040111_rev2.ppt#256,1,Slide1

[e]"Agentur für Luft- und Raumfahrt – MitarbeiterInnen-Verzeichnis." 28 Oct. 2011 http://www.ffg.at/team#AgenturfürLuft-undRaumfahrt

[f]Lionnet, Pierre, Perrier, Paul A., and Moranta, S. "The European space industry in 2010, facts & figures, 15th edition, June 2011." ASD-Eurospace 28 Oct. 2011: 6 http://www.eurospace.org/FANDF2010/eurospace_f_and_f_2011web.pdf

Belgium

Population[a]	10.92 million
GDP[b]	354.4 billion euros
Responsibility[c]	The Department for Space Research and Applications, under the Belgian Federal Science Policy Office, manages Belgian participation in the programmes and activities of the European Space Agency and of other national or international organisations which are responsible for research and scientific public service
Activities[d]	Manages Belgian public participation in R&D activities in the themes of "generic technology, launchers, microgravity, space sciences, Earth observation, telecommunication and navigation", e.g. ESA, EUMETSAT and SPOT. Consulting with other public authorities interested in space, educational activities and popularisation of space research
Budget[e]	2011 ESA contribution of 164.8 million euros
Staff[f]	Department for space research and applications: about 20
Direct employment in Space Manufacturing Industry[g]	1,446

[a]"Eurostat – Tables, Graphs and Maps Interface (TGM) table: Population." 28 Oct. 2011 http://epp.eurostat.ec.europa.eu/tgm/refreshTableAction.do?tab=table&plugin=1&pcode=tps00001&language=en
[b]"Eurostat – Tables, Graphs and Maps Interface (TGM) table: GDP." 28 Oct. 2011 http://epp.eurostat.ec.europa.eu/tgm/refreshTableAction.do?tab=table&plugin=1&pcode=tec00001&language=en
[c]Belgian Science Policy: Space research and Applications." 28 Oct. 2011 http://www.belspo.be/belspo/act/rech/spatres_en.stm
[d]"AMS-02 » Archive » AT HOME, AT LAST." 19 May 2011. AMS-02 31 Oct. 2011 http://www.ams02.org/2011/05/at-home-at-last/
[e]"ESA Budget for 2011." European Space Agency 31 Oct. 2011 http://download.esa.int/docs/DG/ESA_2011_Budget_040111_rev2.ppt#256,1,Slide1
[f]"European space agencies." Czech Republic Ministry of Transport: Space Technologies and Satellite Systems Department 28 Oct. 2011 http://www.spacedepartment.cz/en/european-space-agencies-1/
[g]"NASA – NASA's Hubble Finds Most Distant Galaxy Candidate Ever Seen in Universe." 26 Jan. 2011. NASA 31 Oct. 2011 http://www.nasa.gov/home/hqnews/2011/jan/HQ_11-025_Hubble_Distant_Galaxy.html

Czech Republic

Population[a]	10.53 million
GDP[b]	149.3 billion euros
Responsibility[c]	The Ministry of Education, Youth and Sports is responsible for the implementation of all space-related agreements the Czech Republic has signed, e.g. the agreement with ESA. Additionally, the Czech Space Office maintains direct communication with space-related experts and institutions

	inside the country and at the same time develops relationships and programmes supporting international cooperation
Activities[d]	The key sectors of national space activities include astronomical studies, magnetospheric, ionospheric and atmospheric research, remote sensing, Earth observation, microgravity research experiments, and competitive industry production, including small satellites construction
Budget[e]	2011 ESA contribution of 10.4 million euros
Staff[f, g]	CSO: 13
	Members: 19

[a]"Eurostat – Tables, Graphs and Maps Interface (TGM) table: Population." 28 Oct. 2011 http://epp.eurostat.ec.europa.eu/tgm/refreshTableAction.do?tab=table&plugin=1&pcode=tps00001&language=en
[b]"Eurostat – Tables, Graphs and Maps Interface (TGM) table: GDP." 28 Oct. 2011 http://epp.eurostat.ec.europa.eu/tgm/refreshTableAction.do?tab=table&plugin=1&pcode=tec00001&language=en
[c]"Czech Space Office – Welcome." 28 Oct. 2011 http://www.czechspace.cz/en/welcome-to-czech-space
[d]"NASA – NASA And ESA'S First Joint Mission To Mars Selects Instruments." 2 Aug. 2010. NASA 31 Oct. 2011 http://www.nasa.gov/home/hqnews/2010/aug/HQ_10-181_Joint_Mars_Mission.html
[e]"ESA Budget for 2011." European Space Agency 31 Oct. 2011 http://download.esa.int/docs/DG/ESA_2011_Budget_040111_rev2.ppt#256,1,Slide1
[f]"Czech Space Office – Organisation structure of the CSO." 28 Oct. 2011 http://www.czechspace.cz/en/cso/organisation-structure-of-the-cso
[g]"Czech Space Office – Members" 28 Oct. 2011 http://www.czechspace.cz/en/cso/members

Denmark

DTU Space
Institut for Rumforskning og -teknologi

Population[a]	5.56 million
GDP[b]	234.0 billion euros
Responsibility[c]	The Ministry of Science, Technology and Innovation is responsible for the national space policy and space activities
Activities[d]	In addition to ESA programmes, bilateral cooperation is undertaken with the U.S., Sweden, Russia and France
Budget[e]	2011 ESA contribution of 31.2 million euros
Staff[f]	DTU:145
Direct employment in Space Manufacturing Industry[g]	231

[a]"Eurostat – Tables, Graphs and Maps Interface (TGM) table: Population." 28 Oct. 2011 http://epp.eurostat.ec.europa.eu/tgm/refreshTableAction.do?tab=table&plugin=1&pcode=tps00001&language=en
[b]"Eurostat – Tables, Graphs and Maps Interface (TGM) table: GDP." 28 Oct. 2011 http://epp.eurostat.ec.europa.eu/tgm/refreshTableAction.do?tab=table&plugin=1&pcode=tec00001&language=en
[c]"DTU Space – National Space Institute." 28 Oct. 2011 http://www.space.dtu.dk/English.aspx
[d]"NASA – NASA'S Lunar Spacecraft Completes Exploration Mission Phase." 15 Sep. 2010. NASA 31 Oct. 2011 http://www.nasa.gov/home/hqnews/2010/sep/HQ_10-223_LRO_Success.html
[e]"ESA Budget for 2011." European Space Agency 31 Oct. 2011 http://download.esa.int/docs/DG/ESA_2011_Budget_040111_rev2.ppt#256,1,Slide1
[f]http://www.space.dtu.dk/English/Staff/All.aspx
[g]"NASA – NASA's Hubble Finds Most Distant Galaxy Candidate Ever Seen in Universe." 26 Jan. 2011. NASA 31 Oct. 2011 http://www.nasa.gov/home/hqnews/2011/jan/HQ_11-025_Hubble_Distant_Galaxy.html

Finland

Tekes

Population[a]	5.38 million
GDP[b]	180.3 billion euros
Responsibility[c]	The Finnish Space Committee consists of representatives of all stakeholders and coordinates all of the activities. Tekes is the executive body for space activities and, together with the Academy of Finland for basic research, manages the Finnish participation within ESA programmes and other international projects
Activities[c]	In addition to ESA programmes, Finland has bilateral activities with the USA (TWINS, Mars Science Laboratory, Phoenix, ISS) France (Pleiades), Germany (TanDEM-X and TerraSAR-X) and Japan (ISS), as well as a national space technology programme
Budget[d]	2011 ESA contribution of 20.1 million euros
Staff[c]	Approximately 400 people in Finland and abroad (90 work in regional Centres for Economic Development, Transport and the Environment (ELY Centres)). Twenty programme coordinators from outside Tekes support the work of Tekes' own managers and provide an important contribution
Direct employment in Space Manufacturing Industry[e]	150

[a]"Eurostat – Tables, Graphs and Maps Interface (TGM) table: Population." 28 Oct. 2011 http://epp.eurostat.ec.europa.eu/tgm/refreshTableAction.do?tab=table&plugin=1&pcode=tps00001&language=en
[b]"Eurostat – Tables, Graphs and Maps Interface (TGM) table: GDP." 28 Oct. 2011 http://epp.eurostat.ec.europa.eu/tgm/refreshTableAction.do?tab=table&plugin=1&pcode=tec00001&language=en
[c]"Tekes organisation." 28 Oct. 2011 http://www.tekes.fi/en/community/Organisation/344/Organisation/1288
[d]"ESA Budget for 2011." European Space Agency 31 Oct. 2011 http://download.esa.int/docs/DG/ESA_2011_Budget_040111_rev2.ppt#256,1,Slide1
[e]"NASA – NASA's Hubble Finds Most Distant Galaxy Candidate Ever Seen in Universe." 26 Jan. 2011. NASA 31 Oct. 2011 http://www.nasa.gov/home/hqnews/2011/jan/HQ_11-025_Hubble_Distant_Galaxy.html

France

cnes

Population[a]	65.08 million
GDP[b]	1,932.8 billion euros
Responsibility[c]	The Centre National d'Etudes Spatiales (CNES) is responsible for French space activities. It is under the shared responsibility of the Ministry of Education and Research and of the Ministry of Defence. The Office Nationale d'Etudes et de Recherches Aérospatiales (ONERA) is also responsible for space related research
Activities[d]	In addition to ESA programmes, civil, military and science programmes are undertaken (Pleiades, Helios, Essaim, e-Gorce), as well as bilateral cooperation with the U.S.

	(CALIPSO, Jason 2 and 3) and India (Saral, Oceansat 3, Altika-Argos, Megha Tropiques)
Budget[e, f, g]	Approximately 1.9 billion euros in 2010. 2011 ESA contribution of 751.4 million euros
Staff[d]	CNES: app. 2,400
Direct employment in Space Manufacturing Industry[h]	12,082

[a]"Eurostat – Tables, Graphs and Maps Interface (TGM) table: Population." 28 Oct. 2011 http://epp.eurostat.ec.europa.eu/tgm/refreshTableAction.do?tab=table&plugin=1&pcode=tps00001&language=en
[b]"Eurostat – Tables, Graphs and Maps Interface (TGM) table: GDP." 28 Oct. 2011 http://epp.eurostat.ec.europa.eu/tgm/refreshTableAction.do?tab=table&plugin=1&pcode=tec00001&language=en
[c]"CNES – De l'Espace pour la Terre: About CNES." 28 Oct. 2011 http://www.cnes.fr/web/CNES-en/3773-about-cnes.php
[d]"NASA – NASA Selects OSIRIS-REx as Next New Frontiers Mission." 25 May 2011. NASA 31 Oct. 2011 http://www.nasa.gov/centers/goddard/news/releases/2011/11-037.html
[e]"ESA Budget for 2011." European Space Agency 31 Oct. 2011 http://download.esa.int/docs/DG/ESA_2011_Budget_040111_rev2.ppt#256,1,Slide1
[f]Pagkratis, Spyros. Space Policies, Issues and Trends in 2010/11. European Space Policy Institute: Vienna, 2011: 13, Figure 2.1
[g]"ECB: Euro foreign exchange reference rates." European Central Bank 12 Aug. 2011 http://www.ecb.int/stats/exchange/eurofxref/html/index.en.html. Exchange rate EUR:USD on 31 Dec. 2010
[h]"NASA – NASA's Hubble Finds Most Distant Galaxy Candidate Ever Seen in Universe." 26 Jan. 2011. NASA 31 Oct. 2011 http://www.nasa.gov/home/hqnews/2011/jan/HQ_11-025_Hubble_Distant_Galaxy.html

Germany

DLR

Population[a]	81.75 million
GDP[b]	2,476.8 billion Euro
Responsibility[c]	The German Space Agency within the German Aerospace Center (DLR) is responsible for German space activities. It is under the responsibility of the Ministry of Economics and Technology
Activities[d]	In addition to ESA programmes, Germany has national civil and commercial programmes in Earth observation (RapidEye, TerraSAR-X, TanDEM, EnMAP), Human space flight (ISS, Microgravity experiments), launch services (Eurockot, OHB-Cosmos), associated ground systems and space technologies (such as intersatellite links). Germany is involved in bilateral cooperation with the U.S. (GRACE, Dawn, Sofia) and its military programs include remote sensing satellites (SAR-Lupe radar satellites) and satcoms (Satcom BW)
Budget[e, f]	DLR Overall budget 1047 million euros. Of which 2011 ESA contribution of 713.8 million euros
Staff[c]	DLR for space activities: app. 2,000

Direct employment in Space Manufacturing Industry[g]	6,112

[a]"Eurostat – Tables, Graphs and Maps Interface (TGM) table: Population." 28 Oct. 2011 http://epp.eurostat.ec.europa.eu/tgm/refreshTableAction.do?tab=table&plugin=1&pcode=tps00001&language=en
[b]"Eurostat – Tables, Graphs and Maps Interface (TGM) table: GDP." 28 Oct. 2011 http://epp.eurostat.ec.europa.eu/tgm/refreshTableAction.do?tab=table&plugin=1&pcode=tec00001&language=en
[c]"DLR Portal – DLR." 28 Oct. 2011 http://www.dlr.de/dlr/desktopdefault.aspx/tabid-10002
[d]"ESA – MagISStra – MagISStra mission key data." 20 May 2011. European Space Agency 31 Oct. 2011 http://www.esa.int/esaMI/magisstra/SEMUJ7QOHEG_0.html
[e]"ESA Budget for 2011." European Space Agency 31 Oct. 2011 http://download.esa.int/docs/DG/ESA_2011_Budget_040111_rev2.ppt#256,1,Slide1
[f]"DLR Portal – DLR at a glance – DLR at a glance." 28 Oct. 2011 http://www.dlr.de/dlr/en/desktopdefault.aspx/tabid-10443/637_read-251/
[g]"NASA – NASA's Hubble Finds Most Distant Galaxy Candidate Ever Seen in Universe." 26 Jan. 2011. NASA 31 Oct. 2011 http://www.nasa.gov/home/hqnews/2011/jan/HQ_11-025_Hubble_Distant_Galaxy.html

Greece

Population[a]	11.33 million
GDP[b]	227.3 billion euros
Responsibility[c]	The General Secretariat for Research and Technology (GSRT) of the Ministry of Education is responsible for Greek space activities
Activities[d]	The Greek space activities cover mainly the fields of space physics, ionospheric physics, Earth observation and telecommunications
Budget[e]	2011 ESA contribution of 14.9 million euros
Staff	For space: < 5

[a]"Eurostat – Tables, Graphs and Maps Interface (TGM) table: Population." 28 Oct. 2011 http://epp.eurostat.ec.europa.eu/tgm/refreshTableAction.do?tab=table&plugin=1&pcode=tps00001&language=en
[b]"Eurostat – Tables, Graphs and Maps Interface (TGM) table: GDP." 28 Oct. 2011 http://epp.eurostat.ec.europa.eu/tgm/refreshTableAction.do?tab=table&plugin=1&pcode=tec00001&language=en
[c]"Γενική Γραμματεία Έ ρευναςτΘ και τεχνολογίαςτΘ: αρχική Σελίδα." 26 Aug. 2011 http://www.gsrt.gr/central.aspx?sId=119I428I1089I646I488772
[d]"ESA Portal – ESA astronaut Luca Parmitano assigned to 2013 Space Station mission." 18 Feb. 2011. European Space Agency 31 Oct. 2011 http://www.esa.int/esaCP/SEM5J3PT1KG_index_0.html
[e]"ESA Budget for 2011." European Space Agency 31 Oct. 2011 http://download.esa.int/docs/DG/ESA_2011_Budget_040111_rev2.ppt#256,1,Slide1

Hungary

Population[a]	9.99 million
GDP[b]	97.1 billion euros
Responsibility[c]	The Hungarian Space Office, under the responsibility of the Ministry of Environment and Water, manages, coordinates and represents Hungarian space activities

Activities[d]	Participation in microgravity, Earth observation, life and material sciences and GSTP programs of ESA
Budget	In 2009, Hungary spent about 2 million euros to satisfy its PECS agreements with ESA, and contributed 1.3 million euros to EUMETSAT
Staff[e]	Total space activities: 250

[a]"Eurostat – Tables, Graphs and Maps Interface (TGM) table: Population." 28 Oct. 2011 http://epp.eurostat.ec.europa.eu/tgm/refreshTableAction.do?tab=table&plugin=1&pcode=tps00001&language=en
[b]"Eurostat – Tables, Graphs and Maps Interface (TGM) table: GDP." 28 Oct. 2011 http://epp.eurostat.ec.europa.eu/tgm/refreshTableAction.do?tab=table&plugin=1&pcode=tec00001&language=en
[c]"A Magyar Űrkutatási Iroda – The structure of the Hungarian space activity" 28 Oct. 2011 http://www.hso.hu/page.php?page=215
[d]"NASA – Shuttle Missions – STS-134." 9 June 2011. NASA 31 Oct. 2011 http://www.nasa.gov/mission_pages/shuttle/shuttlemissions/sts134/main/index.html
[e]"ESA – Rosetta – Rosetta triumphs at asteroid Lutetia." 10 July 2010. European Space Agency 31 Oct. 2011 http://www.esa.int/SPECIALS/Rosetta/SEM44DZOFBG_0.html

Ireland

ENTERPRISE IRELAND

Population[a]	4.48 million
GDP[b]	156.0 billion euros
Responsibility[c]	Enterprise Ireland, in association with the Office of Science and Technology of the Department of Enterprise, Trade and Employment, manages and coordinates space activities in Ireland
Activities[d]	Irish space activities are in the fields of software systems and services, precision mechanical components, advanced materials, electronics/microelectronics and telecommunications systems and service engineering
Budget[e]	2011 ESA contribution of 15.6 million euros
Staff[d]	For space: <5
Direct employment in Space Manufacturing Industry[f]	26

[a]"Eurostat – Tables, Graphs and Maps Interface (TGM) table: Population." 28 Oct. 2011 http://epp.eurostat.ec.europa.eu/tgm/refreshTableAction.do?tab=table&plugin=1&pcode=tps00001&language=en
[b]"Eurostat – Tables, Graphs and Maps Interface (TGM) table: GDP." 28 Oct. 2011 http://epp.eurostat.ec.europa.eu/tgm/refreshTableAction.do?tab=table&plugin=1&pcode=tec00001&language=en
[c]"Home – Enterprise Ireland." 28 Oct. 2011 http://www.enterprise-ireland.com/en/
[d]Berger, Brian. "NASA Makes It Official: Orion To Be Multi-Purpose Crew Vehicle." Space News 30 May 2011: 5
[e]"ESA Budget for 2011." European Space Agency 31 Oct. 2011 http://download.esa.int/docs/DG/ESA_2011_Budget_040111_rev2.ppt#256,1,Slide1
[f]"NASA – NASA's Hubble Finds Most Distant Galaxy Candidate Ever Seen in Universe." 26 Jan. 2011. NASA 31 Oct. 2011 http://www.nasa.gov/home/hqnews/2011/jan/HQ_11-025_Hubble_Distant_Galaxy.html

Italy

Population[a]	60.63 million
GDP[b]	1,548.8 billion euros
Responsibility[c]	The Italian Space Agency (ASI), under the Ministry of Education, Universities and Research, manages Italian space activities
Activities[d]	Italian civil space activities include ESA programmes and a national programme. National activities include scientific missions (AGILE, PRISMA, MIOSAT), dual-use Earth observation satellites (Cosmos-Skymed) and military satcoms (Sicral). Italy conducts bilateral cooperation with France (Athena-Fidus) and Argentina (SIASGE)
Budget[e, f, g]	Approximately 820 million euros in 2010. 2011 ESA contribution of 380 million euros
Staff[d]	ASI: app. 200
Direct employment in Space Manufacturing Industry[h]	5,095

[a]"Eurostat – Tables, Graphs and Maps Interface (TGM) table: Population." 28 Oct. 2011 http://epp.eurostat.ec.europa.eu/tgm/refreshTableAction.do?tab=table&plugin=1&pcode=tps00001&language=en
[b]"Eurostat – Tables, Graphs and Maps Interface (TGM) table: GDP." 28 Oct. 2011 http://epp.eurostat.ec.europa.eu/tgm/refreshTableAction.do?tab=table&plugin=1&pcode=tec00001&language=en
[c]"A.S.I. – Agenzia Spaziale Italiana." 28 Oct. 2011 http://www.asi.it/en
[d]"Astrium-Led Group to Build Satellite Center in Britain." Space News 26 July 2010: 8
[e]"ESA Budget for 2011." European Space Agency 31 Oct. 2011 http://download.esa.int/docs/DG/ESA_2011_Budget_040111_rev2.ppt#256,1,Slide1
[f]"ESA – Mars500 – Hatch closed: 18-month Mars500 mission has begun." 3 June 2010. European Space Agency 31 Oct. 2011 http://www.esa.int/esaMI/Mars500/SEMLUB5XT9G_0.html
[g]"ESA Portal – Graduation of Europe's new astronauts." 22 Nov. 2010. European Space Agency 31 Oct. 2011 http://www.esa.int/esaCP/SEMRFLIRPGG_index_0.html
[h]"NASA – NASA's Hubble Finds Most Distant Galaxy Candidate Ever Seen in Universe." 26 Jan. 2011. NASA 31 Oct. 2011 http://www.nasa.gov/home/hqnews/2011/jan/HQ_11-025_Hubble_Distant_Galaxy.html

Luxembourg

Population[a]	511.84 thousand
GDP[b]	40.3 billion euros
Responsibility[c]	Luxinnovation, the National Agency for Innovation and Research, under the responsibility of the Ministry of Culture, Higher Education and Research, coordinates space activities
Activities[d]	Luxembourg focuses mainly on telecommunications, through SES Astra commercial satellite services provider
Budget[e]	2011 ESA contribution of 11.5 million euros
Staff[f]	36 staff members are listed in the 2010 annual report, not all of whom work in space

Direct employment in Space Manufacturing Industry[g]	31

[a]"Eurostat – Tables, Graphs and Maps Interface (TGM) table: Population." 28 Oct. 2011 http://epp.eurostat.ec.europa.eu/tgm/refreshTableAction.do?tab=table&plugin=1&pcode=tps00001&language=en
[b]"Eurostat – Tables, Graphs and Maps Interface (TGM) table: GDP." 28 Oct. 2011 http://epp.eurostat.ec.europa.eu/tgm/refreshTableAction.do?tab=table&plugin=1&pcode=tec00001&language=en
[c]"Luxinnovation, Agence Nationale pour la Promotion de l'Innovation et de la Recherche." 28 Oct. 2011 http://www.luxinnovation.lu/site/index.jsp
[d]"NASA: ICESat." 31 Oct. 2011 http://icesat.gsfc.nasa.gov/icesat/
[e]"ESA Budget for 2011." European Space Agency 31 Oct. 2011 http://download.esa.int/docs/DG/ESA_2011_Budget_040111_rev2.ppt#256,1,Slide1
[f]"2010 Rapport Annuel." Luxinnovation 28 Oct. 2011 http://www.innovation.public.lu/catalogue-publications/acteurs-cles/rapports-activites/Activity-Report_Luxinnovation_2010_FR.pdf: 36
[g]"NASA – NASA's Hubble Finds Most Distant Galaxy Candidate Ever Seen in Universe." 26 Jan. 2011. NASA 31 Oct. 2011 http://www.nasa.gov/home/hqnews/2011/jan/HQ_11-025_Hubble_Distant_Galaxy.html

Netherlands

Population[a]	16.66 million
GDP[b]	588.4 billion euros
Responsibility[c]	The new Netherlands Space Office (2009) is responsible for coordinating international space activities with ESA, NASA and JAXA, as well as for the development of the Dutch national space programme. The Dutch Space Research Organization (SRON) is responsible for national and multilateral space science programmes
Activities[d]	In addition to its contributions to ESA programmes, such as Gaia, Exo-Mars, and GMES, NSO conducts bilateral activities with Japan (SPICA-SAFARI) and is responsible for planning the Dutch National Space Programme. SRON conducts space research in the fields of Earth observation, microgravity and planetary exploration
Budget[e]	2011 ESA contribution of 84.2 million euros
Staff[d, f]	NSO: 26, SRON: app. 200
Direct employment in Space Manufacturing Industry[g]	794

[a]"Eurostat – Tables, Graphs and Maps Interface (TGM) table: Population." 28 Oct. 2011 http://epp.eurostat.ec.europa.eu/tgm/refreshTableAction.do?tab=table&plugin=1&pcode=tps00001&language=en
[b]"Eurostat – Tables, Graphs and Maps Interface (TGM) table: GDP." 28 Oct. 2011 http://epp.eurostat.ec.europa.eu/tgm/refreshTableAction.do?tab=table&plugin=1&pcode=tec00001&language=en
[c]"Netherlands Space Office – Home." 28 Oct. 2011 http://www.spaceoffice.nl/
[d]Brinton, Turner. "STSS Satellites Demonstrate 'Holy Grail' of Missile Tracking." Space News 28 March 2011: 10
[e]"ESA Budget for 2011." European Space Agency 31 Oct. 2011 http://download.esa.int/docs/DG/ESA_2011_Budget_040111_rev2.ppt#256,1,Slide1
[f]"SRON: Netherlands Institute for Space Research – Jobs|Vacatures" 28 Oct. 2011 http://www.sron.nl/index.php?option=com_content&task=view&id=2809&Itemid=2479

[g]"NASA – NASA's Hubble Finds Most Distant Galaxy Candidate Ever Seen in Universe." 26 Jan. 2011. NASA 31 Oct. 2011 http://www.nasa.gov/home/hqnews/2011/jan/HQ_11-025_Hubble_Distant_Galaxy.html

Norway

Population[a]	4.92 million
GDP[b]	311.9 billion euros
Responsibility[c]	The Norwegian Space Centre (NSC), under the Ministry of Trade and Industry, manages Norwegian space activities
Activities[d]	In addition to ESA programmes (in particular Earth observation, telecommunications and launchers), Norway has national support programmes and commercial activities (Telenor). Moreover, Norway operates the Andøya rocket range and the Svalbard ground station. Norway has also a bilateral agreement with Canada on the use of Radarsat 2 data
Budget[e]	2011 ESA contribution of 63.2 million euros
Staff[f]	NSC: 36
Direct employment in Space Manufacturing Industry[g]	293

[a]"Eurostat – Tables, Graphs and Maps Interface (TGM) table: Population." 28 Oct. 2011 http://epp.eurostat.ec.europa.eu/tgm/refreshTableAction.do?tab=table&plugin=1&pcode=tps00001&language=en
[b]"Eurostat – Tables, Graphs and Maps Interface (TGM) table: GDP." 28 Oct. 2011 http://epp.eurostat.ec.europa.eu/tgm/refreshTableAction.do?tab=table&plugin=1&pcode=tec00001&language=en
[c]"The Norwegian Space Centre." 28 Oct. 2011 http://www.spacecentre.no/english/
[d]"ESA – Navigation – Galileo takes its place on top of the world." 20 May 2011. European Space Agency 31 Oct. 2011 http://www.esa.int/esaNA/SEMZICMSNNG_index_0.html
[e]"ESA Budget for 2011." European Space Agency 31 Oct. 2011 http://download.esa.int/docs/DG/ESA_2011_Budget_040111_rev2.ppt#256,1,Slide1
[f]"Norwegian Space Centre – Staff." 28 Oct. 2011 http://www.spacecentre.no/English/About_NSC/Contact_list/
[g]"NASA – NASA's Hubble Finds Most Distant Galaxy Candidate Ever Seen in Universe." 26 Jan. 2011. NASA 31 Oct. 2011 http://www.nasa.gov/home/hqnews/2011/jan/HQ_11-025_Hubble_Distant_Galaxy.html

Poland

Population[a]	38.20 million
GDP[b]	354.3 billion euros
Responsibility[c, d]	Polish space activities are under the joint responsibility of the Ministry for Scientific Research and Information Technology and the Ministry of Economics Affairs and Labour. Within the Academy of Sciences, the Space Research Centre coordinates space activities and hosts the Polish Space Office (PSO)

(continued)

Activities[e, f]	Polish space activities focus on the fields of space science, navigation and Earth observation applications. Poland is participating in several ESA missions, such as Herschel, Planck, Rosetta, ExoMars and BepiColombo
Budget	In 2009, app. 4 million euros (ESA PECS 1.2 million, national 3 million)
Staff[g]	Space Research Centre: 179

[a]"Eurostat – Tables, Graphs and Maps Interface (TGM) table: Population." 28 Oct. 2011 http://epp.eurostat.ec.europa.eu/tgm/refreshTableAction.do?tab=table&plugin=1&pcode=tps00001&language=en
[b]"Eurostat – Tables, Graphs and Maps Interface (TGM) table: GDP." 28 Oct. 2011 http://epp.eurostat.ec.europa.eu/tgm/refreshTableAction.do?tab=table&plugin=1&pcode=tec00001&language=en
[c]"Centrum Badań Kosmicznych." 28 Oct. 2011 http://www.cbk.waw.pl/
[d]"Polskie Biuro do spraw Przestrzeni Kosmicznej." 28 Oct. 2011 http://www.kosmos.gov.pl/
[e]"JAXA | Small Solar Power Sail Demonstrator 'IKAROS' Successful Solar Sail Deployment." 11 June 2011. JAXA 31 Oct. 2011 http://www.jaxa.jp/press/2010/06/20100611_ikaros_e.html
[f]Clark, Stephen. "FAA issues first commercial re-entry license to SpaceX." 22 Nov. 2010. Spaceflight Now 31 Oct. 2011 http://spaceflightnow.com/falcon9/002/101122license/index.html
[g]"Pracownicy etatowi CBK."28 Oct. 2011 http://www.cbk.waw.pl/index.php?option=com_contact&view=category&catid=49&Itemid=82&lang=pl

Portugal

FCT Space Office

Population[a]	10.64 million
GDP[b]	172.8 billion euros
Responsibility[c]	Portuguese space activities are coordinated by the Portuguese Space Office within the Foundation for Science and Technology (FCT), which is accountable to the Ministry of Science, Technology and Higher Education (GRICES)
Activities[d]	Mainly participation in ESA programs (telecommunications systems, technology developments, Earth observation, exploration), as well as space outreach programmes
Budget[e]	2011 ESA contribution of 15.8 million euros
Staff	App. 6
Direct employment in Space Manufacturing Industry[f]	118

[a]"Eurostat – Tables, Graphs and Maps Interface (TGM) table: Population." 28 Oct. 2011 http://epp.eurostat.ec.europa.eu/tgm/refreshTableAction.do?tab=table&plugin=1&pcode=tps00001&language=en
[b]"Eurostat – Tables, Graphs and Maps Interface (TGM) table: GDP." 28 Oct. 2011 http://epp.eurostat.ec.europa.eu/tgm/refreshTableAction.do?tab=table&plugin=1&pcode=tec00001&language=en
[c]"FCT — Space Office." 28 Oct. 2011 http://www.fct.pt/apoios/cooptrans/espaco/index.phtml.en
[d]"ESA News – Next step for ESA's first Moon lander." 16 Sep. 2010. European Space Agency 31 Oct. 2011 http://www.esa.int/esaCP/SEMUV2KOXDG_index_0.html
[e]"ESA Budget for 2011." European Space Agency 31 Oct. 2011 http://download.esa.int/docs/DG/ESA_2011_Budget_040111_rev2.ppt#256,1,Slide1
[f]"NASA – NASA's Hubble Finds Most Distant Galaxy Candidate Ever Seen in Universe." 26 Jan. 2011. NASA 31 Oct. 2011 http://www.nasa.gov/home/hqnews/2011/jan/HQ_11-025_Hubble_Distant_Galaxy.html

Romania

Population[a]	21.41 million
GDP[b]	121.9 billion euros
Responsibility[c]	The Romanian Space Agency (ROSA), under the responsibility of the Ministry of Education and Research, manages Romanian space activities
Activities[d]	Romania has national activities covering space science (space physics and astronomy), space systems (construction of nanosatellites and microgravity experiments) and space applications (telemedicine, Earth observation and navigation space-based services)
Budget[d]	In 2009, app. 2 million euros were spent to satisfy Romania's PECS agreement with ESA. Romania also spent about 12 million for the "Space and Security" programme within Romania's "National Plan for R&D and Innovation"
Staff[d]	ROSA: 36

[a]"Eurostat – Tables, Graphs and Maps Interface (TGM) table: Population." 28 Oct. 2011 http://epp.eurostat.ec.europa.eu/tgm/refreshTableAction.do?tab=table&plugin=1&pcode=tps00001&language=en
[b]"Eurostat – Tables, Graphs and Maps Interface (TGM) table: GDP." 28 Oct. 2011 http://epp.eurostat.ec.europa.eu/tgm/refreshTableAction.do?tab=table&plugin=1&pcode=tec00001&language=en
[c]"Romanian Space Agency (ROSA)." 28 Oct. 2011 http://www.rosa.ro/
[d]De Selding, Peter B. "Inmarsat Secures Ex-Im Financing for Global Xpress Satellites." Space News 10 Jan. 2011: 7

Spain

Population[a]	46.15 million
GDP[b]	1,062.6 billion euros
Responsibility[c]	The Centre for the Development of Industrial Technology (CDTI), under the Ministry of Science and Innovation, is funding and coordinating Spanish space activities
Activities[d]	In addition to ESA programmes, Spain has a national space programme including governmental and commercial programmes, especially in civil and military telecommunications (Hispasat, Spainsat), and Earth observation (SEOSAT/INGENIO, SEOSAR/PAZ, INTAuSAT 1) and is involved in bilateral cooperation projects with France, Canada, the U.S. and Russia. Spain also manages national, ESA and NASA ground facilities
Budget[e, f, g]	Approximately 300 million euros in 2010. 2011 ESA contribution of 201.9 million euros
Staff[h]	CDTI: app. 300
Direct employment in Space Manufacturing Industry[i]	2,526

[a]"Eurostat – Tables, Graphs and Maps Interface (TGM) table: Population." 28 Oct. 2011 http://epp.eurostat.ec.europa.eu/tgm/refreshTableAction.do?tab=table&plugin=1&pcode=tps00001&language=en
[b]"Eurostat – Tables, Graphs and Maps Interface (TGM) table: GDP." 28 Oct. 2011 http://epp.eurostat.ec.europa.eu/tgm/refreshTableAction.do?tab=table&plugin=1&pcode=tec00001&language=en
[c]"Centro para el Desarrollo Tecnológico Industrial." 26 Aug. 2011 http://www.cdti.es/index.asp?TR=A&IDR=2

[d]Berger, Brian. "Rocketplane, Owner File for Chapter 7 Bankruptcy Liquidation." Space News 12 July 2010: 11
[e]"ESA Budget for 2011." European Space Agency 31 Oct. 2011 http://download.esa.int/docs/DG/ESA_2011_Budget_040111_rev2.ppt#256,1,Slide1
[f]"ESA – Mars500 – Hatch closed: 18-month Mars500 mission has begun." 3 June 2010. European Space Agency 31 Oct. 2011 http://www.esa.int/esaMI/Mars500/SEMLUB5XT9G_0.html
[g]"ESA Portal – Graduation of Europe's new astronauts." 22 Nov. 2010. European Space Agency 31 Oct. 2011 http://www.esa.int/esaCP/SEMRFLIRPGG_index_0.html
[h]"ESA – Rosetta – Rosetta triumphs at asteroid Lutetia." 10 July 2010. European Space Agency 31 Oct. 2011 http://www.esa.int/SPECIALS/Rosetta/SEM44DZOFBG_0.html
[i]"NASA – NASA's Hubble Finds Most Distant Galaxy Candidate Ever Seen in Universe." 26 Jan. 2011. NASA 31 Oct. 2011 http://www.nasa.gov/home/hqnews/2011/jan/HQ_11-025_Hubble_Distant_Galaxy.html

Sweden

Population[a]	9.42 million
GDP[b]	346.9 billion euros
Responsibility[c]	The Swedish National Space Board, SNSB, is under the Ministry of Education and Research, and is responsible for national and international activities relating to space and remote sensing, primarily research and development
Activities[d]	Conducting magnetospheric and ionospheric research, astronomical studies, remote sensing, observation of the Earth's atmosphere and environment and specialized industrial competence (technology development), including development of small cost-efficient satellites. Promoting R&D is the main task of SNSB; and special focus is put on *inter alia* on-board computers, telecom equipment, nano-technology and formation flying platforms. Bilateral cooperation is mainly with France. Low-cost, short-term profile projects are represented by the Swedish-multilateral Viking, Freja and Odin scientific satellites
Budget[e, f]	Approximately 700 million Swedish krona. 2011 ESA contribution of 59.9 million euros
Staff[f]	17
Direct employment in Space Manufacturing Industry[g]	760

[a]"Eurostat – Tables, Graphs and Maps Interface (TGM) table: Population." 28 Oct. 2011 http://epp.eurostat.ec.europa.eu/tgm/refreshTableAction.do?tab=table&plugin=1&pcode=tps00001&language=en
[b]"Eurostat – Tables, Graphs and Maps Interface (TGM) table: GDP." 28 Oct. 2011 http://epp.eurostat.ec.europa.eu/tgm/refreshTableAction.do?tab=table&plugin=1&pcode=tec00001&language=en
[c]"Rymdstyrelsen – Home." Swedish National Space Board 28 Oct. 2011 http://www.snsb.se/en/Home/Home/
[d]"Rymdstyrelsen – Space Activities in Sweden." Swedish National Space Board 28 Oct. 2011 http://www.snsb.se/en/Home/Space-Activities-in-Sweden/
[e]"ESA Budget for 2011." European Space Agency 31 Oct. 2011 http://download.esa.int/docs/DG/ESA_2011_Budget_040111_rev2.ppt#256,1,Slide1
[f]"Rymdstyrelsen – About SNSB." Swedish National Space Board 28 Oct. 2011 http://www.snsb.se/en/Home/About-SNSB/

[g]"NASA – NASA's Hubble Finds Most Distant Galaxy Candidate Ever Seen in Universe." 26 Jan. 2011. NASA 31 Oct. 2011 http://www.nasa.gov/home/hqnews/2011/jan/HQ_11-025_Hubble_Distant_Galaxy.html

Switzerland

Schweizerische Eidgenossenschaft
Confédération suisse
Confederazione Svizzera
Confederaziun svizra

Population[a]	7.87 million
GDP[b]	398.9 billion euros
Responsibility[c]	The Swiss Space Office of the State Secretariat for Education and Research of the Federal Department of Home Affairs is responsible for Swiss space activities and cooperates closely with the Swiss Department of Foreign Affairs on that topic. The Federal Commission for Space Affairs (CFAS) is currently preparing a new Swiss space policy. The interdepartmental coordination committee for space (IKAR) is responsible for coordinating activities
Activities[c]	Most of the Swiss activities are undertaken within ESA programmes (space science, human spaceflight, launchers, Earth observation, Prodex and navigation)
Budget[d]	2011 ESA contribution of 96.2 million euros
Staff[c]	SNSB: 16
Direct employment in Space Manufacturing Industry[e]	796

[a]"Eurostat – Tables, Graphs and Maps Interface (TGM) table: Population." 28 Oct. 2011 http://epp.eurostat.ec.europa.eu/tgm/refreshTableAction.do?tab=table&plugin=1&pcode=tps00001&language=en
[b]"Eurostat – Tables, Graphs and Maps Interface (TGM) table: GDP." 28 Oct. 2011 http://epp.eurostat.ec.europa.eu/tgm/refreshTableAction.do?tab=table&plugin=1&pcode=tec00001&language=en
[c]"Staatssekretariat für Bildung und Forschung SBF." 28 Oct. 2011 http://www.sbf.admin.ch/htm/themen/weltraum_de.html
[d]"ESA Budget for 2011." European Space Agency 31 Oct. 2011 http://download.esa.int/docs/DG/ESA_2011_Budget_040111_rev2.ppt#256,1,Slide1
[e]"NASA – NASA's Hubble Finds Most Distant Galaxy Candidate Ever Seen in Universe." 26 Jan. 2011. NASA 31 Oct. 2011 http://www.nasa.gov/home/hqnews/2011/jan/HQ_11-025_Hubble_Distant_Galaxy.html

United Kingdom

UK SPACE AGENCY

Population[a]	62.44 million
GDP[b]	1,700.1 billion Euro
Responsibility[c]	The UK Space Agency, an executive agency of the Department for Business, Innovation and Skills (BIS), is responsible for the strategic decisions on the UK civil space programme and provides a clear, single voice for UK space ambitions
Activities[d]	UK space activities include bilateral cooperation with JAXA, India and the U.S. and within ESA in EO and space exploration (Cassini-Huygens, James Webb Space Telescope, Herschel, and Planck missions)

Budget[e, f, g, h]	The UK Space Agency's current budget is approximately £220 million per year, while other sources list the budget as approximately 520 million euros. 2011 ESA contribution of 265.3 million euros
Staff[i]	UKSA: 16
Direct employment in Space Manufacturing Industry[j]	3,554

[a]"Eurostat – Tables, Graphs and Maps Interface (TGM) table: Population." 28 Oct. 2011 http://epp.eurostat.ec.europa.eu/tgm/refreshTableAction.do?tab=table&plugin=1&pcode=tps00001&language=en

[b]"Eurostat – Tables, Graphs and Maps Interface (TGM) table: GDP." 28 Oct. 2011 http://epp.eurostat.ec.europa.eu/tgm/refreshTableAction.do?tab=table&plugin=1&pcode=tec00001&language=en

[c]"Who we are I BIS." UK Space Agency 28 Oct. 2011 http://www.bis.gov.uk/ukspaceagency/who-we-are

[d]"UK Space Agency I BIS." 28 Oct. 2011 http://www.bis.gov.uk/ukspaceagency

[e]"ESA Budget for 2011." European Space Agency 31 Oct. 2011 http://download.esa.int/docs/DG/ESA_2011_Budget_040111_rev2.ppt#256,1,Slide1

[f]"ESA – Mars500 – Hatch closed: 18-month Mars500 mission has begun." 3 June 2010. European Space Agency 31 Oct. 2011 http://www.esa.int/esaMI/Mars500/SEMLUB5XT9G_0.html

[g]"ESA Portal – Graduation of Europe's new astronauts." 22 Nov. 2010. European Space Agency 31 Oct. 2011 http://www.esa.int/esaCP/SEMRFLIRPGG_index_0.html

[h]"UK Space Funding I Who we are I BIS." UK Space Agency 28 Oct. 2011 http://www.bis.gov.uk/ukspaceagency/who-we-are/how-we-work/uk-space-funding

[i]De Selding, Peter B. "EchoStar Corp. To Acquire Hughes in $2 Billion Deal." Space News 21 Feb. 2011: 4

[j]"NASA – NASA's Hubble Finds Most Distant Galaxy Candidate Ever Seen in Universe." 26 Jan. 2011. NASA 31 Oct. 2011 http://www.nasa.gov/home/hqnews/2011/jan/HQ_11-025_Hubble_Distant_Galaxy.html

Chapter 14
Bibliography of space policy publications July 2010–June 2011

Blandina Baranes

14.1 Monographs

Abeyratne R (2011) Space security law. Springer, Berlin

Alpatov AP et al (2010) Dynamics of tethered space systems. CRC Press, Boca Raton

Ananthaswamy A (2010) The edge of physics: a journey to earth's extremes to unlock the secrets of the universe. Houghton Mifflin Harcourt, Boston

Arbatov A, Dvorkin V (eds) (2010) Outer space: weapons, diplomacy, and security. Carnegie Endowment for International Peace, Washington, DC

Asphaug V, Sorensen E (eds) (2011) Global positioning systems. Nova Science Publishers, Hauppauge

Books LLC (ed) (2010) Space policy of the European Union: Galileo. Life Journey Books, Sharon

Capterou M (2011) Satellites – Dynamiques et applications. Springer, Berlin

Clark S (2010) The big questions: the universe. Quercus, London

Contant JM (2010) Celebrating 50 years of excellence: space for humanity. International Academy of Astronautics, Paris

De Montluc B (2011) A new international strategic context for space policies. Nova Science Publishers, New York

Dench P, Gregg A (2010) Carnarvon and Apollo: one giant leap for a small Australian town. Rosenberg, Lancaster

Dubbs C, Paat-Dahlstrom E (2011) Realizing tomorrow: the path to private spaceflight. University of Nebraska Press, Lincoln

Duggins P (2010) Trailblazing Mars: NASA's next giant leap. University Press of Florida, Gainesville

B. Baranes (✉)
European Space Policy Institute, Vienna, Austria
e-mail: blandina.baranes@espi.or.at

P. Hulsroj et al. (eds.), *Yearbook on Space Policy 2010/2011*,
Yearbook on Space Policy, DOI 10.1007/978-3-7091-1363-9_14,

Flohrer T, Choc R, Bastida B (2011) Classification of geosynchronous objects. European Space Agency. ESA Space Debris Office, Paris

Gomez J (ed) (2011) Kalman filtering. Nova Science Publishers, Hauppauge

Harvey B, Zakutnyaya O (2011) Russian space probes: scientific discoveries and future missions. Springer Praxis, New York

Häuplik-Meusburger S (2011) Architecture for astronauts: an activity based approach. Springer Praxis Books, Wien

Hays PL (2011) Space and security: a reference handbook. ABC-CLIO, Santa Barbara

IFPA (2010) Air, space, & cyberspace power in the 21st-century. The institute for foreign policy analysis international security studies program of the Fletcher School. Tufts University, Washington, DC

Ito A (2011) Legal aspects of satellite remote sensing. Brill, Leiden

Jakhu R (ed) (2010) National regulation of space activities. Springer, Berlin

Kanas N, Manzey D (2010) Space psychology and psychiatry (space technology library). Springer, New York

Kastin JF (2010) How to find a habitable planet. Princeton University Press, Princeton

Kayser V (2010) Launching space objects: issues of liability and future prospects (Space Regulations library). Springer, New York

Kalkhan MA (2011) Spatial statistics: geospatial information modeling and thematic mapping. CRC Press, London

Landfester U, Remuss NL, Schrogl KU, Worms JC (eds) (2010) Humans in outer space – interdisciplinary perspectives (studies in space policy). Springer, Wien

Larned R, Swan C, Swan P (2010) National security space strategy considerations. Lulu.com

Le Moigne J, Netanyahu NS, Eastman RD (eds) (2011) Image registration for remote sensing. Cambridge University Press, Cambridge

Lutes CD, Hayes P (eds) (2011) Toward a theory of spacepower: selected essays. National Defense University Press, Washington, DC

Mix LJ (2009) Life in space: astrobiology for everyone. Harvard University Press, Cambridge

Nardon L (2010) Les réorientations de la politique spatiale américaine en 2009 et 2010. IFRI, Paris

Norris P (2010) Watching earth from space: how surveillance helps us – and harms us. Springer, Berlin

Pekkanen SM (2010) In defense of Japan: from the market to the military in space policy. Stanford University Press, Stanford

Pelton JN, Jakhu R, Sgobba T (2010) Space safety: regulations and standards. Butterworth-Heinemann, Amsterdam

Perozzi E, Ferraz-Mello S (2010) Space manifold dynamics: novel spaceways for science and exploration. Springer, New York

Reg A (2010) Unmanned aircraft systems: UAVS design, development and deployment. AIAA education series. AIAA, Reston

Santos JM (2011) Final call for a European space strategy. IFRI, Brussels

Space Foundation (2010) The space report: the authoritative guide to global space activity. Space Foundation, Colorado Springs

Swinerd G (2010) How spacecraft fly: spaceflight without formulae. Springer, Wien

Tkatchove S (2011) Space-based technologies and commercialized development: economic implications and benefits. Engineering Science Reference, Hershey

Vakoch D (ed) (2011) Psychology of space exploration. NASA, Washington, DC

Von der Dunk F (ed) (2011) National space legislation in Europe: issues of authorisation of private space activities in the light of developments in European space cooperation. Brill, Leiden

Warren DA, Conway BD (eds) (2010) Future of U.S. Human spaceflight: background and issues. Nova Science Publishers, New York

Werth H, Reiter T (2011) Thomas Reiter: Leben in der Schwerelosigkeit. Herbig Buchverlag, München

Wikborg E (ed) (2010) Space tourism issues. Nova Science Publishers, New York

Williams JN (2010) The work of his hands: a view of God's creation from space. Concordia Publishing House, St. Louis

Worden A, French F, Gordon D (2011) Falling to earth: an Apollo 15 astronaut's journey to the Moon. Smithsonian Books, Washington, DC

14.2 Articles

Abdelkhalik O, Gad A (2011) Optimization of space orbits design for Earth orbiting missions. Acta Astronaut 68:1307–1317

Abou Yehia J, Schrogl KU (2010) European regulation for private human spaceflight in the context of space traffic management. Acta Astronaut 66:1618–1624

Acevedo R, Varela F, Orihuela N (2010) The role of Venestat-1 satellite in promoting development in Venezuela and Latin America. Space Policy 26:189–193

Ansdell M, Ehrenfreund P, McKay C (2011) Stepping stones toward global space exploration. Acta Astronaut 68:2098–2113

Aoki S (2010) Japanese law and regulations concerning remote sensing activities. J Space Law 36(2):335–364

Avdeyev S et al (2011) Human space exploration – a global trans-cultural quest. Space Policy 27:24–26

Balogh W et al (2010) The United Nations Programme on Space Applications: status and direction for 2010. Space Policy 26:185–188

Bensoussan D (2010) Space tourism risks: a space insurance perspective. Acta Astronaut 66:1633–1638

Besha P (2010) Policy making in China's space program: a history and analysis of the Chang'e lunar orbiter project. Space Policy 26:214–221

Bhatt BS (2010) Inventions in outer space: need for reconsideration of the patent regime. J Space Law 36(1):1–17

Bini A (2010) The moon agreement in the 21st century. Acta Astronaut 67:496–501

Blount PJ (2010) Informed consent v. ITAR: regulatory conflicts that could constrain commercial human spaceflight. Acta Astronaut 66:1608–1612

Bollweg HG (2010) GNSS-liability by international or European Union law? German J Air Space Law 59(4):551–559

Bouwmeester J, Guo J (2010) Survey of worldwide pico- and nanosatellite missions, distributions and subsystemtechnology. Acta Astronaut 67:854–862

Brachet G, Pasco X (2011) The US 2010 space policy: a view from Europe. Space Policy 27(1):11–14

Bütfering P (2010) Regional convergence platforms in Europe – innovation for space through technology partnerships. Acta Astronaut 66:1520–1524

Buzdugan M (2011) Satellite financing through hosted payloads: benefits and challenges. Air Space Law 36(2):139–160

Cameron BG, Seher T, Crawley EF (2011) Goals for space exploration based on stakeholder value network considerations. Acta Astronaut 68:2088–2097

Caroti S (2011) Defining astrosociology from a science fiction perspective. Astropolitics 9:39–49

Cartier A (2011) Symposium on the regulation of sub-orbital flights in the European context. Air Space Law 36:87–92

Chadda S (2010) U.S. commercial space sector: matured and successful. Journal of Space Law 36(1):19–54

Collins P, Autino A (2010) What the growth of a space tourism industry could contribute to employment, economic growth, environmental protection, education, culture and world peace. Acta Astronaut 66:1553–1562

Cook SB, Druger M, Ploutz-Snyder LL (2011) Scientific literacy and attitudes towards American space exploration among college undergraduates. Space Policy 27:48–52

Coppersmith J (2011) The cost of reaching orbit: ground-based launch systems. Space Policy 27:77–80

Crowther R (2011) The regulatory challenges of ensuring commercial human spaceflight safety. Space Policy 27:74–76

DalBello R, Mendelson M (2011) Private risk management in orbital operations. German J Air Space Law 60(2):218–230

Delagdo LM (2011) When inspiration fails to inspire: a change of strategy for the US space program. Space Policy 27:94–98

De Milliano M, Verhoeven C (2010) Towards the next generation of nanosatellite communication systems. Acta Astronaut 66:1425–1433

Donath T et al (2010) Possible European systems for space situational awareness. Acta Astronaut 66:1378–1387

Dyke G et al (2011) Dream project: applications of earth observations to disaster risk management. Acta Astronaut 68:301–315

Ehrenfreund P, Peter N, Billings L (2010) Building long-term constituencies for space exploration: the challenge of raising public awareness and engagement in the United States and in Europe. Acta Astronaut 67:502–512

Entradas M, Miller S (2010) Investigating public space exploration support in the UK. Acta Astronaut 67:947–953

Feucht U, Gienger G (2011) Launch and early operations of Herschel and Planck. Acta Astronaut 68:1283–1291

Finetto C, Lobascio C, Rapisarda A (2010) Concept of a lunar FARM: food and revitalization module. Acta Astronaut 66:1329–1340

Fong T et al (2010) Assessment of robotic recon for human exploration of the Moon. Acta Astronaut 67:1176–1188

Fukushima Y (2011) An Asian perspective on the new US space policy: the emphasis on international cooperation and its relevance to Asia. Space Policy 27:3–6

Gabrynowicz JI (2011) One half century and counting: evolution of the U.S. national space law and three long-term emerging issues. J Space Law 37:1

Gast MA, Moore SK (2011) A glimpse from the inside of a space suit: what is it really like to train for an EVA? Acta Astronaut 68:316–325

Giannopapa C (2011) Improving Africa's benefit from space applications: the European-African partnership. Space Policy 27:99–106

Gopalaswamy B, Wang T (2010) The science and politics of an Indian ASAT capability. Space Policy 26:229–335

Grabois MR (2011) Apollo: learning from the past, for the future. Acta Astronaut 68:1353–1360

Halstead CB (2010) Prometheus unbound? Proposal for a new legal paradigm for air law and space law: orbit law. J Space Law 36(1):143–205

Hansson A (2011) Refocusing European missile defence. Space Policy 27:116–117

Harrison A (2011) The search for extraterrestrial intelligence: astrosociology and cultural aspects. Astropolitics 9:63–83

Hashimoto T et al (2011) Japanese moon lander SELENE-2 – present status in 2009. Acta Astronaut 68:1386–1391

Häuplik-Meusburger S, Peldszus R, Holzgethan V (2011) Greenhouse design integration benefits for extended spaceflight. Acta Astronaut 68:85–90

Hearsey CM (2011) The nexus between law and astrosociology. Astropolitics 9:28–38

Hempsell M (2010) A phased approach to orbital public access. Acta Astronaut 66:1639–1644

Hertzfeld HR, Baseley-Walker B (2010) A legal note on space accidents. German J Air Space Law 59(2):230–240

Hobe S (2010a) The legal regime for private space tourism activities – an overview. Acta Astronaut 66:1593–1596

Hobe S (2010b) The moon agreement – let's use the chance! German J Air Space Law 59(3):372–381

Jarritt I, Peeters W, Schrogl KU (2011) Space solutions: practical applications for governments and markets. Space Policy 27:113–115

Jiménez-Monroy C (2010) The legal implementation of space cooperation in Latin America. A proposal for a three-phase plan. German J Air Space Law 59 (3):382–390

Johnson PS (2010) The roles of NASA, U.S. astronauts and their families in long duration missions. Acta Astronaut 67:561–571

Jorgenson CM (2010) Space law and policy 2010: the IISL-IAA symposium. Space Policy 26:194–195

Kalery AY, Sorokin IV, Tyurin MV (2010) Human space exploration beyond the international space station: role of relations of human, machine and the 'Earth'. Acta Astronaut 67:925–933

Kaul R (2010) Liability implications of the use of global navigation satellite systems (GNSS) for communication, navigation, surveillance/air traffic management (CNS/ATM) in civil aviation – with special focus on India. Ann Air Space Law XXXV

Kim DH (2010) Proposal for establishing an international court of air and space law. German J Air Space Law 59(3):362–371

Krynitz M (2010) The ground station of the future. Acta Astronaut 67:1252–1254

Kuan S (2010) The legality of the deployment of anti-satellite weapons in earth orbit: present and future. J Space Law 36(1):207–230

Lafleur JM, Saleh JH (2010) Survey of mission evolution and flexibility in the Space Shuttle program. Space Policy 26:236–245

Lee JH et al (2010) An overview of Korean astronaut's space experiments. Acta Astronaut 67:934–941

Lele A (2010) An Asian Moon race? Space Policy 26:222–228

Lempert D (2011) View point and book review. Living in space: cultural and social dynamics, opportunities, and challenges in permanent space habitats. Astropolitics 9:84–111

Lester D, Thronson H (2011) Human space exploration and human spaceflight: latency and the cognitive scale of the universe. Space Policy 27:89–93

Ling Y (2010) Remote sensing data distribution and application in the environmental protection, disaster prevention, and urban planning in China. J Space Law 36(2):435–457

Logsdon JM (2011a) A new US approach to human spaceflight? Space Policy 27:15–19

Logsdon JM (2011b) Change and continuity in US space policy. Space Policy 27:1–2

Lyall F (2010) Who is an astronaut? The inadequacy of current international law. Acta Astronaut 66:1613–1617

Macauley M, Laxminarayan R (2010) The value of information: 'methodological frontiers and new applications for realizing social benefit' workshop. Space Policy 26:249–251

Maggi F et al (2010) Efficient solid rocket propulsion for access to space. Acta Astronaut 66:1563–1573

Marenkov D (2010) New legal framework for foreign investment in the Russian aviation and space industry. German J Air Space Law 59(1):42–59

Masson-Zwaan T, Freeland S (2010) Between heaven and earth: the legal challenges of human space travel. Acta Astronaut 66:1597–1607

Masson-Zwaan T (2010) "Regulation of sub-orbital space tourism in Europe: a role for EU/EASA? Air Space Law 35(3):263–272

McGee BW (2010) A call for proactive xenoarchaeological guidelines – scientific, policy and socio-political considerations. Space Policy 26:209–213

McQuaid K (2010) Earthly environmentalism and the space exploration movement, 1960–1990: a study irresolution. Space Policy 26:163–173

Moore CL (2010) Technology development for human exploration of Mars. Acta Astronaut 67:1170–1175

Murthi KRS, Shoba TS (2010) Technology transfer trends in Indian space programme. Acta Astronaut 67:942–946

Newton EK, Griffin MD (2011) United States space policy and international partnership. Space Policy 27:7–9

Niebuhr SM (2010) Principal investigators and project managers: insights from discovery. Space Policy 26:174–184

Nishida SI et al (2011) Strategy for capturing of a tumbling space debris. Acta Astronaut 68:113–120

Norris P (2011) Developments in high resolution imaging satellites for the military. Space Policy 27:44–47

Pass J (2011) Examining the definition of astrosociology. Astropolitics 9:6–27

Peeters W (2010) From suborbital space tourism to commercial personal spaceflight. Acta Astronaut 66:1625–1632

Pelton JN (2010) The space shuttle – evaluating an American icon. Space Policy 26:246–248

Peoples C (2010) The growing "securitization" of outer space. Space Policy 26:205–208

Pop V (2011) Space exploration and folk beliefs on climate change. Astropolitics 9:50–62

Post MA, Lee R (2011) Lessons learned from the York University Rover Team (YURT) at the University Rover Challenge 2008–2009. Acta Astronaut 68:1343–1352

Raitt D et al (2010) Space expectations: latest survey results. Acta Astronaut 67:1318–1322

Remuß NL (2011) The new space strategy – space as a key driver for innovation improving people's living conditions on earth. German J Air Space Law 60 (2):231–263

Rendleman JD, Faulconer JW (2010) Improving international space cooperation: considerations for the USA. Space Policy 26:143–151

Robinson GS (2011a) United States national security space strategy. German J Air Space Law 59(4):264–279

Robinson J (2011b) Transparency and confidence-building measures for space security. Space Policy 27:27–37

Ruttley T et al (2010) International utilization at the threshold of 'assembly complete' – science returns from the International Space Station. Acta Astronaut 67:513–519

Scholz A et al (2010) Flight results of the COMPASS-1 picosatellite mission. Acta Astronaut 67:1289–1298

Schrogl KU (2011) Is UNCOPUOS fit for the future? Reflections at the occasion of the 50th session of its legal subcommittee 2011. German J Air Space Law 60 (1):93–102

Singer MJ, Musacchio JT (2011) An international environmental agreement for space debris mitigation among asymmetric nations. Acta Astronaut 68:326–337

Sippel M (2010) Promising roadmap alternatives for the SpaceLiner. Acta Astronaut 66:1652–1658

Smirnov NN et al (2010) Impact of debris particles on space structures modeling. Acta Astronaut 67:333–343

Smith MS (2011) President Obama's national space policy: a change in tone and a focus on space sustainability. Space Policy 27:20–23

Som S (2010) An international symbol for sustained exploration of space. Space Policy 26:140–142

Souchier A (2010) Private ground infrastructures for space exploration missions simulations. Acta Astronaut 66:1580–1592

Stubbe P (2011) Conference on "Soft Law" in outer space. The functioning of nonbinding norms in international space law. German J Air Space Law 60 (2):280–285

Su J (2010a) Towards an effective and adequately verifiable PPWT. Space Policy 26:152–162

Su J (2010b) Use of outer space for peaceful purposes: non-militarization, nonaggression and prevention of weaponization. J Space Law 36(1):253–272

Sukol R (2011) Positive law codification of space programs: the enactment of title 51, United States code. J Space Law 37:1

Sun W et al (2011) European small geostationary communications satellites. Acta Astronaut 68:802–810

Szajnfarber Z, Stringfellow MV, Weigel AL (2010) The impact of customer-contractor interactions on spacecraft innovation: insights from communication satellite history. Acta Astronaut 67:1306–1317

Takaya-Umehara Y (2010) TCBMs over the military use of outer space. Acta Astronaut 67:1299–1305

Tomilovskaya ES et al (2011) Effects of long-duration space flight on target acquisition. Acta Astronaut 68:1454–1461

Tronchetti F (2010) The moon agreement in the 21st century: addressing its potential role in the era of commercial exploitation of the natural resources of the moon and other celestial bodies. J Space Law 36:2

Tronchetti F (2011) Preventing the weaponization of outer space: is a Chinese-Russian-European common approach possible? Space Policy 27:81–88

Tsuchida A et al (2011) KIBO (Japanese experiment module in ISS) operations – 2008/3-2009/10. Acta Astronaut 68:1318–1324

Van Fenema P (2010) The international institute of space law in 2010. Air Space Law 35(4/5):337–338

Verhoeven CJM (2011) The origin of satellite swarms. Acta Astronaut 68:1392–1395

Vighnesam NV, Sonney A, Gopinath NS (2010) India's first lunar mission Chandrayaan-1 initial phase orbit determination. Acta Astronaut 67:784–792

Watanabe H (2011) Japanese space policy during the 1980s: a balance between autonomy and international cooperation. Acta Astronaut 68:1334–1342

Webber D (2010) Point-to-point sub-orbital space tourism: some initial considerations. Acta Astronaut 66:1645–1651

Weeden B (2011) Overview of the legal and policy challenges of orbital debris removal. Space Policy 27:38–43

Zhao Y, Bian Y (2011) Export control regime for space items in China: opportunities and challenges in the new era. Space Policy 27:107–112

About the Authors

Blandina Baranes joined the European Space Policy Institute (ESPI) in Vienna in February 2005 and currently holds the position of ESPI Communications Manager. Prior to this, she was the chief librarian of the Jewish Studies Department of the University of Vienna. During the past years she has also worked as a documentalist and librarian for different institutions, such as the Austrian broadcasting corporation, the Der Spiegel magazine and others. She conducted her studies and research in Austria and Israel and graduated with a Masters Degree from Vienna University, Faculty of Philosophy, Department of Social and Cultural Anthropology. She is co-editor of the past two volumes of ESPI's "Yearbook on Space Policy", published several articles and participated as a speaker in a number of space related events on the topic of "Space and Interdisciplinarity".

Daniel Barok is the CEO and owner of BINGO Business Development and Consulting Services, located at Rosh Ha Ha'ayin Israel and established in January 2011. From 2006 to 2010 Daniel served in the Israeli Space Industry as a Director for International Marketing in ImageSat International N.V (the owners and operators of the Israeli-made EROS satellites, as providers of commercial imaging services), and thereafter as Director for Business Development with ELBIT Systems, focusing on space borne remote sensing electro-optical payloads, including advanced hyperspectral systems. At the end of 2005, he retired after 25 years of service in the Israeli Defence Forces (IDF). Daniel started this long career at the Israeli Air Force (IAF), serving in several operational and Intelligence posts. In his last position in the IAF, Daniel was in command of a squadron for remote sensing and imagery analysis. Thereafter, he transferred to the intelligence corps, leading R&D activities at the IDF Geospatial Intelligence Agency (IGA), while he was one of the leaders of the national space programmes, mainly directing the implementation of the national Ofeq satellites capabilities, as well as commercial and other resources among the Israeli intelligence community. In this position, Daniel led a technological unit of several hundred engineers and technicians. Daniel represented the Israeli space community in many international space collaborations, space policy and inter-agency meetings as a professional advisor to Israeli delegations,

P. Hulsroj et al. (eds.), *Yearbook on Space Policy 2010/2011*,
Yearbook on Space Policy, DOI 10.1007/978-3-7091-1363-9,

and as the initiator of part of the agreements. Daniel holds a B.Sc. in Mechanical Engineering and an MBA, both from Ben-Gurion University.

Catherine Doldirina is a post-doctoral fellow at the Institute of Air and Space Law of McGill University, and Co-chair of the Space Generation Advisory Council. She has been conducting legal research relating to space activities since 2005. Her expertise lies in the field of intellectual property law and her current research relates to the legal status of remote sensing data. She has lectured on European competition law, European copyright law and space law at various universities in Canada and Europe. Her professional experience as a lawyer was gained as legal assistant to specialist copyright lawyers (Bremen). She successfully completed an internship at the European Space Policy Institute, Vienna, drafting the study "Case for Space" during this time. She authors work on other aspects of space law, she is engaged in the research activities of the Institute of Air and Space Law, as for example the Space Security Index, and she is a member of the International Institute of Space Law. In 2009 she was awarded the Diderick Verschoor award and prize for the best paper by a young author for the Colloquium on Space Law of the International Institute of Space Law.

Peter Hulsroj pursued his legal studies at the University of Copenhagen from 1974 to 1979. After military service, Peter Hulsroj did post-graduate studies at Harvard Law School in the USA, where he earned a Master's Degree in 1981. Returning from the USA, Mr Hulsroj joined a large commercial law firm in Copenhagen, where he became a full member of the Danish Bar. After three and half years of private practice in Copenhagen, Mr Hulsroj became a Contracts Officer in ESTEC, and thereafter served for almost 14 years as the Head of Contracts and Legal Affairs at Eumetsat in Darmstadt, Germany. Before rejoining ESA, Mr Hulsroj was the Legal Adviser of the Preparatory Commission for the Comprehensive Nuclear-Test-Ban Treaty Organization in Vienna. Peter Hulsroj was appointed ESA Director of Legal Affairs and External Relations (D/LEX) in May 2008 and served as such until he became the Director of ESPI in 2011.

Alexander Kolovos holds a doctorate degree in European security and international relations and teaches space applications at the Hellenic Air Force Academy. He joined the Hellenic National Centre for Space Applications (HNCSA, Ministry of Defence) in 1985 and he has been continuously associated with it since, both in the fields of space policy and satellite imagery exploitation. An acknowledged expert on Space policy issues, he proposed and managed inter alia: the participation of the Hellenic Republic to the BOC initiative (Besoins Opérationnels Communs), and to both Helios 2 and MUSIS programmes (2003–06); the formulation of the "European Security and Defence Policy and Space" study, from which the importance of space applications and functions was recognised by the EU Council; the participation to the European Commission's Green Paper process on the future of the EU in space (2002); and the formulation of the Western European Union's Space policy (1998). His contribution to the EU first and second pillars was acclaimed by EU Research Commissioner Phillipe Busquin and EUROSPACE (Association of European Space Industry). As an acknowledgement of his capacities, the Hellenic Republic proposed in 2009 his candidature for the post of the Director of the European Union's Satellite Centre (EUSC). Brig.Gen. (ret.) Kolovos has written one book, which is included in the recommended reading list of two universities, and two monographs on space applications. He has also written over thirty papers, hosted in prestigious international publications, such as Space Policy, Space News, and the ESPI Perspectives' series. References to his work can be found in EU documents, publications of the EU Institute for Security Studies, as well as in international magazines and newspapers such as Space News, Defence News, and Aviation Week and Space Technology.

Sa'id Mosteshar is Director of the London Institute of Space Policy and Law, Professor of International Space Law, Senior Associate Research Fellow at the Institute of Advanced Studies of London University, and member of the UK Space Leadership Council, formerly Professor of Global Communications at the University of California. He leads the Institute's ESA funded Study, "Evidence from Space", examining the utility of Earth Observation information as evidence in judicial and administrative proceedings at national and international levels. He has written extensively on space law and policy topics and is a frequent speaker at international conferences. He recently contributed to the National Regulation of Space Activities, winner of the 2011 Social Sciences Book Award of the International Academy of Astronautics. Sa'id is a former Chairman of the Space Law Committee of the International Bar Association, and member of the California Bar's International Law Section. His professional memberships include the International Institute of Space Law. Sa'id holds degrees in Physics and Econometrics. He is a member of the English Bar, as well as that of California and U.S. Federal Court, he was admitted to Fellowship of the Institute of Chartered Accountants in England and Wales, and is a member of the Chartered Institute of Arbitrators.

Julia Neumann is Attorney-at-Law (Rechtsanwältin) with the international law firm White & Case LLP in Berlin, Germany, where she advises private and public entities on public law and regulatory issues. She was admitted to the Bar in 2009. From 2007 to 2009 she performed her legal training (Referendariat) at the German Federal Ministry of Transport, Building and Urban Affairs in Bonn, Germany, at SES ASTRA in Betzdorf, Luxembourg, at ESA Headquarters in Paris, France, and in the telecommunications practice of the international law firm Freshfields Bruckhaus Deringer LLP in Cologne, Germany. From 2003 to 2007, Julia Neumann was (senior) research assistant with the Institute of Air and Space Law at the University of Cologne. During that period, she provided research with regard to a great variety of space law issues, including the legal regime for suborbital flights, and participated in various international conferences on space law. Her doctoral thesis 'Bridging the Digital Divide' is concerned with legal aspects of bridging the international digital divide with satellite telecommunications, and will be published in late 2011. Moreover, she has published several articles on space law. She is a member of the International Institute of Space Law (IISL) since 2006.

Spyros Pagkratis currently holds the position of Resident Fellow at the European Space Policy Institute in Vienna, which he joined in 2010. His fields of expertise include space policy, security and economy. He was the study leader of two annual ESPI Reports on "Space Policies, Issues and Trends" that provide a comprehensive overview and analysis of developments in the field of international space policies on a yearly basis. These ESPI Reports are among the Institute's leading publications and they are presented as an information document to ESA member states' delegations. He is also part of the editorial team and co-author of ESPI's "Yearbook on Space Policy" annual flagship publication, as well as the editor of the "ESPI Perspectives" series. Prior to joining ESPI, he briefly worked as a space policy analyst with the Western European Union Assembly and EADS Astrium Space Transportation in Paris. He holds a degree in history from Athens University and two Master's degrees, one in International Relations' History from Sorbonne University, and the other in Defence Policy and Armament Procurement Planning from Panthéon/Assas University, in Paris. He has published articles in the field of space policy and economy, participated in international conferences and workshops on the subject, and reviews articles for Acta Astronautica.

Rosa Rosanelli is currently a post-graduate student at the 2011–2012 Safran promotion of the Master 2 on Space Law and Telecommunications of Paris-Sud University. Prior to this, she worked as a Research Intern at the European Space Policy Institute (ESPI) in Vienna, where she focused on the Galileo/GMES flagship European space programmes and the role of space in support of the Europe 2020 strategy. She was also an Intern at the Security and Defence Department of the Istituto Affari Internazionali (International Affairs Institute/IAI) in Rome, where she conducted research on European Security and Defence Policy. She holds a BA in Political Sciences and International Relations, and a MA in International Relations from the "Sapienza" University of Rome. Her final thesis, on the subject of "Space Activities and Security and Defence Policies" was published as a "Quaderno IAI". She also studied at the Law Faculty of Nantes and attended the ECSL Summer Course in Lisbon, in 2009.

Matxalen Sánchez Aranzamendi is an Associate Fellow at the European Space Policy Institute (ESPI), based in Brussels. Before, she has worked as Resident Fellow for Space Legislation and Regulations from 2008 to 2011 at the ESPI seat in Vienna. Since joining ESPI she has authored two studies on the effects of National Space Legislation and of Space Regulations on the development of commercial space activities in Europe and she has coedited two ESPI Reports on Third Party Liability for GNSS in Europe and legal Issues of Remote Sensing. In addition to this, she has published conference proceedings and articles in the German Journal of Air and Space Law (ZLW) and in International Institute of Space Law (IISL) Colloquia proceedings. Matxalen Sánchez Aranzamendi is a member of the International Institute of Space Law (IISL) since 2010. She has delivered guest lectures on National Space Legislation at Leiden University, SIOI Rome and the ECSL summer course 2010 in Jaen and has participated as a speaker in international conferences, such as the IISLECSL Symposium at the UNCOPUOS Legal

Subcommittee in 2010, the EU Council Presidency conference on space governance in 2010 in La Granja, or the IAC 2010 in Prague. She has also provided advice in National Space Legislation drafting processes, i.e. for the Austrian National Space Legislation, and the drafting of a proposal for the ILA Model Law for National Space Legislation. Previously she was an intern at the European Commission Directorate for Space Policy, GMES and Security (2007–2008), and she worked as a Junior Policy Adviser at the Basque Delegation in Brussels. Matxalen Sánchez Aranzamendi is a graduate from Leiden University (the Netherlands) and Deusto University (Spain), where she obtained her Advanced LLM degree in European Business Law and her degree in Law and Economics respectively.

Nicolò Sartori is a Researcher at the Istituto Affari Internazionali (International Affairs Institute, IAI) in Rome, within the Security and Defence Department. He works on the evolution of technologies in the public safety and security domain, and on the potential convergences between the security and defence industrial sectors in Europe. He is also an Energy Research Fellow at the Italian Military Centre for Strategic Studies (CeMiSS), where he contributes to the activities of the Strategic Observatory and of the Quarterly in the energy domain. He is a Ph.D. student at the Department of Politics and International Relations at the University of Kent, preparing a doctoral thesis on the concept of energy security in the international theory debate. He holds a BA in International and Diplomatic Studies from the University of Bologna (Forlì Campus) and an MA in International Relations (Bologna Campus). He served as faculty advisor to the NATO Defence College in Rome, where he also conducted research on NATO's role in energy security. His research interests include Russia's defence and security policy and international energy security, with a particular focus on Central Asia and the broader Caspian region.

Hideaki Shiroyama is director of the Policy Alternatives Research Institute, and a professor of public administration at the Graduate School of Public Policy and the Graduate School of Law and Politics, at the University of Tokyo. His fields of expertise include international administration, science/ technology and public policy, and public policy process. He led the effort to set up a Space Policy Programme at the Graduate School of Public Policy in 2010. His publications include: "Transformation of Political Space and Policy Innovation 1 Political Theory of Policy Innovation" (University of Tokyo Press 2008), "The Structure of International Aid Administration" (University of Tokyo Press 2007), "Governance of Science and Technology" (Toshindo 2007), "The Harmonisation of Automobile Environmental Standards between Japan, the United States and Europe" Pacific Review vol. 20–3, "Administrative Reorganization and Public Sector Reform in Japan" in The Public Sector in Transition: East Asia and the European Union Compared (Nomos 2007), "Technology Innovation and Diffusion for Environmental Protection," in Energy Market Restructuring and the Environment (The University Press of America 2002). He also served as a chairman of the Planning Committee of New Initiatives for Humanities and Social Sciences Programme at the Japan Society of Promotion of Science from 2003 to 2008, as president of PI forum, an NPO for consensus building in Japan from 2005 to 2008, and as a member of various advisory councils on the use of the international space station, nuclear safety, food safety, scenarios for climate mitigations, industrial policies and higher education.

Anna Clementina Veclani is a Junior Researcher at the Istituto Affari Internazionali (International Affairs Institute, IAI) in Rome, within the Security and Defence Department. She mainly covers European research projects on space-related affairs and aerospace/defence industry. Previously, she was a research assistant within the Political Committee of the WEU-ESDA Assembly in Paris and a trainee within the Political Office of the Italian Embassy in Washington, DC. She holds an MA in International and Diplomatic Studies obtained at the University of Bologna (Forlì Campus) and one in International Studies and Military Strategy from the University of Milan, in the framework of the Joint Services Command and Staff College, Rome. She also studied abroad, at Sciences-Po in Paris and at York University in Toronto.

Cornelius Vogt works as Programme Officer for the "Berliner Forum Zukunft" of the Research Institute of the German Council on Foreign Relations in Berlin, Germany. He studied Political Sciences and Human Geography in Germany, France and the United States. He graduated with distinction with a study on "Strategic Aspects of German Space Security Policy". Amongst others, he worked and interned at the German Bundestag and the Federal Ministry of Defence. He gained additional practical training and study experience in Singapore and Moscow. His areas of specialisation are: German foreign, security, and defence policy; reform and operations of the German armed forces; and German space security policy. Cornelius Vogt received scholarships from the German National Academic Foundation, the Fulbright-Commission and the German Academic Exchange Service.

Joanne Wheeler is a partner at CMS Cameron McKenna of London, and in particular at the Technology, Media and Telecommunications team. Joanne specialises in satellite and telecommunications regulatory and policy matters and commercial contracts in the communication sector. She represents operators, manufacturers, communications companies, financial institutions, governments and international institutions in all aspects of domestic, European and international communications and spectrum regulations and procedures and in all types of commercial contracts, including in the negotiation and drafting of procurement contracts. She has been involved in various studies undertaken by governmental and intergovernmental agencies, covering a variety of subjects from space debris issues to public-private partnerships in satellite communications. She also advised the UK Government's Space Innovation and Growth Team on regulatory, legal and policy issues in the space and satellite sector; this work is ongoing. In addition to this, she has advised a government body on recognised spectrum access and spectrum trading. Apart from her private practice experience, Joanne has extensive and unique experience from her three years at the European Space Agency (ESA) in Paris and at the UK Office of Communications (Ofcom), where she was involved in drafting Ofcom's satellite filing procedures. She participates in various panels and committees, including the British Standards Institute Panel on space debris and the Advisory Board of the University College London satellite project, while she is a member of the International Institute of Space Law. Joanne is also the Secretary of the UK branch of the European Centre for Space Law and a Fellow of the Royal Astronomical Society. She has also been recently appointed to represent the UK satellite industry at the Legal Subcommittee of the UN (UNCOPUOS). Joanne has lectured on satellite and communications law topics at Queen Mary College, University of London and the International Space University in Strasbourg. She is a regular speaker at international space and communications conferences, has published several articles in international journals and has contributed chapters for several books on space and satellite topics. She is admitted to practice law in England and Wales and also in Scotland, where she is a notary public. Joanne holds degrees from the Universities of Aberdeen, Oxford and Rotterdam and a diploma in telecommunications from Queen Mary College, University of London.

Index

P. Hulsroj et al. (eds.), *Yearbook on Space Policy 2010/2011*,
Yearbook on Space Policy, DOI 10.1007/978-3-7091-1363-9,

MIX
Papier aus verantwortungsvollen Quellen
Paper from responsible sources
FSC® C105338
FSC www.fsc.org

If you have any concerns about our products,
you can contact us on
ProductSafety@springernature.com

In case Publisher is established outside the EU,
the EU authorized representative is:
Springer Nature Customer Service Center GmbH
Europaplatz 3, 69115 Heidelberg, Germany

Printed by Libri Plureos GmbH
in Hamburg, Germany